KV-638-325

THE BASES OF ECONOMIC GEOGRAPHY

To Renée for the summer of '76;
to Dorothy Wilson who can and does

RONALD R. BOYCE

To Sarah who can't and to
Sally who won't

ALAN F. WILLIAMS

First British Edition

THE BASES OF ECONOMIC GEOGRAPHY

Ronald R. Boyce
SEATTLE PACIFIC UNIVERSITY

Alan F. Williams
UNIVERSITY OF BIRMINGHAM

HOLT, RINEHART & WINSTON
London · New York · Sydney · Toronto

Holt, Rinehart and Winston Ltd: 1 St Anne's Road
Eastbourne, East Sussex, BN21 3UN

Printed in Great Britain by Clarke, Doble and Brendon Ltd, Plymouth and London

ISBN 0–03–910198–3

Last digit is print No: 9 8 7 6 5 4 3 2 1

Contents

PART 4 ECONOMIC ASPECTS OF URBAN LOCATION

PART 5 GEOGRAPHICAL PROSPECTUS

Preface

'My principal Design, in publishing the following Treatise, is, To present the younger sort of our Nobility and Gentry, with a Compendious, Pleasant and Methodical Tract of Modern Geography, that most useful Science, which highly deserves their Regard. . . . That Geography doth merit the Title of Science in several Respects, and that the Knowledge thereof is both pleasant and useful to Mankind, is a Truth so universally granted, that 'twere altogether needless to enter a Probation of it.'

Preface to *Modern Geography*, Patrick Gordon, 1754.

In the 2nd Edition of this book, Ronald Boyce explained the purpose of *The Bases of Economic Geography*: to explore the spatial characteristics of people and their economic activities. He aimed at a simple and direct essaylike format, selecting those topics which would provide a meaningful basis—hence his title. He felt strongly that the student attracted to economic geography needs to be taught to swim before being thrown into the deep waters of an ever-growing subject. He also made the important point that the book made no pretence to cover, even superficially, all the commodities, products, goods or services that might be relevant to economic geography. Instead, he would use selected phenomena as examples of the ordered location of things; by proceeding in a graduated way through a selected series of activities, he aimed to develop a framework for geographical appreciation and analysis.

These aims and purposes are retained in this new British Edition, the emphasis being firmly rooted in basic principles, concepts and theories. These have been further developed where it is recognized that the needs of British or other English-speaking students may be different from those of freshmen in American colleges or universities, and we have added a wide range of case studies and examples mainly, but by no means exclusively, based on experience in Britain and Europe.

The text is divided into five Parts. Part 1 examines population and its distribution on the earth and discusses the nature of economic geography. Part 2 contains several chapters which discuss the bases for understanding the spatial patterning of various activities. Part 3 examines the location of economic activities. Here, we progress from those activities found among primitive and traditional economic systems, through the extractive activities and on to agricultural and industrial location problems and patterns in Europe, North America and elsewhere in the world. In Part 4 we treat economic aspects of urban location, with an emphasis on those activities which relate to matters considered in earlier chapters. Finally, in Part 5 we undertake a brief review of the earth's crises of the present and problems for the future.

Overall, we have tried to provide a synoptic view of economic geography which can be used safely as a point of entry for A-level candidates or first year university students. But we have in mind, too, the needs of college and professional course students who are likely to have a relatively modest geographical background. The particular requirements of the various Examining Boards vary widely and syllabuses are subject to changes which reflect new trends. This is because economic geography has fast-developing theoretical-quantitative branches which are exciting to explore. But these branches are often tackled from inadequate foundations, so that they become a minefield for the ill-prepared or unwary. That this is so is regularly confirmed by teacher and lecturer, while even as this Preface is being written, in the current issue of 'Teaching Geography' (January, 1979) a student records her plea for experienced geographers to provide a 'fairly simple explanation of the various geographical theories concerned with population, agriculture, industry, transport and urban geography'.

Sometimes, where foundations are provided, they cannot be comprehended without continuous recourse to long lists of supplementary reading which are by no means available everywhere. Here, we are aware that the influences of time, distance and cost all tend to be deterrents to ready reference, which is only practicable within reach of the major libraries of college or university. And even there, economic geography is but one of the many necessary foundations required of the student within the broader field of modern geography. For these reasons, we have kept the chapter bibliographies within limits consistent with the need to provide alternatives.

The reader will find, therefore, that he can proceed smoothly through the text unless he wishes to increase the depth or breadth of his understanding by reference to the occasional classic, the well-written 'standard' text, or a recent contribution. The index provides a useful guide back to important topics and a glossary offers a checklist of essential terms. Metric measures are used throughout except where the text relates to Figures unchanged from the earlier American editions; an appendix gives conversion factors.

In the compilation of the British Edition, undertaken entirely within the University of Birmingham, special thanks are due to Professors David Thomas, Rowland Moss and Paul H. Temple, whose work in many allied fields of geography has indirectly, but importantly, influenced the British author. Very specially, however, he would like to thank his colleagues with whom he has shared the teaching of a foundation course in economic geography, namely Dr Robert Gwynne, Dr Martin Parry, Dr Denis Shaw and Mr Michael Tanner. All these colleagues offered valuable criticisms of parts of the manuscript. I am also indebted to Visiting Professor Tatsuo Kimura, University of Kobe-Gakuin, Japan, who checked over many of the new Figures so ably drawn by Mrs Jean Dowling. Finally, for the publishers, Tom Perlmutter exercised the greatest patience and was a most effective progress-chaser.

To all these people we owe so much. But despite the many individuals involved in development and production, this author takes full responsibility for the organization and specific content of this edition. Any errors that may be found are mine.

Birmingham, England
January 1979 ALAN F. WILLIAMS

Preface to the Second American Edition

This text explores the spatial characteristics of people and their economic activities. The presentation given here is a personal one: I have written this book in an essaylike format and have selected those topics which I felt best provided a meaningful geography of the earth occupancy pattern.

The book makes no pretence to cover, even superficially, all the commodities, products, goods, or services that might be relevant to economic geography. Rather than try to explain the locational character of all that is relevant, this book simply uses selected phenomena as examples of the ordered placement of things. By proceeding in a graduated way through a selected series of activities, a framework for geographical appreciation and analysis is developed.

In some respects this work parallels the intention of Hendrik Willem Van Loon, who said in his preface to *Van Loon's Geography* (New York: Simon and Schuster, 1932): 'No, I don't merely want a new geography. I want a geography of my own, a geography that shall tell me what I want to know and omit everything else . . .' Likewise, I have written an economic geography that tells me what I would like to know about the subject and nothing else.

The reader will find that this text, unlike most other textbooks, is not interrupted by overly detailed maps, tables, and other peripherally related materials. Distracting, 'filler-type' items are also conscientiously avoided. This book does not attempt to be a substitute for atlases, the US Census, dictionaries, or encyclopedias. Nor does this book contain a summary of current research articles. Only those articles that are relevant as supplementary reading are included in the bibliographies. In some cases older works are preferred to more recent ones because they are better written.

In this edition photographs and a glossary have been added. The photographs have been carefully chosen to illustrate, either directly or indirectly, the principles discussed in the text. A glossary has been included to define those technical terms and expressions that are either not found in a dictionary or are used here in a slightly different geographical sense.

Eminent geographers who helped me develop the manuscript for the first edition of this book and to whom I am much indebted are Dr Preston James, formerly of Syracuse University; Dr John Borchert, University of Minnesota; and Dr James Lindberg, University of Iowa.

Many persons have been of inestimable aid in the development of the revised second edition. I would especially like to thank Dr Nico Sheepers of Rand Afrikaaners University in Johannesburg who was my constant research companion for a year. I also owe a great debt to Dr Douglas Fleming, Dr William Beyers, Dr Richard Morrill, Dr Edward Ullman, Dr Phillip Bacon, and Dr John Sherman—all at the University of Washington at the time—and others who provided me with many insights and possibilities for improvement.

The illustrations have been substantially revised and improved. Many of these revisions and the cartography were done by Eleanor Mathews, with assistance from John Sherman and Carl Youngmann.

My daughter Renée was primarily responsible for obtaining the photographs used in this book. These were supplied by many nations and states, among them the states of Georgia, Colorado, and Idaho and the nations of Israel, the USSR, and South Africa. I would particularly like to thank Dr Yehuda Hayuth at the University of Rhode Island, on leave from the University of Haifa, Israel, for his photographs of Israel.

The editorial staff at Holt, Rinehart and Winston has been not only gracious and helpful throughout but has added greatly to the quality and accuracy of this text. Ms Rosalind Sackoff worked with me from the beginning of the revision to add relevant sections and otherwise update the book. The general reorganization of the text is in a large part due to her suggestions and insights. One can hardly say enough about the general expertise and careful work of Mrs Jean Shindler. I have worked with many editors but none so meticulous, single-minded, and efficient as she. I owe both these editors a tremendous thanks.

Numerous changes in the book are the result of thoughtful comments and analysis by various reviewers. In particular, I wish to thank Dr Russell Capelle, University of Vermont; Dr Robert A. Young, California State University, Fullerton; Dr John McGregor, Indiana State University; Dr William Cheek, Southwest Missouri State University; Dr Prentice Knight III, University of Kansas; Dr Emmett F. Stallings, University of Southern Louisiana; and Dr Kathleen M. Brown, University of Michigan.

I am also grateful to many of my former students who have used my book in their classrooms and who have passed on important suggestions for improvements. Among these are Dr Roger Crawford, San Francisco State University; Dr Harvey Heigis, San Diego State University; Dr Siim Soot, University of Illinois; Dr Ronald Schultz, Florida and East Atlantic University; Dr Marjorie Rush, University of Texas, Houston; and Dr David Johnson, Southwestern Louisiana State University.

Finally, special thanks are due to Mrs Dorothy Wilson, administrative secretary for the School of Social and Behavioral Sciences at Seattle Pacific University. Her heroic efforts at every stage of the manuscript are tremendously appreciated.

Despite the many individuals involved in the development and production of the text, I take full responsibility for the organization and specific content of this book Any errors that may be found are mine.

Seattle, Washington RONALD R. BOYCE
June 1977

Part One

PERSPECTIVE ON POPULATION AND PLACE

'Yes, but how about geography? No, I don't merely want a new geography. I want a geography of my own, a geography that shall tell me what I want to know and omit everything else . . .'

Hendrik Willem Van Loon, *Van Loon's Geography* (New York: Simon and Schuster, 1932), p. vii.

1

World Population Distribution

In the study of almost any subject, people come foremost. In the study of economic geography, the initial question is, Where are people and why are they so situated? The second question is, What do they do there? Such a query automatically brings up other considerations such as the nature of people as producers, consumers, or simply as inhabitants of various parts of the earth. Closely related is the question of the so-called standard of living of populations in various places. Such differing economies, in turn, are directly related to the needs of the earth's resources, the extent of the transportation and interconnection systems, and a myriad of other features that are relevant to the understanding of people and their arrangement on earth.

If we were to stand back from the earth like astronauts (something we may simulate by studying satellite images taken on the Gemini, Apollo and other space missions) we would be struck not by the beauty, order and life found on earth, but by its seemingly 'shipwrecked' condition. Two-thirds of the earth is covered with water. The majority of the land portion is either covered with ice (Antarctica), lacking sufficient moisture for crops (deserts), distinguished by an extremely limited growing season (far northern and southern latitudes, plus high elevations), or otherwise characterized by major occupancy problems (for example, swamps, mountains, barren rock-laden areas, lakes, volcanoes and related lava flows, tropical jungles and sandy wastes). Only a few places are truly appropriate for year round occupancy. Spaceship earth is more like an ecological disaster than a Garden of Eden.

Such a view of our planet is further supported by an examination of the distribution of people upon it. Human beings have attached themselves to places selected for survival in an otherwise hostile world. People are concentrated in only three or four parts of the earth, except for a few isolated places far from the concentrated strongholds of civilization. The remainder of the land has less than one person per square kilometre. Within the three major centres, or nodes, of population in Eurasia are found some three-quarters of the world's population, yet these three centres account for less than 10 per cent of the earth's area.

Two great clusters are at once noticeable: (1) an enormous massing of population in southern and eastern Asia; and (2) a sizable concentration of population on either side of the North Atlantic Ocean, in Western Europe and in North America. About one-half of the world's population is found in such nations as China (over 800 million), India (about 590 million), Pakistan, Japan, and Indonesia (each with well over 100 million). Another one-fifth of the world's population is found in the Western European countries. A much smaller population cluster (about 120 million persons) is the northeastern quadrangle of the United States and the Ontario peninsula of Canada. Together these areas account for another 3 per cent of the earth's inhabitants. Thus these two great clusters of population contain almost three-quarters of the earth's population in an area that comprises no more than 5 per cent of the land's surface (Figure 1–1).

The remaining one-quarter of the earth's population is widely scattered. However, the more densely populated nodes are represented by metropolises near coasts, for example, along the Pacific coast of the United States, in the southeastern part of South America, and southeastern part of Australia, and about West Africa. Other small nodes are found inland, as in Central Mexico and the East African highlands about Lake Victoria, or sporadically situated near major river estuaries.

Today the population of the world is estimated to be slightly over 4 billion. Table 1–1 shows that nearly two-fifths of the world population lives in the Southeast Asian perimeter. However, there are also vast areas of China and much of India and Pakistan that contain little population. Therefore, there is extreme concentration of population in only a small fraction of the earth's surface.

If the other nations of the world were occupied at the same density as these heavy population nodes, and if water area were eliminated from the calculations, the population of the earth would contain over 16 billion persons. But would the earth be able to support that many persons at the present stage of technological development? The rest of the land area of the world is not as well-endowed physically as is China and the Indian subcontinent. Even so, at the ecomonic level of people in this area, the earth could certainly support two or three times the present population without any further technological development.

Table 1–1 Population importance of Southeast Asian perimeter

Country	Population (millions)	Percentage of world population	Land area (thousands of square miles)	Percentage of world area (including water)
China	800	20.0	3692	1.87
India	586	14.7	1230	0.62
Bangladesh	71	1.8	55	0.03
Sri Lanka (Ceylon)	13	0.3	25	0.01
Burma	30	0.8	262	0.13
Subtotal	1500	37.6	5264	2.66
World total	4000+		197,667	

It is instructive to determine what the population of the earth might be if it were populated at the same density as the United States. The population of the United States is about 215 million, or 5.4 per cent of the world's population. The land area of the United States is 3.6 million square miles or about 6.4 per cent of the land area of the world. Therefore, with the density average for the United States being about 58 per square mile overall, the population of the world would be about 3.4 billion persons. Given the combination of open spaces and cities, the United States is a fairly average representation of the world overall, at least as far as concentration and density are concerned.

But there are some areas of the world that are very scantily and unevenly populated. Australia, about the same area as the US not counting Hawaii and Alaska, (2,968,000 square miles), contains only some 13.5 million people, some 80 per cent of whom concentrate in a small number of cities. The vast majority of the continent is very lightly populated with only some 4.5 persons per square mile. At the same density as that of Australia, the earth would contain only about 250 million people or only 16 per cent of its present amount. The world would contain a population of about that actually found in the USSR.

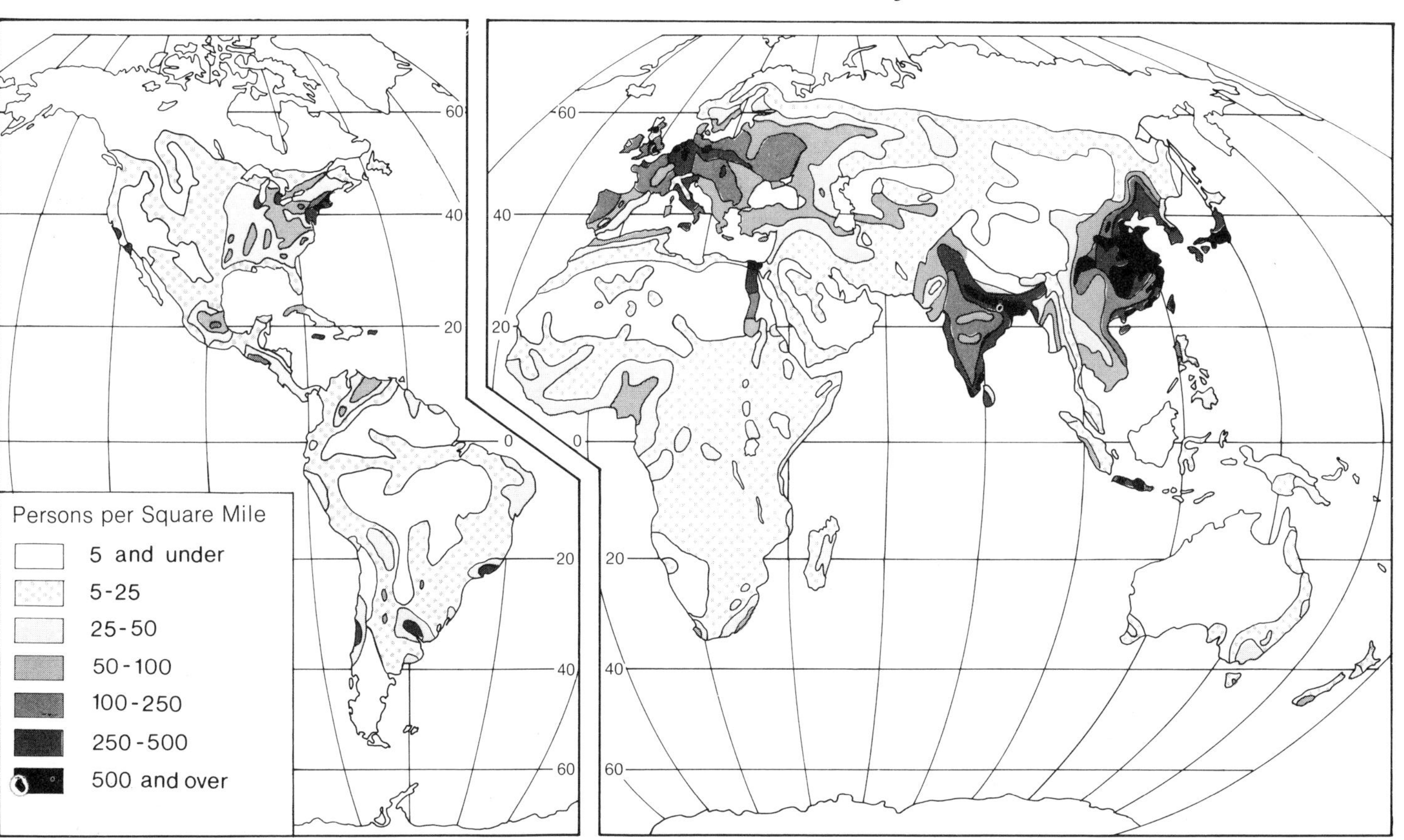

Figure 1–1 World population distribution.

CHARACTERISTICS OF HIGH DENSITY AREAS

We can see that the world's areas of greatest population density are in the Northern Hemisphere and generally in middle latitudes. All are near the sea and most are characterized by relatively low elevations. Because of the proximity to rivers and oceans they are called riparian civilizations. The concentration of high-density population is strikingly coincident with those parts of the earth that are most arable and that have the greatest capacity for grain-type agriculture, for example, corn, wheat, rice, and millet. In short, it would appear that the majority of the world's population is close to the food supply source.

The earliest civilizations were in somewhat similar environments; they were in the Northern Hemisphere in midlatitude areas, on major rivers, and astride corridors of good soil. The most fundamental difference is that the earlier civilizations were associated with exotic rivers, that is, rivers that flowed through desert (or foreign) environments. Such riparian civilizations include those in northern China along the Hwang Ho, in western India and Pakistan along the Indus, and most important of all, those in northern Africa and Southwest Asia in the areas of the Nile and the Tigris and Euphrates rivers. These were the first places where people became truly fixed on the earth, where they could break out of a nomadic or semisedentary lifestyle and settle down in one place.

These exotic streams provided people with their greatest invention, the city. Here was a year-round growing season, plenty of water and sunshine, an almost inexhaustible soil (periodically replenished by river flooding), a means of transporting foodstuffs to the city by the great rivers, and an easy means to control nature by means of canals for irrigation and transportation. Given the technology of the time, these areas were physically ideal. Today the Nile Valley, the lower Tigris-Euphrates, the Indus, and the Hwang Ho are still densely settled areas, but they are no longer the centres for the major populations of the earth.

The nodes of population concentration are economically diverse. In southern and eastern Asia, except Japan, there is a generally low-income, agriculturally-dependent, fast-growing population. In this area are some of the major zones of periodic starvation, the booming large cities with their perimeters of squatter squalor, and the less advanced technologies. For much of the population the economy is not appreciably above a subsistence level, whereas for others, a rarified elite minority, it is superior in lifestyle to the economies of Western Europe and the United States. This 'dual society', with its great poverty and squalor adjacent to great affluence and elegance, is one of the characteristics of this population node.

The other population nodes of the earth, although slightly smaller in population, are much more developed in urbanization, industry, agriculture, and trade. They contain the urban-industrial communities; here are the population nodes where almost 70 to 80 per cent of the population live in urban places. The northwest European and eastern North American population nodes form one great trading link. The North Atlantic route bears the greatest volume of ocean shipping in the world; manufacturing products, often of the same type, flow both ways. Here are the world's most affluent peoples, created largely by the industrial and transportation revolutions. These revolutions so mechanized agriculture that millions were freed, or forced, to leave the farms and seek the plentiful industrial jobs in the cities.

It is also the European and North American population nodes with which the rest of the world trades. Crops such as tea, coffee, bananas, and cacao come from tropical areas; summer harvests come from the Southern Hemisphere during Northern winter; the industrial giants reach out to far corner areas for mineral resources: petroleum, iron ore, copper, and many other items. The earth is being scoured to find the needed resources to keep the cities and factories viable.

Indeed, many of the minor population nodes throughout other areas of the world are merely outposts for gathering produce. Nowhere is the pattern more evident than in the coastal areas of South America, Africa, and Asia, with their many port cities. In some respects, this pattern is a remnant of European colonialism and mercantilism; in other respects, it is a fair reflection of the nature of population outposts in a world where the population is gathered in a few strategic areas.

CHARACTERISTICS OF LOW DENSITY AREAS

It is fairly easy to understand why many areas of the earth are scantily populated or generally uninhabited. Yet many of the truly severe places are occupied in outpost fashion by people attempting to gain particular resources or advantages peculiar to these areas (for example, strategic military installations, exploratory resource groups, and huddled bands of a native population). Mineral resources have long brought settlement to remote and difficult areas. The North Slope of Alaska is today invested with oil-related construction projects. A good part of the Northwest Territories of Canada is likewise being combed for potential resources. Desert areas, where they contain sizable mineral deposits, are also occupied for such purposes. Nonetheless, these are spotty and extremely minor nodes of usually temporary population. Other non-utilitarian parts of the earth may be but seasonally visited for recreational or scientific purposes.

As we examine the map of world population density further, we are struck by the sizable areas of the earth's surface that are occupied by less than five persons per square mile. These are the lands on the margins of the good earth. In some of them the population survives fairly well economically, but most such areas are characterized by marginal economies and a concomitant low standard of living. The predominant occupations are of a nomadic nature, for example, shifting cultivation, herding of animals over a wide area, or some system of hunting and gathering.

Such population distribution results in still other problems for peoples in different parts of the earth. There are problems peculiar to areas where there is much material wealth and problems of a different nature in those areas where there is little wealth. Such spatial differentiation results in three general types of difficulties: (1) those pertaining to economic characteristics of areas as a result of population inequality; (2) problems pertaining to ecology and environment; and (3) a broad assortment of human problems as the result of the population distributions themselves.

The nature of such occupancy-based features is reflected in all manner of public policy questions. For example, what should be done in regional development for 'underdeveloped' areas? How might the standard of living and the general welfare of people be raised, independent of raising the economic level of the areas themselves? How might environmental conditions relate to possibilities for encouraging population and/or economic growth in a particular area?

As might be anticipated, the major problems of people are most pronounced in the lagging or underdeveloped parts of the world—the Third World, as it is called. On the other hand, environmental problems are most concentrated in the highly urbanized and industrialized places. Thus some problems result from areas having too little economic activity in comparison with their population, and some problems stem from areas having perhaps too much economic activity. The first directly affects the population in its standard of living, the second affects the environment.

CAUSES OF REGIONAL INEQUALITY

The reasons for spatial variation of particular activities over the earth's surface will be presented later. The purpose here is to gain some appreciation for the general cause of differentiation of areas. The general concept of core, central, and outlying areas is most useful in this regard. It forms a framework that provides an explanation for regional differences, past, present and future. Moreover, the concept is operable on various territorial scales—at world, national, regional, and even local levels.

Simply put, the concept is based on the idea that for any given territory there is a prime region for carrying on most economic activities. In the economic life of the rising nation-states of Europe, certain 'core areas' became crucial (Figure 1–2). They were often the best agricultural lands and favoured by ruling dynasties who granted fairs and markets. Thereafter, they became lands of continuous increment. They benefited first from innovations to gain maximum urban and industrial development, major local markets and good access to all other places. Because of their importance and dominance, the core areas are at the focus of major transportation routes. The core area thus has first choice in the manufacture and marketing of most products. Other areas are primarily limited to a raw material supply operation whereby they feed the core area with needed commodities.

The classic historical case was the British Empire. A great many areas of the world were supplying Britain with raw materials for her industry and foodstuffs for her population. In Britain value was added to these commodities through manufacturing, after which some of the products were returned to other areas at much increased cost. The difference between the costs of buying and selling was used by Britain to supply herself with additional means of production, with increased quantities of raw materials and foodstuffs, and, in general, to enhance her economic development. The result was that the core area, Great Britain, got richer faster than the poorer colonial areas.

Somewhat similar conditions are found on a world scale today. For example, the core area of the world is Western Europe and the northeastern quadrangle of the United States. This region includes—technologically, industrially, and politically—the most advanced nations on earth. Most other areas of the world, with the exception of the Soviet bloc and Japan, supply this core area with raw materials needed to operate its huge urban, industrial, and military complex. In return the noncore areas receive manufactured goods and technical assistance. Thus while both areas benefit, the core area appears to be benefiting most. The gap in economic development and standard of living between the two areas has grown so appreciably that the core area nations are often referred to as the 'have' nations and the other nations of the world as the 'have not' countries.

Whether the urban-industrial core areas of the world will be able to maintain their historic economic superiority is highly debatable in light of the commodity cartels now being formed among resource-rich nations. The Organization of Petroleum Exporting Countries (OPEC) founded in 1960 is a classic case. The core areas are finding that they must pay the price asked for basic energy and manufacturing raw materials. Moreover, it is now evident that the advanced countries of the world are no longer able to dictate trading terms for the price of manufactured goods.

Core area superiority still operates within nations. For example, the core area of the British Isles, once to be defined as the agricultural-commercial lowlands of Southeast England centred on the Thames Valley, became enlarged in the industrial era to include great manufacturing cities from London to Lancashire. This core area today includes over 60 per cent of Great Britain's total population, industrial plant, commercial, tech-

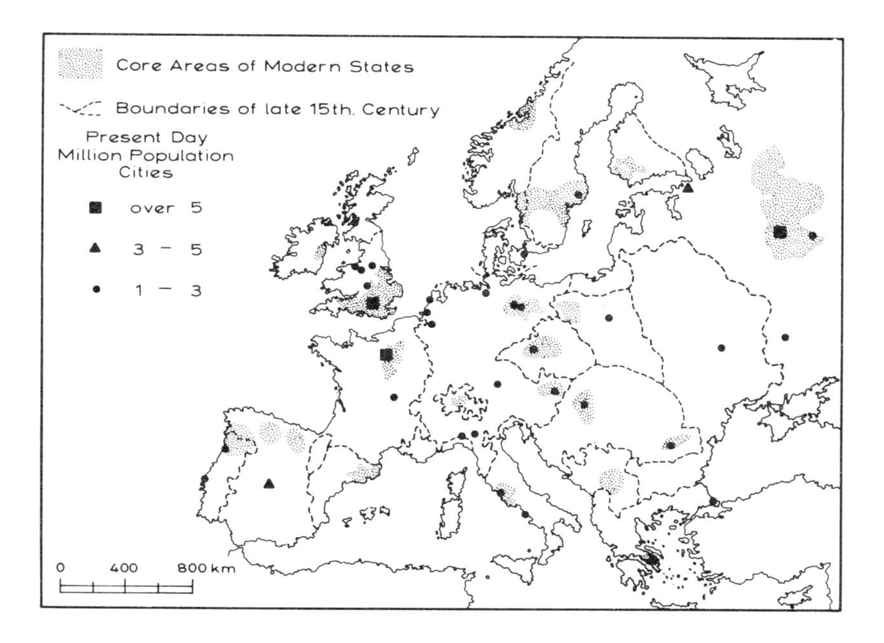

Figure 1–2 (a) Core areas in the historical context of the rise of European nation states (adapted from Pounds and Ball, 1964; see Bibliography). Many present-day 'millionaire' cities suggest the confirmation and extension of early core zones.

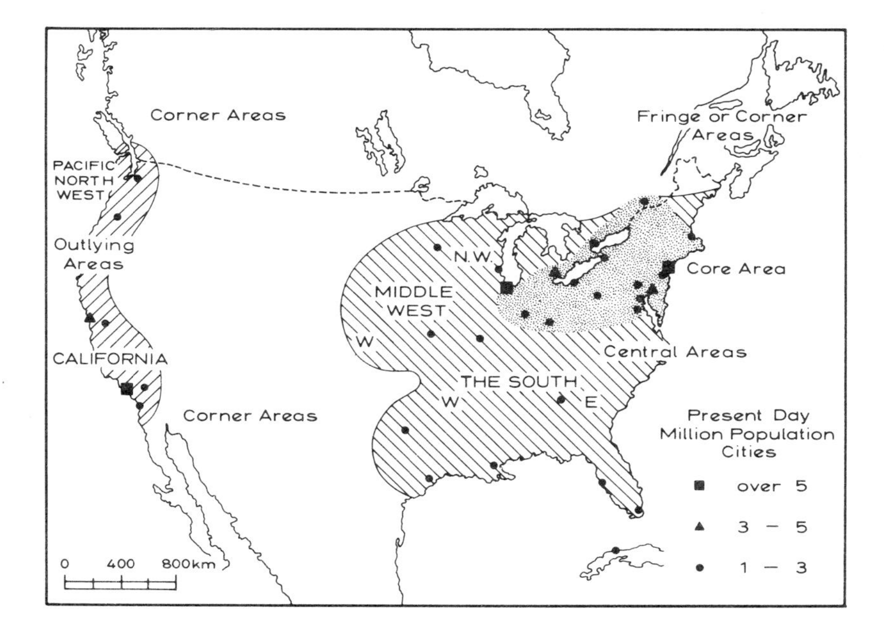

Figure 1–2 (b) Core, central and outlying areas of the contemporary North American economy. In addition, stranded areas (see text) may occur within either central or outlying areas. Present-day 'millionaire' cities also shown.

nical and scientific establishments, and seats of learning. The core area in the United States is also a great manufacturing belt. It is an area (Figure 1–2) at the focus of almost all national transportation routes—in fact, it arranges them that way. The core area of the United States includes less than 10 per cent of the country's total area but contains over one-half of the nation's population and much greater proportions of the nation's manufacturing employment. It is a crucible of brainpower and innovation, and the communications, financial, and administrative centre of the nation.

Why this has happened is the subject of much speculation. Perhaps it was due to an early start. Perhaps it was the result of having great quantities of such industrial raw materials as coal and iron ore fortuitously close at hand at the right time. Perhaps it was the result of having great agricultural resources. Perhaps it was based on early development of international trade. It is likely that all of these factors played an important, if unequal, part. As with other core areas in the world, the economic superiority of these areas is one not easy to change.

We can suggest, too, why other areas might have some difficulty competing with the core area. In the next best position are those areas adjacent to it called the central areas. These central areas are often in a fairly good position for developing their economic activity if they have natural resources—resources that the core area lacks. Barring that, the central areas do at least get second choice in the production and distribution of products and they are able to manufacture for a national market if their products are high in value compared with the costs of transportation. Manufacturing plants here are often characterized by having a high weight-loss ratio. Central areas also have the possibility of intercepting many flows from distant areas en route to the core.

The next tier of regions is called the outlying areas. Because these areas are farthest from the core, they are less able to compete for national markets. Most of the outlying areas are restricted to the production of (1) raw materials for the industries of core or central areas, (2) foodstuffs for the general population, or (3) resources that cannot be found in comparable quantity and quality closer to the core market.

Thus to compete effectively, outlying areas must have a very superior natural base. This generally consists of utilizing a very specialized resource like timber, minerals, soil, or even sunshine. The Pacific Northwest and California, as outlying areas in the United States, specialize heavily in supplying commodities not readily available closer to the market, for example, timber, agricultural products, certain minerals, and a climate suitable for retirement living and recreation. A somewhat similar situation may be observed within Western Europe, where the core areas of the United Kingdom, the Low Countries and West Germany are provided with agricultural commodities from Eire and Denmark, minerals from Spain and Poland, paper and timber from Sweden and Finland, while the sunshine fringe from Portugal to Greece provides for recreation. Australia and New Zealand, which are outlying areas on a world scale, specialize in commodities that cannot be supplied more cheaply, or in sufficient quantities, by countries between them and the European core.

Industrial production in outlying areas is generally geared to local markets. With such limited markets, activities here, in contrast to those found in core and central areas, can profit very little from large-scale production. In fact, many outlying areas have a population or market so small that profitable production cannot be undertaken. On the other hand, there are such exceptions as southern California and southern France (the Midi) where the concentration of population affords a regional market large enough to support the production of many locally consumed items.

In one sense, outlying areas are becoming better off as transportation improves and cuts down on the problems of transferability to the core area. On the other hand, the trend has been to improve methods of transporting raw materials versus finished products so that many of the raw materials are now moving directly to core and central areas rather than being processed in outlying areas.

Considerable distinction should be made among outlying areas. Some are so far removed from central areas that they are best described as fringe or corner areas. Their tendency, even within the most economically advanced nations, is to be left in the backwash of development. As examples, we may note the Western Hebrides in the United Kingdom and most of the Province of Newfoundland and Labrador in Canada.

At this stage some distinction should also be made between outlying areas and stranded areas. Although underdevelopment in outlying areas may occur for a number of reasons, the primary cause is their spatial position relative to other places and materials. Other places having better opportunities for growth thus substitute for the growth that might otherwise occur in outlying areas. In contrast, the stranded area may occur within either central or peripheral areas. Characteristically, it is an area that was formerly in tune with major development but was not able to keep up. This stranded condition may result from a multitude of factors including a depletion of resources or general inabilities to compete with other places having greater opportunities.

Thus it can be seen why regional disparity is the rule. Vast differences in economic advancement at almost every territorial scale are the result. On a world scale the advanced and the underdeveloped nations exemplify these differences. Within nations some regions are fairly well off while others lag or are underdeveloped by comparison. Such regional disparity is found within nations, both rich and poor.

Within the underdeveloped nations of the world, large cities often stand out in contrast to the backward rural areas around them. The disparity here occurs within a very short spatial distance. The growing and rapidly industrializing cities stand in stark contrast to the primitive economy and largely illiterate and unskilled people in the countryside. Such a dual society is especially evident in much of Southeast Asia, where cities like Bombay, Calcutta, Bangkok, Saigon, and Hanoi stand out as pillars of affluence and growth in a sea of rural backwash. (Of course, there are also many poor in such cities so that there is a tremendous local disparity within them.)

Regional disparity is also the rule in more economically advanced nations. Note the variation in economic activity between the northern and southern parts of Japan, northern and southern Italy, northern and southern Argentina, or between the north and the south in the United States. (In all but Japan the northern areas are the more highly industrialized and urbanized.)

A more careful examination of regional disparity, however, reveals that a simple, geographical dichotomy between growing and lagging areas is too crude a formulation. In the United Kingdom, for example, various studies report at least three different types of problem areas. Firstly, there are the sparsely populated hill and pasture regions of considerable extent but few local employment opportunities; central Wales and the Scottish Highlands and Islands are examples, both of which have experienced strong out-migration of people. Secondly, there are underdeveloped regions in various rural 'corners', like Southwest England and Southwest Scotland, distant from the major industrial areas. Thirdly, there are several depressed urban areas where basic industrial activities—like coal-mining, shipbuilding, cotton manufacture—are no longer the profitable mainstays of the regional economy.

Thus the reasons for disparity are multiple. In the United States, the Appalachians, from southern Pennsylvania to northern Florida, are suffering as a result of poor farmland and a technology that requires fewer workers in coal production. Other places, both in Western Europe and the United States, have found that their resources are no longer competitive with those available elsewhere. Still others find that because of transportation improvements or industrial and technological developments elsewhere, they have been left with little competitive potential. As might be expected, many of the farming areas are in difficulty simply because agriculture requires fewer people and fewer areas today than it did formerly.

PROBLEMS RELATED TO REGIONAL DISPARITY

If the distribution of people conformed with the economic opportunities of areas, few problems resulting from spatial inequality would exist. To the contrary, at least on a world scale, the areas with the fewest activities and the least hope for economic development also contain the greatest populations. Classic examples in this regard are India, China, and other countries in Southeast Asia. Unfortunately, even in economically backward areas with few people such limited population often has a low standard of living.

A hasty solution comes to mind: encourage migration from areas of inferior location to those with superior actual or potential development. This is not much practised, however, primarily for political reasons. Migration between countries is tightly controlled. Thus most peoples of the world are restricted to their own nations—whether or not they can earn a livelihood in them. To improve their economic status, their own area must be improved. Even within nations, the facilitation and encouragement of large-scale migration is rarely practised.

Migration is not the simple solution it appears to be at first glance for yet more reasons. Poor people in lagging areas may be socially strong and their cohesion sadly weakened by a selective migration which leaves behind the young and the aged. Poor people, too, are also found in advanced areas. These people are the less educated and the less skilled—characteristics that apply in abundance to most of the population in underdeveloped areas.

Thus the primary solution consists in efforts to improve the economic potential of lagging and depressed areas. As the concept of core, central, and outlying areas demonstrates, however, this is very difficult to achieve. In fact, despite major area development programmes within and between nations, the gap between the rich and the poor and the have and have-not areas has been widening rather than narrowing. Near starvation is a recurring fact of life in many of the underdeveloped countries of the world. Their agricultural resources are simply not adequate to support the population properly. It is therefore commonly—albeit superficially—suggested that the problem is overpopulation relative to local food resources.

Yet, as a whole, the world is capable of feeding its inhabitants. It has been argued, therefore, that the primary problem is one of equalization of wealth and distribution of foodstuffs. Because of the many political and institutional problems involved, however, this equalization has not occurred. Consequently, even in an affluent world, many people are without jobs, without food, and without hope. The fact is that a minority of people in the world have an overabundance of material wealth whereas most others have only a scanty minimum for survival. Moreover, the gap is growing, with those on the lower end of the economic scale becoming more and more discontented.

In this regard three types of countries are commonly differentiated: (1) the Western countries, (2) the Communist or Soviet-bloc countries, and (3) the Third World countries. The first category commonly includes the industrialized countries of Western Europe, the United States, Canada and such outliers of European colonization as Australia, New Zealand, and South Africa. (Japan, although highly industrialized and urbanized, cannot be considered 'Westernized'.) The second category, the Soviet-bloc countries, are usually technologically advanced but lack an individual standard of living comparable to that of the West European and North American nations. Finally, the Third World includes the countries of practically all other areas, particularly Asia, Africa, and South America. Incidentally, the South American nations might be considered still another bloc. They are true outliers of European colonization but are nonetheless primarily raw-material-oriented in terms of their level of economic development. Internal political problems and the fact that these countries have been used as a source of raw materials and as a field for investment by the Western countries have perhaps prevented them from attaining a higher economic level.

It is the people of the Third World countries who have been most affected by problems of underdevelopment. These countries suffer from a serious imbalance between a rapidly growing population and available resources, an imbalance that grows progressively more serious. Their technology is inferior. Their knowledge and use of their own resources are inadequate. Their transportation remains largely primitive. As might be anticipated, much of the population is plagued by problems of malnutrition, illiteracy, and lack of the skills needed for urban industrial purposes.

REGIONAL DEVELOPMENT DILEMMAS

Our primary concern with regard to the effects of unequal development of areas should, of course, be the people who are its victims. In practice, however, most of the emphasis, reflected in the programmes for change, has been placed on regional improvement rather than on human improvement. Regional economic improvement of the territory through industrialization appears to be a logical way, although perhaps not the best way, of improving the quality and standard of living for the population.

A better, people-based solution in some places might be improving agriculture, education and job training, and encouraging and aiding people to move out of areas that have little potential (or that are over-populated in terms of opportunities) and into more advantageous areas. Nonetheless, most government programmes are focused on improving employment within lagging areas. Very little consideration is given to direct improvement of the people if this involves moving them out of the area.

One reason that areas rather than people are emphasized is that the various countries of the world and regions within them, are vying with each other to improve the economic status of their territories. Every small country wants to improve its economic status. Specifically, most want to develop a major iron and steel industry. Obviously, this is unreasonable for many places in the world, not only because of the lack of a raw material base but also because of limited local markets. Nevertheless, many underdeveloped nations pursue such a deliberate industrialization policy in the hope that by this means they can more quickly reach a 'take-off' stage in economic development.

Similar strategies for improving regional growth within economically advancing nations also are much in evidence. The policy has been primarily to bring about more industrialization in the underdeveloped places. One strategy is to develop urban growth poles in these outlying areas. For example, noncoastal settlement in Australia was fostered by placing the national capital in Canberra. Brazil has attempted to develop much of its inland area by building its national capital at Brazilia. To date, the results have been mixed at best. In most of the world's outlying areas, however, it seems much more efficient to transport the resources available there to some other place. Oil from the North Slope of Alaska, for example, is being transported to refineries in the coterminous United States, metallic ores are shipped from West Africa to the smelters of North Sea Europe, coal and iron ore reach Japan from Western Canada and Australia. Thus little major or long-term employment is created in many outlying places. Similar kinds of events are evident in the large, mineral-supplying areas of Peru, Venezuela, Chile, and the Middle East.

ECOLOGICAL AND ENVIRONMENTAL PROBLEMS

Although they exist throughout the world and are usually associated with widespread extraction of resources in outlying areas, environmental and ecological problems are most acute in the heavily populated core areas because they affect greater numbers of people. Industrial activities have contaminated many riverine areas. In Canada, many lakes in Ontario are polluted with the waste products of mining; in Europe, the River Rhine has been described as an 'open sewer' and parts of the Mediterranean Sea are declared unsafe for bathing.

In the United States, Lake Erie is badly polluted as are rivers like the Hudson, the Ohio, the Illinois, and many others. As in the core areas of other industrial nations, industrial effluent, sewage and run-off from agricultural land on which chemical pesticides and fertilizers have been used are most responsible. Air pollution, although primarily associated with large cities, blankets wider areas downwind of heavy industrial activity. In the United Kingdom, the old plague of urban smog has almost disappeared with the introduction of strictly-enforced smoke-control zones; yet the industrial 'haze' remains, and far away across the North Sea, the Scandinavian countries complain that their air is polluted by the smoke of British industry. In Japan, where the bulk of recent industrial growth has been oil-based, excessive densities of sulphur dioxides and petrochemical smog are the most widely occurring types of air pollution. Should the present rate of pollution growth continue, the environment for living will continue to deteriorate dramatically.

In other areas different kinds of environmental problems occur. The classic case was the erosion problem in the South during the heavy period of cotton growing. The dust bowl in Oklahoma is another tribute to man's attempt to exploit the earth at all costs. Great sores on the surface of the earth exist where people have almost wantonly exploited the mineral resources. This is particularly evidenced in Appalachia, in northern Minnesota, in Bingham, Utah, and in other areas where strip mining is widely practised. In major oil-producing regions, the area is blanketed with oil wells and other paraphernalia, which disfigure it and deny other uses.

Another example is the destruction of considerable forest areas in the United States. In northern Michigan and Minnesota forests were cut out in such a way as to render the area largely unusable for any other purpose. Even the redwood forests of northern California are quickly falling under the modern chainsaw. Such resources are almost irreplaceable, given the time it takes to grow new redwood trees. But vast areas of forest exploitation occur throughout the North American West. Although transplanting is now becoming more common, the destruction of the ecological environment is severe. Not only are many areas ruined for recreational and other purposes, but flooding, the destruction of wildlife, and other harmful consequences commonly occur.

The role of people in changing the face of the earth has in many cases been disastrous. Generally, little attention has been given to the future longevity of areas. Conservation has been much talked about but very little practised. To say that many areas of the world have become ugly and dilapidated through man's intervention is an understatement.

The classic endeavour of human beings has been to change the natural environment. However, their actions have treated the environment more as an enemy than as an integral part of their existence. Fortunately, we are entering a new age of consciousness and concern about how people occupy and utilize the land. Much more attention is being given to questions concerning proper and wise use of resources. (In fact, a number of prime resources will soon be depleted unless more careful attention is paid to the way in which they are used.) The big consumers, as might be expected, are the core areas of the world. The central and outlying areas are primarily engaged in feeding this massive giant. However, some nations have begun to rebel and to suggest that it is not to their benefit, regardless of price paid and profits gained, to deplete their resource base. Others, however, look upon the sale of their resources (a case in point is Venezuela and its petroleum) as a proper way to get funds to finance other aspects of their regional economy.

BIBLIOGRAPHY

Bernarde, M. A. *Our Precarious Habitat*. New York: W. W. Norton, 1970.
Berry, B. J. L. *Strategies, Models, and Economic Theories of Development in Rural Regions*. Agricultural Economic Report no. 127. Washington, D.C.: U.S. Department of Agriculture, 1967.
Boyce, R. R. 'Man the freak-seeker: reflections on a visit to Mount Rainier.' *Landscape*, Autumn 1976, p. 49.
Chisholm, M., and Manners, G. *Spatial Policy Problems of the British Economy*. Cambridge University Press, 1971.
Cipolla, C. *The Economic History of World Population*. London: Penguin, 1970.
Clayton, K. M., ed. *Pollution Abatement*. Newton Abbot: David & Charles, 1975.
Demko, G. J., ed. *Population Geography: A Reader*. New York: McGraw-Hill, 1970.
Detwyler, T. R., and Marcus, M. G. *Urbanization and Environment*. Belmont, Calif.: Duxbury Press, 1972.
Estall, R. C. *New England: A Study in Industrial Adjustment*. Praeger, 1966.
Grigg, D. *The Harsh Lands: A Study in Agricultural Development*. Macmillan, 1970.
Holdren, J. P., and Ehrlich, P. R., eds. *Global Ecology*. New York: Harcourt Brace Jovanovich, 1971.
James, P. E. *All Possible Worlds: A History of Geographical Ideas*. New York: Bobbs-Merrill, 1972.
Landsberg, H. H., Fischman, L. L., and Fisher, J. L. *Resources in America's Future: Patterns of Requirements and Availabilities, 1960–2000*. Baltimore: Johns Hopkins Press, 1963.
Leighton, P. A. 'Geographical aspects of air pollution.' *Geographical Review*, April 1966, pp. 151–174.
Perloff, H. S., Dunn, E. S., Jr, Lampard, E. E., and Muth, R. F. *Regions, Resources, and Economic Growth*. Baltimore: Johns Hopkins Press, 1960.
Phelps, E. S., ed. *The Goal of Economic Growth*. New York: W. W. Norton, 1960.
Pounds, N. J. G., and Ball, S. S. 'Core-areas and the development of the European States System'. *Annals of the Association of American Geographers* 54 (1964), pp. 24–40.
Pyle, G. F., ed. 'Human health problems: spatial perspectives.' *Economic Geography*, April 1976, pp. 95–102.
Rostow, W. W. *The Stages of Economic Growth*. Cambridge University Press, 1963.
Thomas, W. L., ed. *Man's Role in Changing the Face of the Earth*. Chicago: University of Chicago Press, 1956.
Trewartha, G. T., ed. *The More Developed Realm: A Geography of its Population*. Oxford: Pergamon Geographies, 1978.
Ullman, E. L. 'Amenities as a factor in regional growth.' *Geographical Review*, January 1954, pp. 119–132.
Zelinsky, W. *A Prologue to Population Geography*. Englewood Cliffs, N.J.: Prentice-Hall, 1966.
Zelinsky, W., Kosinski, L. A., and Prothero, R. M. *Geography and a Crowding World: A Symposium on Population Pressures Upon Physical and Social Resources in Developing Lands*. Oxford: Oxford University Press, 1970.

2

The Nature of Economic Geography

Geography is the discipline that attempts to explain the uneven locations, distributions, extents, and uses of selected things over the surface of the earth. Economic geography is concerned with explaining the locational characteristics of things of utilitarian value to people—particularly things people consider valuable enough to move from one part of the earth to another. By focusing on the spatial dimensions of items of economic value, partial explanation is provided for the distribution of people themselves.

Sometimes, economic geography is wrongly described as simply a fusion of economics and geography. But to regard it as such would be no more true than to view political geography as a combination of political science and geography or social geography as a combination of sociology and geography. It has also been stated, more truly, that economic geography needs more economics; however, the two definitions offered above will suggest very firmly that economic geography is strongly rooted in the field of academic geography. As such, it has a strong core bias and focus. That bias is spatial: what concerns geographers are the locational elements of things.

Thus the economic geographer is marked out among behavioural scientists by his preoccupation with the ways in which economic activities are spatially arranged on or near the surface of the earth. He is notably concerned not only with 'where?' and 'how where?' but also with 'why where?' This is a way of suggesting that he is not only concerned with spatial arrangements but also with the processes which lead to the spatial arrangements.

In the course of succeeding chapters, it will be appreciated that the processes may be highly complex, so that in his quest for understanding the economic geographer must interact with economists, sociologists and with specialists in many other areas. Indeed, advances in many fields of study may inform him and it frequently happens that his explanations for the spatial positioning of things must necessarily come from a wide assortment of inputs not always obvious or clearly defined. For economic geography, however, these most often include physical, cultural, historical, institutional, and certainly not least, economic factors. These critical inputs, considered in Chapter 3, are called the bases of economic geography.

If man and his activities were distributed evenly over the earth, there would perhaps be little interest in economic geography. But in Chapter 1 we have already drawn attention to the uneven distribution of people and their activities. The major premise in economic geography, as in all geography, is that despite an uneven location and pattern, people and their activities are distributed in an orderly and therefore understandable manner. However, such underlying orderliness is usually neither so simple that it can be explained by a single factor such as environment, nor so complicated that it defies explanation except through approximation, probability, and randomization procedures. Those who argue that single factors offer satisfactory explanation assume far too much spatial orderliness and give people too little response choice. Those who argue that the distribution of things can be approached only through randomization and probability assume far too little spatial orderliness. The truth lies somewhere in between and, as such, involves an interplay among a number of variables.

If we accept the premise that people and their activities within the economic system are distributed in a logical and orderly manner, then such distributions can be understood through the use of concepts, theories, and general spatial principles. It is not necessary, therefore, to spend a great amount of time learning the distributions of thousands of different items. Instead, through various conceptual principles and insights, the causes for a particular pattern can be readily understood.

Thus a careful blend of factual and conceptual information is necessary for geographical understanding. Theory without facts, or with only a few facts calculated to verify the theory, often leads to blind acceptance of presumed spatial arrangements—arrangements that may exist only in the mind of the theoretician. A person who relies solely on theory for geographical understanding actually knows nothing for certain about the real earth. On the other hand, a person who relies solely on facts for geographical understanding probably knows little of meaningful or lasting value. Indeed, without a conceptual base, the 'facts only' advocate will not be able to differentiate the general from the unique and, unless constantly renewed with new facts, will soon find himself hopelessly out of date and irrelevant.

There are also certain inherent dangers in attempting to achieve a proper blend of theory and facts, which is one of the goals of this book. The factually based and theoretically based approaches are more formally labelled the inductive and the deductive. In the inductive approach one takes facts and then tries to 'make sense' of them. In the second approach one tries to deduce, based on certain logical premises or theories, what the spatial effect will be. Both methods are sound when properly used and properly tested—one against theory, the other against facts.

A common pitfall of the factually-based inductive approach involves the correlation of one spatial distribution with another. For example, wheat yield might be correlated spatially with the degree of field slope in a particular area. Such spatial covariation is often valuable in providing insights and in removing incorrect assumptions. Nonetheless, such spatial covariation does not necessarily provide explanation in any causative way. Such a procedure often gives the deceptive appearance of having explained something when it merely showed that it relates to other similar patterns—patterns that may have no direct causal relationships.

An equally serious fallacy in the pursuit of a proper blend of fact and theory is to use only selected facts to 'verify' a particular theory. In geography a misapplication often results when a particular area, in which things operate much as the theory predicts, is purposely selected. The tendency here is to seek out only those facts and those areas that coincide with the theory in question. By so doing, one fails to see how well, or how poorly, the theory works to provide any general explanation. As a fair test of the theory, data from a number of areas should be examined.

Aside from a judicious selection of facts and theories, this book will attempt to build spatial knowledge through the use of the ceteris paribus, or 'other things being equal', approach. Often it will be stated that given certain postulates and premises, such and such should follow, other things held constant. Such use of ceteris paribus is extremely common in economics, and its use here will follow in that tradition. In fact, most statements implicitly contain other-things-being-equal arguments. For example, one might say that a structure should have a physical life of, say, 50 years if fire, earthquake, or other forces do not destroy it. This makes the problem of verifying or denying theoretical statements by facts extremely complex and difficult. The tests required for verification are often quite complex so that only a few will be demonstrated here. Therefore some faith, based on logical (deductive) proof, will be required.

Nonetheless, the most important principle used here is common sense—an uncommon commodity—applied in a spatial, or geographical, context. It is hoped that, once explained, the reasons behind any particular spatial response will become fairly clear, and in some cases, even seemingly obvious. Most of the material here, although unfamiliar to the average person, is not complicated and does seem rather simple and obvious once it is discussed. This is truly the nature of most discovery and surely an attribute of common sense.

Geography is sometimes defined in a tongue-in-cheek manner as consisting of what geographers do. By the same token, economic geography might be defined as consisting of what is included in the economic geography texts. This may not provide a very satisfactory definition, but it will surely reveal a great deal of the philosophy of what economic geography has been and ought to be. If anything is abundantly clear, it is that the content of economic geography texts has changed drastically in the past several decades.

The first major college-level economic geography text to win favour in the English-speaking world was written by J. Russell Smith in 1913 for Henry Holt & Company, the forerunner of the publisher of this work. In terms of organization and approach, Smith's book was largely limited to those activities directly associated with the physical features of the earth. Discussion was largely on farming, fishing, forestry, mining, and general trade patterns.

During the 1930s and 1940s manufacturing was gradually added to economic geography texts. The inclusion of these secondary activities greatly strengthened the coverage of and gave greater coherence to a body of knowledge. Still the general focus was on particular food crops, materials for industry, and the manufacturing of basic industrial materials such as iron and steel. Transportation was also included, particularly international movements of commodities and minerals. Thus these economic geography texts emphasized the locational characteristics of critical components of production, particularly those types that are highly correlated with certain physical characteristics (for example, climates, soils, natural vegetation, geology) of the earth. Consequently, by today's standards these texts gave only passing attention to economic principles affecting the location of things.

It was also during the pre-World War II period that a general theory of economic geography was developed. In 1950, Clarence Jones summed up this 'new' viewpoint of economic geography as follows:

> '*Not all types of work are included in the field of economic geography. Many people—doctors, teachers, ministers, politicians, bankers, writers, musicians, etc.—obtain their living through other types of work. Economic geography deals with the productive occupations. . . . The statement that the classes of goods—food, clothes, shelter, fuel, tools, and materials of industry, and luxuries —have physical bases, called the factors of the natural environment, is axiomatic*'. (Clarence Fielden Jones, *Economic Geography* (New York: Macmillan, 1950), p. 7).

By this he meant that only economic activities with direct physical connections had a legitimate place in economic geography. Many types of manufacturing were excluded, along with people in cities, their location and activities.

For the past twenty years or so, these and many other activities have been usually included in standard textbooks on economic geography. And in about the same period of time, the geographer's approaches to an enlarged subject-matter have come closer to those of the economist. Why this did not happen sooner was suggested by the British economic geographer Michael Chisholm:

> '*Two considerations seem to have been dominant. The first is the lack of attention given by economists to spatial matters, largely because problems of spatial organisation have not until recently seemed to be as important as other issues. This has been partly because of the history of the subject and partly a result of the complexity of the problems of price, value, rewards to factors of production, etc., in which economists have been mainly interested. The second important reason for the estrangement of the two subjects lies in the character of geography as a discipline. With its roots in exploration, geography tended to emphasize the role of the physical environment in conditioning the geographical distribution of phenomena. With this emphasis has gone the habit of taking each phenomenon in turn and relating it to the physical environment, whereby the mutual relationships between the phenomena have been underemphasized.*' (Michael Chisholm. *Geography and Economics* (London: Bell, 1966), pp. 23–4).

Today 'the mutual relationships between the phenomena' have become central for many economic geographers. Also, it might be said that the distribution of anything that has utilitarian value is considered a legitimate candidate for economic-geographic inquiry. This would include anything that men judge to be worth buying, begging, borrowing, selling, bartering, stealing, or moving. Such a stance also means that there are more topics than can possibly be treated in any one economic geography text. But from earlier remarks it will have become clear why comprehensiveness of topic and detail can be subordinated in favour of greater attention to the ways in which ideas and information are handled. To underline this point we may draw more formal attention to the geographer's need for theories, models and systems.

In paragraphs above, 'theories' were contrasted with 'facts'. We are exposed to many kinds of theory in our everyday lives—to political theories, for example, within which we may 'envelope' our particular ideas on how we should govern the state or our local community. Models, too, either held in the hand or in the mind, are part of our learning process. In Science, theories are very carefully erected or formulated. Here, we may think of them as structures within which our information and observations are organized and integrated. We may think of models as any construction which reduces the complexity of the real world to a meaningful size for experiment. They may be built physically or mentally, and expressed in words, mathematical symbols, graphs or maps; but the important thing about them is that they can be tested. The 'other things being equal' approach, already noted, plays a vital part in modelling. It enables us to hold some features or elements constant while we examine the behaviour, under differing conditions or circumstances, of some other variable which we perceive to be critical.

Many relatively simple spatial models are introduced in the course of this book. Some, like those which relate to agricultural location (Chapter 9) and industrial location (Chapter 11) are deductive models which begin with a set of assumptions and develop predictions about behaviour. Others, like the gravity model (Chapter 6) are built up from the observation of real world events and from simplifying generalizations; that is, they are inductive models. Some models combine both deductive and inductive approaches. It is important to observe that no model can exactly represent reality. But insofar as they enable us to gain insights into otherwise hidden operational processes, they may help us to frame the most logical questions at the frontiers of our knowledge about spatial patterns in the real world.

To theories and models we may add 'systems'. In common use, we may refer, for example, to the 'school system', or the 'railway system'. In both of these we may envisage not only concrete objects (like classrooms and railway stations) but also many visible and invisible links necessary to make them function. To further the point, to be effective a 'library system' must link people and books with each other in many ways, including purchase, classification, indexing, storage, retrieval, borrowing facilities and procedures, and much more. In science, engineering and industrial organizations, a systems approach is commonly adopted in order to unravel complicated relationships and procedures. It is an approach which stresses not only a set of objects and their attributes but also the dynamic relationships between them. The economic geographer's 'objects' are all those activities, like fishing, mining, manufacturing, wholesaling and shopkeeping, that have a role to play in the operation of the economy. The relationships between these activities—the ties by which they are connected—are the flows of materials, products, information and people.

So here we are concerned most of all with the economic system, its major sections—which we shall call activity components—and their working relationships with each other.

THE ACTIVITY COMPONENTS

Six major sectors, or activity components, can be recognized in any study of economic geography. These components include the (1) procurement, or primary sector, (2) production, or secondary sector, (3) marketing, or tertiary sector, (4) servicing or quaternary sector, (5) consumption, and (6) transportation, a connecting link among them. (Figure 2–1). On the other hand, many economic geographers include both goods and services as part of the tertiary sector. This third sector is commonly labelled the service sector. Others separate the exchange component (retailing) from the service component (here called the quaternary sector). The quaternary sector, when used by others, almost always includes the office type functions and white collar workers. Some authors

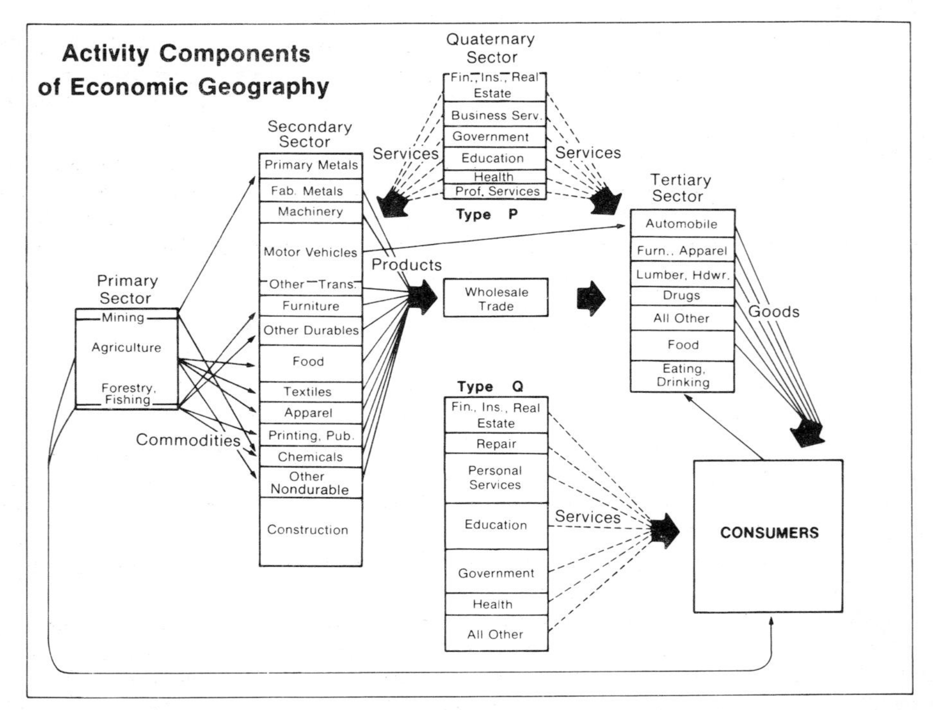

Figure 2–1 Activity components of economic geography. The height of the various columns represents the proportionate numbers of workers in each sector. Commodities produced in the primary sector are moved through the secondary and tertiary sectors to the consumer. The quaternary sector is divided to differentiate schematically those services which aid the flow of material items from those services which are primarily consumer-oriented.

have adopted a much more restrictive range of functions in the quaternary sector than is here used. Types of services commonly associated with business centres, for example, are labelled as being in the tertiary sector.

The Primary Sector

In the primary production sector, commodities are obtained in natural form from the earth's land, seas, mines, forests, and farms. Such resources are taken directly as found, or as grown on the earth in collaboration with human beings.

The major activities in this category include (1) primitive hunting and gathering of wild plants, animals, and fish, and gathering of certain minerals, (2) commercial exploitation of such products, (3) augmentation of nature's bounty through farming methods, (4) 'private' sport use of such items. In commercial economies the first three include fishing, forestry, mining, and agriculture. The recreation activity of hunting and fishing, however, is another major activity which utilizes such extractive procedures.

Primary production often provides the first link in a vast chain of production before these commodities are consumed as goods in the market-place. Usually it is only in very primitive societies that such goods are consumed without going through some of the other major activity sectors.

In the United States only a small percentage of the total labour force is employed in the primary activities (actually less than 10 per cent today). Nevertheless, the primary sector might be thought of as a basic catalyst for the addition of employment in other sectors. Paradoxically, the smaller the proportion of workers in the primary sector, the more advanced the economy. How many of the nations of Europe have 'moved up' in this regard can be illustrated by the case of Sweden. This country ended the nineteenth century with over 77 per cent of its population living by mining, forestry and agriculture, reducing to 40 per cent in 1940 and 15 per cent in 1977.

Until the 1970s it was generally assumed that the primary sector was simply a response sector to the consumer demands as reflected in the manufacturing arena. Thus the primary sector was viewed as primary simply because it was first in the chain of material movement from the state found in nature to the moulded form desired by people. The concern over an increasing population when coupled with finite food resources, the environmental and energy crises, and the growing organization of the primary producers has caused this sector to become primary in the classic meaning of the word. It is now becoming appreciated that the finite nature of the primary sector (food, energy, and the materials for industry) is the critical control for the further development of material standards of living.

The Secondary Sector

In the secondary production sector, materials produced in the primary sector (commodities) are changed in form through the process of manufacturing (products). Thus manufacturing is the intermediate step between initial acquisition of materials from nature and their consumption by people in the retailing or tertiary sector. These are more appropriately called consumer products as contrasted with producer products (that is, products used in the production process).

The manufacturing sector is a very complex one. Only rarely does material move in a one-step process from the primary to the tertiary sector. In most cases a series of steps within the secondary sector is involved (see Figure 2–1). One firm's 'finished product' often becomes another's raw materials. These are commonly labelled secondary transfer products.

The major locational forces for secondary activities are of three types: (1) the need to be near the source of raw material, that is, near the primary sector; (2) the need to be near the market, that is, near the tertiary sector; (3) the need to be near other things that are critical in the production process, such as labour, power, transportation, and so forth. Thus it is evident that the choices available for the location of manufacturing are more numerous than for activities in the primary sector.

The Tertiary Sector

There is much confusion among economic geographers as to what constitutes the tertiary sector. Historically, it was simply added as a third sector to complement the primary and secondary sectors; it included all other activities, but particularly those associated with cities. Over time the tertiary sector has been referred to as the 'service' sector. Consequently some present-day authors take this literally and exclude from the tertiary sector retailers, wholesalers and handlers of goods. Others, however, have made a distinction among kinds of services and have labelled most of the office, governmental, educational, and health services as the quaternary sector and have left business services such as barber shops and lawyers in the tertiary sector. Still others have limited retailing and wholesaling to an exchange category—yet exchange is more a facet of transportation than anything else.

It is therefore appropriate to define the term in a way that is most meaningful in the light of the flow of material from the primary sector to the consumption sector. Clearly, if manufacturing is the secondary sector and produces goods to be shipped to whole saling and retailing establishments, the latter should be properly labelled the third stage, or tertiary sector. Business establishments dealing not in products but in services should not be labelled as tertiary. Thus tertiary activities are simply the third step in a triad of sectors ranging from the first stage (primary), where raw commodities are taken, to the third stage (tertiary), where products are displayed for purchase by the final customer.

The Quaternary Sector

As here defined, the quaternary sector includes the provision of all types of services to the productive sectors (primary, secondary, and tertiary) and to the consumer sector. These services are logically divided between those concerned with the extraction, movement, manufacturing, and selling of material items (type P) and those provided to the consumer (type Q) (see Figure 2–1). The former are labelled producer services and the latter consumer or personal services. In practice, however, this two-fold division may be difficult for the reason that most such quaternary activities have dual purposes within the same establishment. For example, a legal firm may serve many kinds of big business, manufacturing or otherwise, as well as many private consumers. It is even more difficult to make a hard-and-fast separation between such services as communications, education, public administration, finance, and insurance.

Quaternary activities often have locational variables similar to those of tertiary activities and are often found in conjunction with tertiary activities. For example, in a typical 'retail business district' there might be a post office, a telephone exchange, a lawyer, a doctor, a housing office, a barber and beauty shop, an insurance company office, several financial institutions, plus various 'business repair' services—all quaternary activities. None of these deals directly in products or goods. All perform services, either for businesses or for consumers.

All together tertiary and quaternary uses cover no more than one-hundredth of one per cent of the land surface on the earth. On the other hand, the territory used for primary activities alone covers more than one-tenth of the earth's land surface. Tertiary and quaternary activities are highly market-oriented, but they are not necessarily simple to understand. Paradoxically, their locational choice is so restricted that there is considerable competition for sites among the various commercial activities. Within this limited territory, various firms bid for sites, and presumably those best able to pay the highest rents receive the best sites. But this is an oversimplification of the problem and understanding must await detailed treatment later.

The Consumer Sector

The consumer sector includes the total population: workers in all sectors, all their dependents, and unemployed persons. Basically, this is a sector to which goods and services are supplied. In the western world's 'market economy' the consumer sector is the initiating demand sector for the many activities distributed over the earth. By contrast, the primary sector is the initial response, or supply, sector. Of course, the nature of the response will be influenced by transportation conditions, by the various governmental features that differentiate one economic system from another, as well as by many other factors, such as national trade agreements.

This use of land for residential purposes, while utilizing less than one per cent of the land surface, is the home of the consumer, the ultimate user of material and service products, who is clearly a key to the spatial understanding of the other activities.

We might quickly argue that people live near their work and hence merely follow their work. This may be true, but they only live near their work in most cases. Moreover, their choice of residence has direct impact on the location of the tertiary and quaternary employment sectors—sectors that account for the majority of the employed population in postindustrialized countries such as the United States, Great Britain, France or West Germany.

The Transportation Sector

The transportation component consists of several major linkage types (Figure 2–1). First it supplies a series of material flow links among the primary, secondary, tertiary, and consumer sectors. In order to distinguish among the types of materials moved among these sectors, different terminology is usually employed, namely, commodities, products, and goods. The material items physically transferred from the primary to the secondary sector will be called commodities; those transferred from the secondary to the tertiary sector will be called products; and those transferred from the tertiary sector to the consumer will be called goods.

These increase in value as they proceed through each sector not only because of the cost inputs like transportation, storage, and labour but because things are worth more in some areas than in others. As things are conveyed along the road to the consumer, they often change in form (manufacturing) and in ownership (exchange). Thus transportation is the agency whereby value is increased along the path from point of initial production to final consumption.

There is also the flow of people who travel from their residences or places of work to purchase goods in the tertiary sector. The dashed lines in Figure 2–1 depict the flow between the consumer and eating and drinking establishments where the good is consumed on site. Thus the dashed lines represent people. The solid lines, however, represent material items—goods—transported by the consumer himself from the tertiary sector. Connections from the quaternary sector to other sectors are also mostly 'people flows'.

The term transportation pertains only to the physical flows of commodities, products, goods and people. But the services of the quaternary sector, of course, also include the flows of communication and information (not distinguished in the diagram). There is also, importantly, the flow of ideas, energy, money, credit and know-how, which are considered under the broader topics of communication and circulation.

It is informative to rank the sectors of economic activity according to their proximity to the sector of final consumption. In the case of primitive economies and sports-recreation, production and consumption are direct and often immediate. For most other activities, however, there is a decreasing series of steps between initial production and final consumption. In the case of fishing there are three intermediate steps before the fish reaches the final consumer: processing, wholesaling, and retailing. In mining too, many processing and manufacturing activities take place between the initial point of production and final consumption.

Within this connectivity system, the question might be raised as to which is the critical, or catalytic, sector in the flow. Although this is difficult to determine, it is clear that the final consumption sector, which triggers the demand through which all other activities directly or indirectly respond, is the prime sector. People are clearly the key to locational understanding. All goods and services are ultimately aimed at them, albeit in a complex web. In terms of the consumer, residential land use is the only sector connected locationally with all the sectors.

In the analysis of any particular location problem, however, a complex feedback situation occurs. One approach used in this book will be from the vantage point of these linkages. Such linkages among the procurement, production, service, exchange, and consumption sectors provide a key element of spatial understanding.

THE LOCATIONAL PROBLEM BY SECTOR

The degree of complexity in the locational problem varies greatly among the various sectors of economic activity. In general, the degree of locational freedom (and hence the difficulty of spatial understanding) increases as one proceeds from the primary through the other sectors. Thus it will be noted that there is only one freedom of locational choice for extractive activities (at the resource site), two degrees of choice in agriculture (because of slippage between land-based production costs and accessibility to market), three degrees of choice in the secondary sector (raw material, market, or some

Table 2–1 General degree of complexity in the locational problem.

Type of activity	One location fixed or given	Major locational component variables[a]							Rank of locational complexity	Degrees of locational choice
		(a)	(b)	(c)	(d)	(e)	(f)	(g)		
Primitive economies	X	X	X					–	A	1
Primary: extractive	X	X	X	X				–	A	1
Primary: agriculture		X	X	X				–	B	2
Secondary (manufacturing)		X	X	X	X			–	C	3
Tertiary (retailing)		X	X	X	X	X	X	–	D	Many
Quaternary (services)		X	X	X	–	X	X	X	D	Many
Consumer sector		X	X	X	–	X	X	X	D	Many

[a] Locational component variables are: (a) accessibility, (b) site quality, (c) regulation and control, (d) on-site procurement costs, (e) site competition for strategic sites on a microscale, (f) site affinities and disaffinities among activity types, and (g) a host of additional variables not otherwise identified. X indicates a locational component of major importance. The dash indicates other variables not otherwise specified.

intermediate location), and a great number of locational choices among the other sectors (Table 2–1).

The number of major components that provide general explanation for the location of each activity also generally increases as choices in location become more numerous. Note that for primitive, noncommercial economies and the extractive activities, the location of the site of production is given—for production to occur at all, it must be at the location of the resource in question. There are choices among resource sites, however, based on quality of the resource and accessibility to points of consumption. Moreover, the primary activities often have a regulatory variable imposed. The secondary activities are even more complicated in locational understanding because of additional variables pertaining to raw material procurement costs. Finally, the tertiary and quaternary sectors are most complicated of all inasmuch as they are affected in location not only by most of the variables of the other sectors, but also by intensive competition for particular sites within cities.

THE REALM OF ECONOMIC GEOGRAPHY

The realm of economic geography consists of four activity sectors, plus the services and transportation sectors. As we proceed, we will attempt to understand the spatial dimensions of each. In one sense, of course, understanding them fully involves tracing present

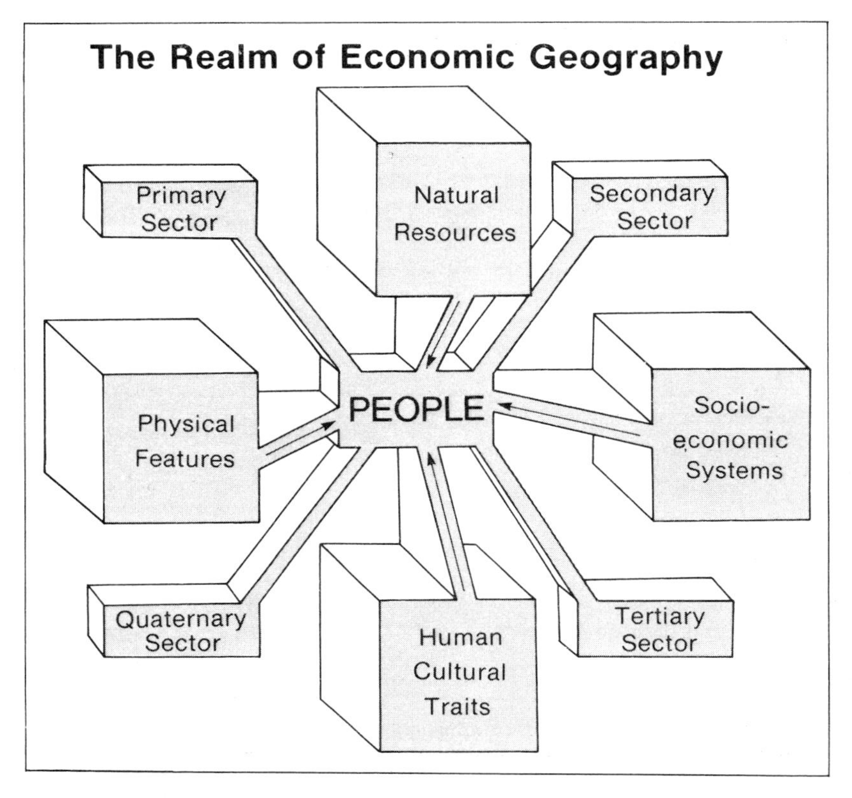

Figure 2–2 The realm of economic geography. Economic geography is arbitrarily differentiated from connecting influences.

patterns back to the dawn of time and space. But this approach would take us out of the arena of economic geography and into the wide expanse of general philosophy. Thus we shall be content to start and to finish on a closer and firmer standing base—a base that will be accepted as given (Figure 2–2).

To be more specific, four major outside influences will be taken as they are found, with no attempt on our part at explanation—unless such a pursuit provides some direct and necessary light, otherwise unavailable, on the activity under investigation. These four givens are: (1) the physical features of the earth; (2) the natural resources of the earth; (3) the human and cultural-trait features and patterns as they vary over space; and (4) the respective socio-economic and institutional systems among different peoples over the earth. Explanation of these outside influences belongs more properly to disciplines other than economic geography—to physical geography, geology, anthropology, psychology, sociology, economics, and many others.

Thus our purpose will be to employ existing factors for the locational explanation and understanding of still other factors, specifically those comprising the various sectors of economic geography. And from this analytical thrust we shall examine other factors that will provide even greater understanding of the location of people and their activities on earth. We will focus on the general types of locational patterns and problems that are encountered as one progresses from the primary to the demand, or consumer, sector. Part Two lays much of the groundwork by which the subjects later discussed can be better understood. These so-called bases include physical features, economic location principles, institutional activities, and the connecting links of transportation.

BIBLIOGRAPHY

Selected texts in economic geography.

Most of the following contain materials that will be of use throughout this book.

Abler, R., Adams, J.S., and Gould, P. *Spatial Organization*. Englewood Cliffs, New Jersey: Prentice-Hall, 1971.

Alexander, J. *Economic Geography*. Englewood Cliffs, N.J.: Prentice-Hall, 1963.

Dohrs, F. E., and Sommers, L.M., eds. *Economic Geography: Selected Readings*. New York: Thomas Y, Crowell, 1970.

Eliot Hurst, M. E. *A Geography of Economic Behaviour*. Englewood Cliffs, N.J.: Prentice-Hall, 1972.

Fosberg, F. R. 'Geography, ecology, and biogeography', *Annals of the Association of American Geographers*, March 1976, pp. 117–128.

Fryer, D. W. *World Economic Development*. New York: McGraw-Hill, 1965.

Griffin, P. F., Chatham, R. L., Singh, A., and White, W. R. *Culture, Resources, and Economic Activity: An Introduction to Economic Geography*. Boston: Allyn & Bacon, 1971.

Highsmith, R. M., Jr., ed. *Case Studies in World Geography: Occupance and Economy Types*. Englewood Cliffs, N.J.: Prentice-Hall, 1961.

Highsmith, R. M., Jr., and Northam, R. M. *World Economic Activities: A Geographic Analysis*. New York: Harcourt, Brace & World, 1968.

Hodder, B. W., and Lee R. *Economic Geography*. London: Methuen, 1975.

Jones, C. F., and Darkenwald, G. G. *Economic Geography*. New York: Macmillan, 1965.

Lloyd, P. E., and Dicken, P. *Location in Space: A Theoretical Approach to Economic Geography*, London: Harper & Row, 1977.

McCarty, H. H., and Lindberg, J. *Preface to Economic Geography*. Englewood Cliffs, N.J.: Prentice-Hall, 1966.

Morrill, R. L. *The Spatial Organization of Society*. Belmont, California: Wadsworth, 1970.

McNee, R. B. *A Primer on Economic Geography*. New York: Random House, 1971.

Roepke, H. G., ed. *Readings in Economic Geography*. New York: John Wiley & Sons, 1967.

Smith, R. H. T., Taaffe, E. J., and King, L. J., eds. *Readings in Economic Geography*. Chicago: Rand McNally, 1968.
Thoman, R. S., Conkling, E. C., and Yeates, M. H. *The Geography of Economic Activity*. New York: McGraw-Hill, 1968.
Van Royen, W., and Bengston, N. A. *Fundamentals of Economic Geography*. Englewood Cliffs, N.J.: Prentice-Hall, 1964.
Yeates, C. I. *Man's Economic Environment*. New York: McGraw-Hill, 1976.
Zimmermann, E. W. *World Resources and Industries*. New York: Harper and Brothers, 1951.

Other sources

Berry, B. J. L. 'Further comments concerning "geographic" and "economic" economic geography', *The Professional Geographer*, January 1959, pp. 9–12.
Chisholm, M. *Geography and Economics*. London: Bell, 1966.
Commission on College Geography. *A Systems Analytic Approach to Economic Geography*. Washington, D.C.: Association of American Geographers, 1968.
Ginsburg, N. *Atlas of Economic Development*. Chicago: University of Chicago Press, 1961.
Ginsburg, N., ed. *Essays on Geography and Economic Development*. Chicago: University of Chicago Press, 1960.
Lukermann, F. 'Toward a more geographic economic geography,' *The Professional Geographer*, July 1958, pp. 2–11.
McCarty, H. H. 'Toward a more general geography,' *Economic Geography*, October 1959, pp. 283–289.
McNee, R. B. 'The changing relationships of economics and economic geography,' *Economic Geography*, July 1959, pp. 189–198.
Taaffe, E. J. 'The spatial view in context,' *Annals of the Association of American Geographers*, March 1974, pp. 1–16.
Toyne, P. *Organisation, Location and Behaviour: Decision-making in Economic Geography*. London: Macmillan, 1974.
Webb, M. J. 'Economic geography: a framework for a disciplinary definition,' *Economic Geography*. July 1961, pp. 254–257.

Part Two

ELEMENTS OF ECONOMIC GEOGRAPHY

> '. . . *explanations for the spatial positioning of things must necessarily come from a wide assortment of inputs not always obvious or clearly defined. For economic geography, however, these most often include physical, cultural, historical, institutional, and certainly not least, economic factors. These critical inputs are called the bases of economic activity.*'

Chapter 2, 'The Nature of Economic Geography', p. 16

3

Physical Bases

As suggested at the close of the last chapter, in economic geography some things must be taken as given—that is, we will not try to explain their causes but only their consequences with regard to the location of economic activities. This is especially true with regard to the physical features of the earth such as climates, soils, land forms, mountains, and rivers. Our purpose is rather to discuss and explain the general influences of such physical bases on economic geography.

The physical features of the earth are, in one sense, fundamental to the understanding of the distribution of people and their activities upon it. The nature of the land and water relationships alone would dictate a fairly predictable general response. Likewise. the major climatic zones provide a condition for occupance which cannot be ignored. On the other hand, these physical features must be taken by people as they are given to them; people have not been able to modify them appreciably, at least not until very recently. People, moreover, react to such physical features within their own cultural and technological conditions. Rarely do the physical features play a deterministic role; they are only some of the many factors that relate to people's choices of occupancy and use.

The natural environment, complex as it is, is sometimes given exaggerated attention: in some cases human beings are surely working against the grain of nature, in other cases in co-operation with nature. With regard to agriculture and related activities, we must accept that the natural environment strongly *conditions* the activity response. Nonetheless, the environment does not *determine* what people can or cannot do in any given place.

SOME EXAGGERATED VIEWS OF NATURE

Given the obvious fact that many of the relatively unoccupied parts of the earth have foreboding environments—for example, the polar ice caps and tundra, the high plateaus and mountains, the major deserts, and various jungle and tropical areas—economic geographers early postulated the 'too' concept. According to the developers of this concept, people have avoided the extreme environments but have sought out the 'good lands'. The avoided areas are thereby considered too hot, too dry, too cold, too steep, or too something else for major settlement and land use.

The positive version of the 'too' concept is that people have sought out the not too hot, not too dry, cold, steep or not too something else: the good lands, the areas most suitable for agriculture and, subsequently, for cities and industry. Some 90 per cent of the earth's population do live and work on less than 10 per cent of the land surface. Many of the select places are in midlatitude areas and climatically temperate zones; they are mostly accessible to the seas and most, too, are characterized by relatively level terrain or rolling plains. But it does not follow that the physical bases of the earth provide the primary explanation.

Yet the 'too' concept, developed in extremis, led an earlier generation of economic geographers to assert that what people do is determined primarily by their environmental conditions. This concept was called environmental determinism. Although geographers have long since formally rejected it, some still persist in placing too heavy an emphasis on environmental factors, coining the neodeterministic phrase environmental possibilism. They hypothesize that the environment sets the stage for a range of possibilities for economic activity, and that people select, through their culture, from among these limited choices. This concept, while generally correct, also commonly overstates the role of the physical bases in the location and pattern of economic activities. Close observation of these patterns reveals that people have more choices open to them than such deterministic or possibilistic concepts suggest. Other variables in land use, such as location and culture, may also play a restrictive role, causing otherwise good land to be left unused or very poor land to be heavily used.

Even if physical factors were determining elements, there would still be different optimum environments for different activities. For example, optimum environmental conditions for bananas are not the same as for cotton, and those for cotton are not ideal for wheat. In fact, an impossible environment for one crop may actually be an optimum environment for another. Thus, even in agriculture, it makes little sense to apply a general label of 'good' to some lands and 'bad' to others.

This is further the case because people have substitution capabilities whereby they may play off costs of production and yields. Even where an optimum physical condition can be found (Figure 3–1), yield and cost substitution factors enter the picture. There

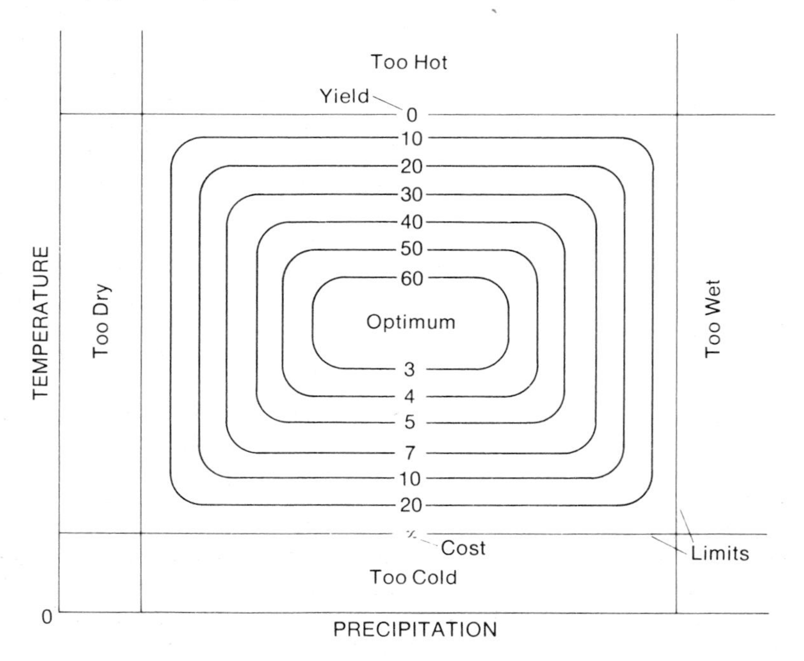

Figure 3–1 A simple physical restraint model. This model demonstrates the effects of varying combinations of temperature and precipitation on crop yields and production costs. (From McCarty, H. H., and Lindberg, J. B. *A Preface to Economic Geography*, 1966, p. 61. By permission of Prentice-Hall, Inc., Englewood Cliffs, New Jersey.)

is a limit beyond which a crop cannot be produced because of zero yield at infinite costs. To find an optimum as illustrated in Figure 3–1, it must be tailor-made to a specific crop. And the limits of profitability are wide, indeed, for most crops.

Therefore it should be evident that physical features do not limit production of a particular crop only to that location which is physically optimum. Instead, production often occurs in 'marginal' areas. Here the yields may be lower, but if the marginal area is close to the final market, savings in distribution costs may more than offset lower yields or higher production costs. On the other hand, it is also possible to obtain yields even higher than those at the physically optimum location if higher input costs (purchasing fertilizer, for example) are made. Lastly, the optimum physical space may be usurped by some other, more profitable crop. For these reasons, any given crop is rarely produced at its most optimum physical location. The true optimum location is where greatest profitability exists, not necessarily where physical features are most advantageous.

Finally, people often modify physical features to suit their needs. Agricultural crops are much affected by climate and soil variables, yet people have many ways of overcoming such natural obstacles: they can fertilize poor soil, drain wet soil, irrigate arid areas, and through the application of chemicals overcome such problems as soil acidity and alkalinity. Given steep slopes, where soil erosion would normally prevent agriculture, they may terrace land to make it usable. In rolling topography, the land may be levelled to make it suitable for irrigation. Such modifications of nature may be precluded because they are too costly; but the demand for the use of areas near markets often makes modification more feasible than using better-endowed areas farther from the markets.

CLIMATE, SOILS, VEGETATION, AND ECONOMIC ACTIVITY

Possibly the most important physical features to be considered in their relationship with economic activities are climate, soils, and vegetation. On a world scale, these are interconnected. Climate sets the stage for general vegetation, and vegetation strongly affects the soil types. For quick analysis, the nature of an area's wild vegetation provides the first strong clues to crop potential in that area. Tropical areas, for example, with their large, evergreen broadleaf trees, indicate a suitable physical environment for plants that require much precipitation and high, even temperatures throughout the year, and for those that can survive on heavily leached soils. On the other hand, grassland areas may have potential for growing grains like wheat and flax. In dry areas, only plants with special, deep-root systems and leaves that allow little transpiration can survive. The date, palm, olive, and certain nut trees, for example, do well in the drought-stricken summers characteristic of Mediterranean climates. The temperate area of mixed forest and grasslands provides a wide array of crop possibilities. Indeed it is in this area that the greatest variety of commercial crops is found.

On a detailed area basis, however, the correlation between climate, vegetation, and soils quickly breaks down. Wild vegetation varies considerably among similar climatic areas, not only because of the succession and diffusion of plants over time but also because of differing physical conditions. For example, soil is a product of vegetation, parent material, slope, drainage, and local factors, all operating over time. Moreover, actual vegetation, often the crops raised by man, further alters the original soil conditions.

Drastic soil differences within climatic regions are caused by two of earth's major features: mountains and river valleys. Volcanic soils and alluvial river-valley soils are extremely rich. In these cases the parent material often outweighs other factors in producing a particular soil. For example, civilization first developed along the banks of the

Tigris-Euphrates, the Nile, the Indus, and the Hwang Ho rivers, which flooded each year and replenished the soil. Thus, unlike the inhabitants of most other areas of the earth, the people of these river regions did not deplete the soil. They could thus remain in one place rather than migrate to new soil areas.

Volcanic soils are possibly the only other soils on earth that are so intrinsically rich that they are almost impossible to deplete. Perhaps the most densely populated rural area on earth is found in Java, where the soil is volcanic. Hawaii, Japan, Southern Brazil, and river valleys in much of Oregon, Idaho, and Washington also have similar soils. Examination of these areas with others nearby reveals tremendous variation in intensity of land use and density of population. Sumatra has the same climate and physical location features as Java. Yet, by comparison, Sumatra has a low density of population. The major difference between them is a volcanic soil on Java and a basaltic soil on Sumatra. In Oregon, Washington, and Idaho a fortuitous combination occurs: volcanic soils and rivers for irrigation.

Yet, if we temporarily ignore the two distorting features, mountains and river valleys, it is safe to generalize further on the spatial variation of climates, vegetation, and soils. In this regard geographers generally recognize five broad types of climate and associated soil and vegetation types (Figure 3–2). The climatic types, for lack of a more detailed name, might be labelled as follows, from equatorial areas to the poles: (1) tropical, (2) dry, (3) temperate, (4) snow, and (5) ice. Trees can grow in all but the very dry and the ice types. The ice climates have little soil in the traditional sense and only limited, tundralike vegetation. Grasses grow primarily in semidry and temperate areas.

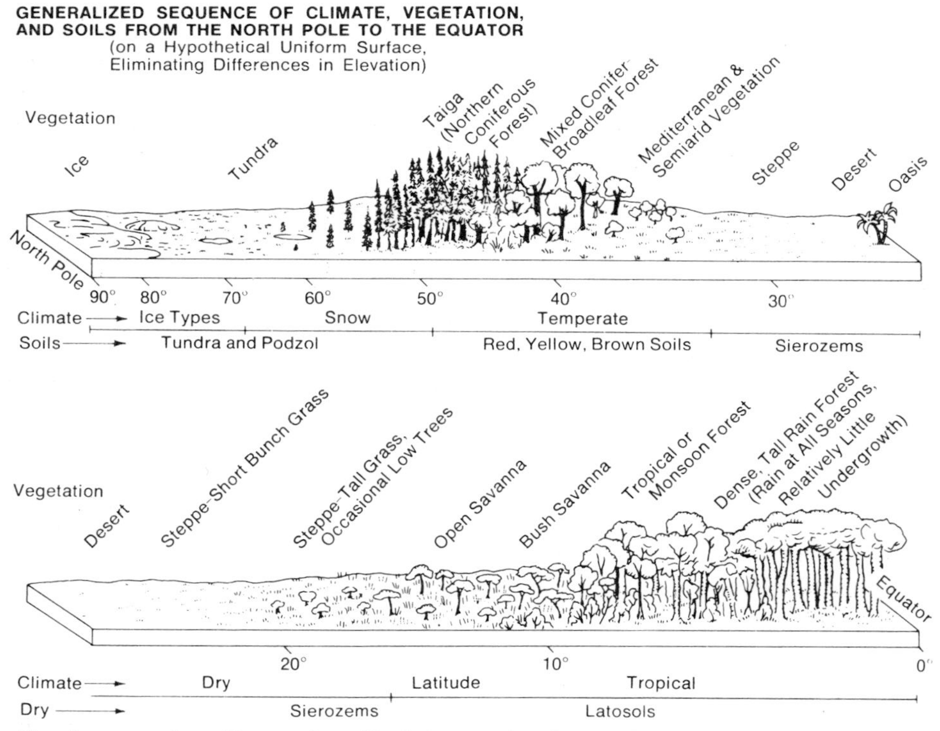

Figure 3–2 Relationships among climate, vegetation, and soils for any latitude. (Adapted from Rhoads Murphey. *An Introduction to Geography*, 1961, p. 82. By permission of Rand McNally College Publishing Company, Chicago.)

The soil types, in the same sequence as above, are (1) latosols, (2) sierozems, (3) a wide variety of red, brown, and yellow soils, (4) podzols and tundra, and (5) none.

As is evident from Figure 3–2, the dividing lines between the five major climatic types and their associated soil and natural vegetation zones are arbitrary. In fact, there is a general transition in all of these from the equator to the poles. The transition zone outward from the tropical zone is the savanna, a zone of seasonal dryness characterized by a more open pattern of vegetation and a less leached soil. The transition zone around deserts is usually called the steppe, a zone of grasslands and more productive soils. A transition zone on the west coasts of continents is a marine type climate affected by ocean currents. The tundra is a transition zone between the northern coniferous forest (called the Taiga in Eurasia and the Boreal forest in North America) and the ice caps.

Wet Tropical Lands and Economic Activity

The tropical areas are generally found about 10 to 15 degrees on either side of the equator. In this area, monthly mean temperatures are above 64.4°F (18°C). Precipitation is exceedingly high, well distributed throughout the year, and generally over 60 inches annually—more than enough for almost any kind of crop. On the outer edges of the climatic type are two areas that are alternately wet and dry, depending on the season of the year. These are the savanna climates, which have a wet season during high-sun times of the year and a relative dry period during low-sun periods, and the monsoon climate, which is related to coastal areas but especially to places where mountains bring about orographic precipitation part of the year.

The lateritic or latosolic soil in the tropics is very deep, but most of the soluble parts have been leached out. They are, however, heavy in iron hydroxides and various hydroxides of manganese and aluminium. This gives the surface soil a reddish or yellowish appearance. Organic material decomposes very rapidly under the high temperatures and heavy rainfall, so that this soil type is not very fertile. Nevertheless, it can support very heavy tree growth and the predominant economic activities in this area are based on trees. The broadleaf trees include rubber, teak, ebony, and mahogany. Coffee, cocoa, cassava, abaca, and kapok are also important products, providing employment in areas accessible to world trade. Important also are forest products such as rubber, nuts, and various other parts of the tree.

Because trees do not generally grow in stands as they do in midlatitude and northern areas, gathering activities require considerable labour. In some parts of the tropics, the development of plantations has remedied this problem. But much of the area is occupied only by subsistence-economy peoples, who eke out a living by gathering tropical products or by a primitive, slash-and-burn agricultural system (see Chapter 7).

Thus the tropical areas are generally among the most sparsely settled on earth. Nonetheless, people in the industrialized areas have long been concerned with the possibilities of greater use and development of the tropics. It is thought that with better health and sanitation methods, better transportation facilities, and the ability to 'farm' many of the areas, the tropical areas will increase in importance.

Dry Lands and Economic Activity

The dry areas of the earth, where evaporation exceeds precipitation, represent another major paucity in world population distribution (Figure 3–3). There is little doubt that this lack of population is directly related to physical factors. One of the foremost requirements for intensive use of the area is, of course, the provision of water for irrigation and other uses. This has been possible in those dry areas traversed by exotic rivers like the Nile, the Colorado, and the Columbia, or where oasis-type conditions occur.

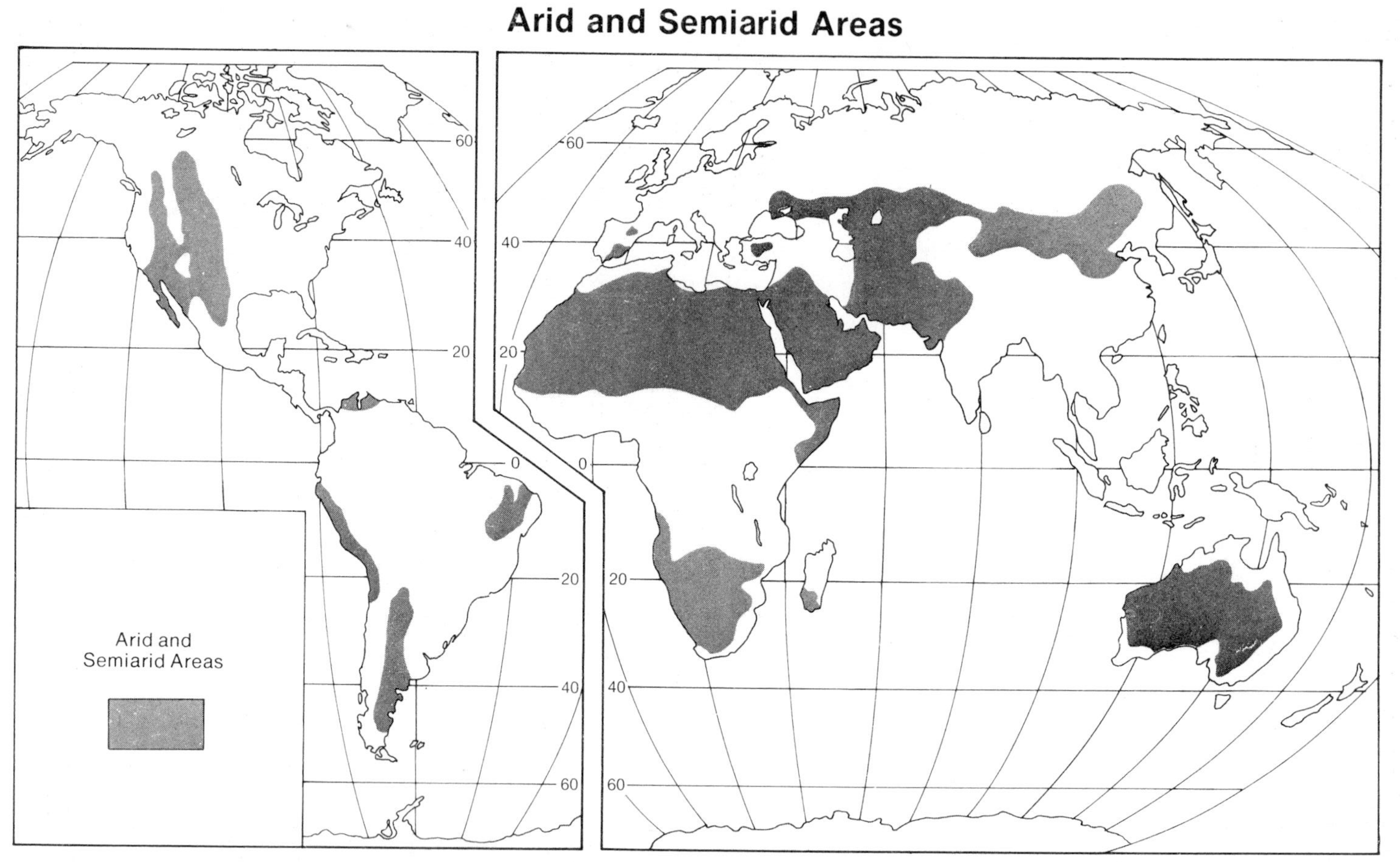

Figure 3–3 Arid and semiarid areas of the world. The dry areas of the world are primarily the result of descending air masses in subtropical high pressure zones. Other causes include the rain-shadow effects of mountains, off-shore, cool ocean currents, and extreme continentality.

Desert soils are usually of the sierozem type and are grey in appearance. Like the soils in the tropics, they contain little humus, in this case, however, because there is little vegetation to decay. There is no leaching of the soil, as is characteristic of tropical climatic areas, because of lack of precipitation. In fact, the opposite effect occurs, so that a lime crust, that is a calcium carbonate deposit, is often found within a foot or so of the surface. Such soil is suitable for cultivation only when finely textured and, of course, where water is available.

With the exception of exotic stream areas and oasis settlements, such as Salt Lake City where adjacent mountains are adequate in precipitation, most of the dry parts of the world are practically uninhabited. These include about one-fifth of the earth's surface. The few inhabitants in these areas are often migratory, moving seasonally with their animals to places with better vegetation.

Again, we have had great hopes for deserts, but the results to date have been disappointing. In some desert areas we have found rich mineral resources, particularly petroleum in the Middle East, and in others we have been able to develop an intensive agriculture based on irrigation. There have been great hopes too that desert areas near the seas could be utilized through desalinization of sea water. Desalinization has indeed developed to serve urban settlements, but converted sea water is still too costly to be used on a large scale for irrigation.

The Warm Temperate Lands and Economic Activity

It is difficult to discuss the temperate areas and economic activity briefly because of their great physical and economic diversity. Some temperate areas, such as the southeastern portions of the United States and China, are 'humid subtropical'; some, such as the 'mediterranean' climates, are characterized by summer drought. Some enjoy moist, mild 'marine' climates on the west coasts of continents; others, such as the Midwest in the United States, are 'continental'. Yet, all are temperate (lowest monthly mean temperature over 26.6°F (−3°C)) and are suitable for the growing of many crops (Figure 3–4).

It is in these areas where the most productive agriculture and the most industrialized and urbanized parts of the world are found. It is in these temperate midlatitude areas where the majority of the world's population resides. In fact, most of the world population lives in the midlatitude forest area—a natural vegetation type characterized by deciduous trees intermixed with pines and coniferous trees. Much of the northwestern part of the United States, Western Europe, northern India, and Japan are characterized by such a vegetation type.

People are not evenly spread over these 'choice' lands, however. Instead, they are clustered only in selected areas within it. The most attractive physical features for urbanization and industrialization appear to be (1) coastal and river valleys, and (2) the flatter, plainlike areas at the lower elevations. Nonetheless, the primary reasons for the particular sitings of urban-industrial development in this area appear to be not physical but economic. These will be discussed in Chapter 4.

Cold Lands and Economic Activity

At about 50 degrees north and south latitude and continuing to the poles are the severe winter climates. These cold lands may generally be divided into three types: (a) A continental, warm-summer, cold-winter type, where the growing season is about three months long and sufficient for many crops. In terms of natural vegetation, coniferous trees like fir and spruce are the rule; (b) The tundra area, where the growing season in so short and insolation so little that only mosses, lichens, and very limited grasses

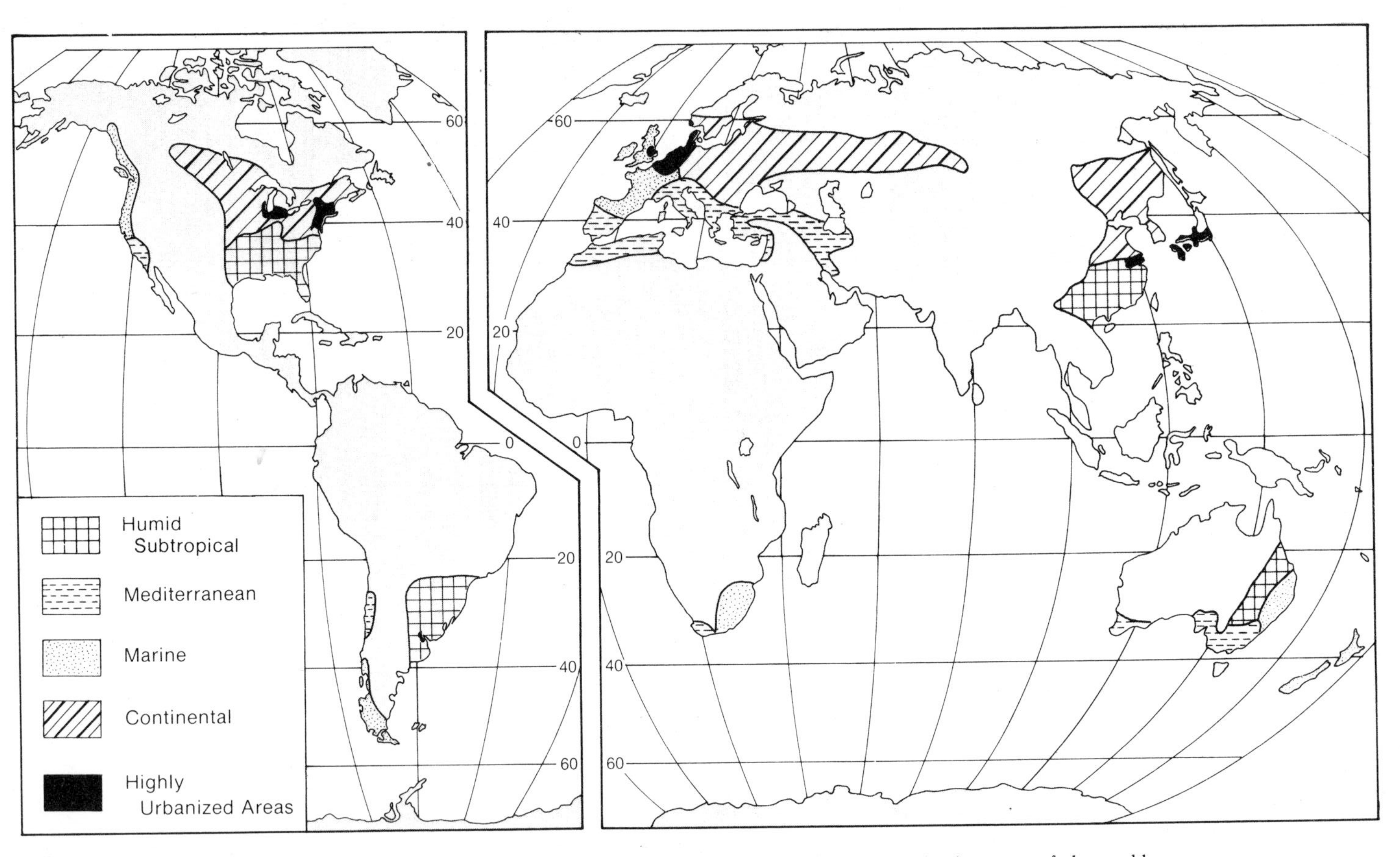

Figure 3–4 Warm temperate areas of the world. These are the highly productive areas of the world. Note their relationship to urban areas and industrialization.

grow. This area often has a permanent layer of frozen soil (permafrost) beneath the thawed-out summer surface and is thereby often boggy and waterlogged. Clearly, agriculture in this area is almost nonexistent; (c) The ice caps, where no vegetation or soils exist.

The continental climate, which has a podzol soil and generally a coniferous-tree natural vegetation, has the greatest (though limited) potential for economic activities in the cold lands. These activities, however, are of an extensive nature, thus making population in these areas limited. Lumbering is by far the most important activity here. As summers are too short for broadleaf hardwood trees, softwood conifers like spruce are predominant. The main use of such wood is for the pulp and paper industry.

Podzol soils, in contrast to the alkaline desert soils, are characteristically acid. Another distinctive feature of podzol soils is a strongly leached horizon that is whitish grey or ash grey. The leached colloids are removed only to the layer below where they often form a clay layer that becomes impervious. These soils are low in fertility also because of the coniferous tree vegetation. Conifers need little colloid material such as phosphorus, potassium, calcium, and magnesium, and so they do not bring these to the surface of the ground. Instead, they remain as a 'hardpan' layer in a lower soil horizon. Of course, this acidic condition and the problem of leached bases can be corrected by drainage, deep ploughing and soil treatment. But only limited areas in this zone are usable agriculturally, because glaciation has left it with a highly rock-infested soil and many morainal swamps and lakes. Because of peculiar geological conditions, however, minerals are often widespread and rich here, and mining has become a major activity.

MOUNTAINS AND ECONOMIC ACTIVITY

Mountains do not fit well into either a foreboding land concept or a good land concept. Certainly, they are sparsely inhabited; but inasmuch as there is an altitude counterpart in climate, vegetation, and soils that corresponds to the latitude-related patterns discussed above, mountains must be considered a mixture of possibilities.

The positive aspects of mountains are numerous and important. First, they provide a diversity of vegetation types over a small distance. Affording the opportunity to move animals from lower to higher elevations in the summer (transhumance) is one way in which mountains might benefit a rural activity. In many areas such a climatic change is a life-giving feature. Here mountains often act as humid islands, inasmuch as they receive much more rain than the surrounding country. Streams from such mountains often provide water for nearby settlements, industry and agriculture. This is the case with the Cumbrian Mountains in Northern England (which supply Manchester) and the Trossachs in the Southwest Highlands of Scotland (which supply Glasgow). Climatic and vegetation changes allow forests to occur in areas that would otherwise be devoid of timber: the Sierra Nevadas in California provide a growth of timber in an otherwise dry and grass-laden area. Mountains are a major source of timber in many of the Alpine parts of the world, with steep slopes often an asset rather than a deficit in getting the material to market. Conversely, mountains also make deserts out of much of the earth that would otherwise be humid and potentially highly productive. Most of the deserts in midlatitude areas are the result of a rain-shadow effect produced by mountains on their leeward or eastern sides.

Paradoxically, the climatic differences created by mountains also create possibilities for spatial interaction. As the environment is often different on either side of the mountain, such areas have differing agricultural and industrial potential. Where irrigation can be applied on the leeward side of the mountains, as in eastern Washington and

British Columbia, tremendous interaction may occur. Thus mountains often set up the physical bases of area differentiation, and hence the potential for trade. They then act somewhat as barriers to such interaction because of the difficulty of transportation over them.

The diversity in climate and vegetation of mountains, as distinct from the homogeneity of the surrounding countryside, also makes them attractive as recreational areas for the nearby population. They may be cooler in summer, like the hill-stations at Simla above the plains of India, and where the nearby lower-elevation lands have mild winters —for example, along the coasts of western North America and western Europe—the coldness of mountains makes them excellent for skiing. Mountains often provide unusual and picturesque landscapes highly valued as National Parks, like Yellowstone in the USA, Banff in Canada, and the English Lake District. However, the recreational value of mountains can be assessed only in terms of the accessibility of the mountains to those demanding the resource.

Undoubtedly the most valuable asset of mountains is their mineral wealth. The folding, faulting, and other tectonic forces are particularly favourable to the formation of rich mineral deposits. Some resources strongly correlated with mountains include coal, iron ore, copper, many alloys, and precious metals. This in no way implies that such mineral resources are found only in mountains, but most minerals are more closely associated with mountains than with any other single physical feature.

Mountains are also often valuable for the generation of water power. The steep river gradients, deep valleys, rock-fill for dams and possibilities for tunnelling make for considerable water potential. Here again the degree to which this potential is actually used will depend largely upon the proximity of such potential to demand sites and the alternatives available for the generation of power.

It will be noted from the above that the use of mountains is often highly seasonal. Thus they have little permanent population. Activities that may operate on a year-round basis, such as mining, forestry and the generation of hydro-electric power, employ few people, are short-lived, and consequently lead to few permanent settlements. Probably no other physical feature of the earth has so many assets and generates so little permanent settlement. Some mountain communities are isolated, backward, and have little or no economic potential; others are tourist, mining, and recreational centres, rich and with great future prospects.

RIVERS AND ECONOMIC ACTIVITY

Rivers, like mountains, are in some cases assets and in other cases deficits. Where they have associated alluvial soils they provide some of the most productive agricultural land on earth. Rivers also provide primary channels of inland transportation, not only directly in the form of navigable waterways but indirectly through their valleys where railways and other routes are numerous. The attraction of man and his activities to river valleys, however, has also led to some of the world's worst flood disasters. Rivers periodically and naturally flood unless dams, levees, and other flood-control devices are constructed. Even then, the greater intensity of use following installation of such flood-control features often means that a smaller flood will cause even greater hardships and damage.

Thus it is truly difficult to generalize about rivers. Their valleys contain the most dense and concentrated settlements in the world, as evidenced by the valleys of the Yangtze, the Hwang Ho, and the Ganges in Asia, the Nile in Africa, the Danube in Europe, and the Mississippi in the United States. Some of the other great rivers, however, are little used, and their valleys contain almost no population. In this respect note the Congo in Africa, the Amazon in South America, and the Ob, Yensie, and Lena in Asia; all are without significant nearby population. Aside from their great value in desert areas where they operate as exotic streams, rivers appear to be of little use for agriculture and its consequent activities unless they are found in midlatitude areas.

Elsewhere, however, greater rivers and their associated valleys and plains have many different uses. The rivers themselves can be used for a variety of activities, so many in fact that there may be conflicts between irrigation, navigation, recreation, hydropower, domestic water consumption and industrial waste disposal. In particular, such conflicts may arise where great rivers flow through many countries. The classic case in Europe is the Rhine. This great river not only became the boundary between France and Germany, but also flows onwards to the sea through the Low Countries. The Rhine could only become a great artery of international trade by agreement between all the nations along its course from Switzerland to the sea. The great port cities of the Netherlands, now dominated by Rotterdam, owed much of their commanding position in West European commerce to their location near the lowest reaches of the Rhine. Indeed, so many great cities of the world were established where fresh water meets salt: in the United Kingdom, we can instance London on the Thames, Glasgow at the head of the Clyde, Liverpool on the Mersey. This is a circumstance which underlines the modern importance of river estuaries, which in the developed countries offer up the study of interfaces. They are increasingly used for great dock systems, as avenues of transportation (by road, rail, river, or canal), as sources of water, as channels for disposal of wastes arising from urban life and industrial activity. Such developments frequently conflict with the role of estuaries as the natural habitat of shellfish, of residential and migratory populations of wildfowl, and as gateways for the passage of migratory fish like salmon. Finally, of course, they may have considerable amenity value for leisure, recreation and aesthetic enjoyment.

It is evident from the above that the standard statistics generally given about rivers, such as 'longest river in the world', are of little value in assessing a river's potential use or uses. Instead, at least seven features should be examined: (1) location, which is by far the most important; (2) flow or discharge, which affects possibilities for most uses; (3 and 4), depth and width, which are clearly critical for navigation and sometimes for recreational purposes; (5) grade, or gradient, which is important for hydroelectric potential and perhaps for recreational use; (6 and 7) direction of flow and circuitry of stream, which have particular importance with regard to transportation potential.

Other river features, such as water quality, temperature, dendritic pattern, and character of the drainage basin slope, may also prove critical for various purposes. For example, a river may have good attributes for hydropower—discharge, gradient, and proximity to market—but may not have economic possibilities because a dam site cannot be found for reservoir impoundment.

Critical deficiencies of some of the world's 'great' rivers will perhaps best illustrate the nature of these various elements. Undoubtedly, the Congo River has the greatest power potential of any river in the world. The tremendous flow and steep fall from the African plateau to the coast give it great potential as a hydropower source, but distance from market makes the river unusable. The waterfall character of the river also greatly decreases its transportation potential. On the other hand, the Amazon River has the largest flow of any river in the world and has no steep gradient to interfere with navigation. But the Amazon is likewise little used—in fact, it has much less traffic than most small river canals in Europe. Again, location is the deterrent factor. The major rivers in Northern Asia run in the 'wrong' direction. They flow northward, where the estuaries become icebound in winter, instead of westward, toward the heart of the Soviet Union. In this regard one might say that the Mississippi River in the United States also runs in the wrong direction. Nevertheless, given the cheapness of water transportation relative to road or rail, it still has some advantages. There are many exotic rivers in the world with potential irrigation value, but it takes tremendous amounts of money to build dams and irrigation networks. It also requires that there be a market for agricultural goods. Undoubtedly, many of these rivers, like the Indus in Pakistan and the Nile in Egypt, will be put to more use with the building of large dams. In many dry areas, however, there is a scarcity of river water. This is particularly the case in the western part of the United States, where Utah and Colorado need more water from the Colorado River and the Rio Grande but must by law release much of it to California, Arizona, and Mexico. So critical is the water problem in many highly industrial, dry areas such as Los Angeles, that water is brought in from hundreds of miles away. There is considerable pressure, too, to bring water from the Columbia River in Washington to Southern California.

BIBLIOGRAPHY

Mackay, J. R. 'The World of underground ice', *Annals of the Association of American Geographers*. March 1972, pp. 1–22.

McCarty, H. H., and Lindberg, J. B. *A Preface to Economic Geography*. Englewood Cliffs, N.J.: Prentice-Hall, 1966.

Mihalyi, L. 'Electricity and electrification of Zambia', *Geographical Review*, January 1977, pp. 63–70.

Miller, D. H. *Water at the Surface of the Earth*. London: Academic Press, 1977.

Strahler, A. *Introduction to Physical Geography*. New York: John Wiley & Sons, 1970.

Ullman, E. L. 'Are mountains enough?' *The Professional Geographer*, July 1953, pp. 5–8.

Ullman, E. L. 'Rivers as regional bonds: the Columbia-Snake example,' *Geographical Review*, April 1951, pp. 210–225.

Wolfanger, L. A. 'The great soil groups and their utilization.' *Conservation of Natural Resources*. Ed. Smith, G.-H. New York: John Wiley & Sons, 1958, pp. 37–67.

4

Economic Bases

In Chapter 3 it was noted that the physical 'stage' was not neutral but consisted of certain parameters of choice for location, parameters which in some cases were highly suggestive of response and others that gave wide latitude for experimentation, adjustment, and development. Its content thus initiated a many-sided framework for analysis of any given distribution or territory. In order to widen the horizon of spatial understanding, certain economic factors or bases in economic geography will be examined in this chapter.

It may appear highly peculiar to treat the economic bases as a single chapter in a book dealing with economic geography. And so it is, in the sense that economic principles are emphasized throughout. The first purpose here is to provide a brief, skeletal outline of certain pervasive economic forces that mould and shape the location of things on the earth's surface. The second purpose of this distinct treatment of economic bases is to emphasize the fact that economic considerations, although extremely important and even paramount in many cases, are nonetheless but one of a series of bases necessary for a full comprehension of spatial reality. The reader will already be mindful that the word 'economic' is used here as an adjective to describe the kinds of activities to be considered from a geographical perspective. More than economic principles from the field of economics is necessary to appreciate why these kinds of activities are distributed in particular places and patterns. Among these additional principles are those pertaining to the physical bases, as already outlined, and those pertaining to the cultural and institutional bases, to be discussed later.

In reality, all of these bases operate in a highly complex, interrelated, and contagious fashion. For pedagogical and organizational purposes, however, these components have been separated into rather arbitrary and isolated frameworks. This separation makes it easier to see how each of these forces, or bases, operates and combines to form any particular pattern. The procedure for focusing on a particular relationship, while holding others in abeyance, is a hallmark technique of economics. It has proven highly instructive in solving economic problems which involve otherwise hidden operational processes and its use is equally valuable in geographical analysis. Another technique borrowed from economics which will be heavily employed is that of simple graphics to illustrate abstract theories and models. This chapter will thus introduce the reader to such devices in a manner whereby they will be more fully appreciated and utilized in the forthcoming and more advanced discussions.

Although the analytical techniques of economics are valuable and necessary to our spatial understanding, there are also certain pitfalls in their use. Indeed, some economic devices, if not judiciously applied, can lead to conclusions as naïve as conclusions based on strictly environmental features. Nonetheless, if the reader keeps in mind that principles and forces are being discussed, not the real earth, little misinterpretation and misapplication should result.

A further cautionary note: the economic principles discussed here are those that bear heavily and directly on locational problems. For this reason, the student who is competent in economics may find much of the discussion quite different from the typical economic approach. The focus is deliberately distorted to gain a view of geographical content. The point to remember is that we are trying to understand the effect of such forces on the distribution of people and their activities. Our goal is not to understand the reasons for or behind the economic principles themselves; these, like mountains, are taken as given.

As with transportation, economic principles will constitute a continuing theme; the blend between these economic principles and transportation principles is involved in many of the theories later discussed. Only the more general and simple principles will now be presented. Specific applications of these principles to agriculture, industry, or land use within cities will be discussed under those subject categories in later chapters.

A PRINCIPLE OF COMPARATIVE TERRITORIAL ADVANTAGE

A principle of comparative territorial advantage, in part, means that areas having the greatest advantage for a particular activity will specialize in the activity that brings them the greatest return. Areas less favourable for producing various items—whether physically poorer, less accessible to markets, having less productive labour, and so on—must take second or third choice in the things they do. In a simple territorial context this means that the better areas get first choice of production, and will specialize in producing the more profitable crop or product, but that less fortunate areas will be left with residual choices.

At first this sounds rather bleak for less-endowed areas, but it is in fact the saving grace for the many disadvantaged and remote areas of the earth. Let us take the example of area A, which is fertile, close to market, and well endowed in all respects. Let us assume that in area A the profit per hectare on potatoes is £200, on wheat, £100, and for grazing purposes, £50. Now compare these with the profits per hectare in the less endowed area B and area C as shown below:

Area	Profit per hectare on Potatoes	Wheat	Grazing
A	£200	£100	£50
B	100	50	25
C	50	25	5

It is evident that area A has an absolute advantage in growing all crops. Therefore why would there be anything left for areas B and C to do? In some cases, there is not. But the general rule is for each area to specialize in the activity that gives it the greatest advantage and to import from other areas items produced at a comparatively lower cost. Under this system, area A would grow potatoes (first choice), area B would grow wheat (second choice), and area C would be used for grazing (third choice). Incidentally, if the profit from grazing in area C were to be £0, and these were all the options open to it, then area C would be unused. According to the principle, area A would import wheat from area B and cattle from area C even though, theoretically, it could produce such things more cheaply itself. By producing such things, however, it would make less than by specializing in the activity that brings it the greatest profit.

Another saving grace for poorly endowed areas is that they may be closer to the market than the physically well-endowed area. If transportation costs are greater than the difference in costs between the two, then the closer but poorer area will have a comparative advantage over the better endowed but more distant area. Such a phenomenon was particularly evident during the last century in Great Britain, where intensive and highly profitable garden-type agriculture was found near almost every major city, even though such territory was often otherwise inferior to areas more distant from the market. The general effect of improved transportation, however, is to allow the areas with the greatest advantage to do the producing. For example, a considerable proportion of Britain's needs for temperate-area garden produce, field vegetables and fruit is now met from beyond the major island of the United Kingdom: suppliers include Eire (potatoes), the Channel Islands (tomatoes), Brittany (cabbage, onions), Provence (apples), Israel (early potatoes), to be joined in the market-place with a further range of farm produce from even greater distances. The result of this is to cause nearer, but intrinsically poorer, land to be put in a 'second choice' position. The traditional market-gardening industry, long-established in such places as the Lea Valley, near London, the Vale of Evesham near Birmingham, and in the valley of the Clyde near Glasgow, is in new difficulties because of this situation.

Over the world as a whole, where artificial restraints on trade are very much in effect, such as trade agreements, tariffs, and outright embargoes, a different condition holds. (These institutional types of economic factors are considered in Chapter 5.) Nevertheless, the general principle is still applicable. It is succinctly stated by Towle as follows:

> *'Just as a given community, or nation, may increase its real income and the incomes of its individual citizens by abandoning individual self-sufficiency for individual specialization, so may the world increase its real income and the income of the individual nations by forgoing national self-sufficiency for national specialization. Nations, like individuals, are endowed by nature with special facilities; and nations, like individuals, acquire special skills. Minerals can be produced only in those nations where nature has stored them: nickel in Canada; tin in Bolivia, the East Indies, and Malay States; manganese in Brazil and Russia; petroleum in the United States, certain Latin American countries, Russia, Iraq, and the Dutch East Indies [Indonesia]. Nations not endowed by nature with stores of these minerals can acquire them only through exchange with nations more favourably endowed. Many agricultural products require for their cultivation a special climate. Coffee, tea, rubber, and silk, for example, demand a warm or tropical climate, while most of the cereal crops do much better in the cooler lands of the temperate zones. Consequently, the United States imports its coffee, tea, rubber, and silk, while tropical countries import from countries in the temperate zone most of their cereals'.* (Lawrence W. Towle, *International Trade And Commercial Policy* (New York: Harper and Brothers, 1956), pp. 10–11).

The most distinctive feature of the world distribution of economic activities is the tremendous specialization of areas. In fact, places having a mixed farming system—like Midland England or the American Midwest, for example—or cities having a widely diversified industrial base are exceptions. The general rule is one of great territorial specialization. Thus within the United States monoculture reigns: the Great Plains area stands out as specializing in wheat, Wisconsin is primarily a dairying region, and the Mississippi Valley of Arkansas shows a concentration of rice. Territorial specialization is so prominent that in North America we can label certain areas 'the corn belt' or 'the cotton belt'; in western Europe, too, we speak of the 'manufacturing Midlands' or the 'industrial belt of the Ruhr'. Specialization with regard to minerals is equally prominent: coal, copper, iron ore, and many other minerals are found in many areas, but they are heavily exploited in just a few. The reason is comparative advantage. In fact, it would make an instructive study to contrast the reserves of various minerals with the pattern of those actually used; the discrepancy is truly astounding.

Unfortunately, such exchange and specialization do not benefit all areas equally, and perhaps this general rule, when allowed to run its natural course, may not bring about the best of all possible worlds. It should also be pointed out that the comparative advantage model presented above is highly simplified and has been modified to serve the general purposes intended. As fully presented, it involves a rather complicated comparison of ratios of advantage from one place to another and in agriculture it commonly results in an optimum mix of crops rather than one crop in each area.

PRICE AND THE PRINCIPLE OF SUPPLY AND DEMAND

In its most simple and general form, price is the result of an interplay between the forces of supply and demand. These two forces are considered contradictory and run in opposite directions (Figure 4–1). From the supply side, people would opt to supply

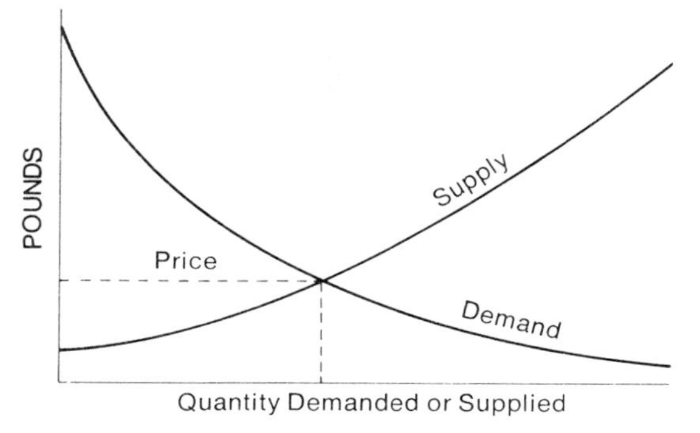

Figure 4–1 Price as a compromise between supply-and-demand curves. Price is a reflection of two contradictory relationships: (a) the supplier wants to supply considerable merchandise at high prices; (b) the consumer wants to pay low prices and is willing to purchase much more of the merchandise at low prices than at high prices. Thus there is a direct relationship between market price and the amount of merchandise that will be demanded.

great quantities of any item at a high price but very little, if any, at an extremely low price. Running counter to these desires is the demand of any items. If those demanding the item had their way, they would prefer to purchase a great deal at a low price but very little, if any, at an extremely high price. Thus the price actually used represents a kind of compromise between these two forces.

The specific shapes of these supply-and-demand curves for any given activity, the degrees of elasticity in them, and various other conditions are the rightful subject of economics. For our purposes we will simply take such an interplay as given and make note that price is generated at the place where these two curves cross. Also note that there is always an interrelation between price and quantity produced, or demanded. Thus price is in large part a reflection of the abundance or scarcity of an item and is always a function of (that is, dependent upon) quantity.

In Figure 4–2 it is shown that the quantity of wheat demanded and consumed is

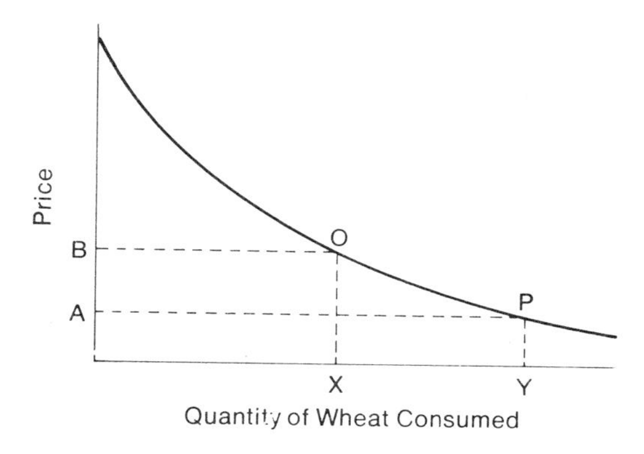

Figure 4–2 Market price and quantity of wheat consumed. Note that at price B only OX amount of the wheat will be demanded, but that at lower price A, PY amount of wheat will be demanded. This relationship generally holds for most items although price will have small effect on amount of food demanded. This latter inflexibility is the result of inelasticity of demand.

based on price. If the price is B, then only X units of wheat will be consumed (B, O, X). On the other hand, if the price is over, say, A, then Y units of wheat will be consumed (A, P, Y). Thus, such a price-quantity relationship is achieved from the general shapes of the supply and demand curves as shown in Figure 4–1.

It is evident that price, and hence profit, is not a simple matter of general cost plus some markup but occurs in a complicated marketing system. For example, a change in the supply or demand of an item will affect its price, and a change in price will in turn affect its supply and demand curves. To see how this operates, note that an increase in the price of a commodity such as wheat, due perhaps to an increase in the demand or a temporary decrease in supply, will cause more wheat to be produced the coming year. With more wheat then being produced, however, the consumer will be in a better bargaining position (that is, a changed supply curve) and the price should fall. This price decline must occur if the wheat then available is to be purchased. Theoretically, assuming demand elasticity, the amount of wheat produced would not be purchased at the old price.

Any quantitative change in production of an agricultural crop, perhaps as the result of weather conditions, may have a considerable effect on price. Thus farming is a fairly risky pursuit compared with manufacturing and retailing. In these latter cases the demand curve can be manipulated by adjustments in production. This is especially so if a firm has a monopoly on the manufacture or sale of a product through patents or if it is the only firm in a particular area. In such cases it is in a supreme bargaining position with the demand sector and can charge considerably more than would be the case under a fully competitive situation. Therefore it is much more profitable to grow some crops than to grow others, to manufacture certain products, to sell some kinds of goods, to render certain services.

Such relationships also have spatial implications. For example, all locations do not have equal opportunities for production of various items. Given the principle of comparative advantage, that area having the greatest advantage gets first choice. The less favourably endowed areas will have to take second best.

It is also evident that there is only so much demand for various goods, regardless of price. Thus even if the earth were one vast, physically perfect area, not all of it would be put to use; nor would it, because of accessibility differences, be put to equal use. There still would be vast areas of good lands left vacant. For example, let us assume that tomatoes were the most profitable of all agricultural crops, and that all the earth's land was equally fertile and climatically suited for tomato production. Clearly, people would not demand solely tomatoes. Also, if only tomatoes were produced, there would be such a glut on the market that their price would drop. Consequently, tomatoes would be an unprofitable crop in many areas. Indeed, they would no longer be a very profitable crop at all.

At this stage the reader should be somewhat perplexed. First, even if this supply-and-demand principle is operating to determine some proper mix of commodities produced, how is it that some crops are still more profitable than others? Why does this principle not also operate to bring about equal profit for all crops? In one way it does. The degree of profit to be made on any given crop varies from year to year, and changes in rank often occur. But this explanation is still unsatisfactory.

The real explanation is a geographic one. The earth is not homogeneous. All areas are not equally good for tomatoes. The amount of land for some crops is limited indeed. For example, olives can be grown only in a mediterranean climate. This geographic inequality places a premium on certain lands and on the value of certain crops that must use such lands. Thus other things being equal, it would follow that those crops with the most exacting physical demands would bring higher prices than others. Such crops also would have the first choice of suitable land.

However, there are a number of ways to 'beat the system', so to speak. It is possible to make high profits in marginal locations by judiciously substituting land, labour, and capital inputs. For example, wheat is typically a frontier crop, and theoretically, because it is less profitable per hectare than many other crops, it is squeezed out at some distance from the market and from better physical areas. (Storage capabilities and ability to grow in physically marginal areas are other reasons for its being a frontier crop.) But a wheat farmer can also make considerable amounts of money by increasing the land scale of operation. Instead of farming a small unit of 100 hectares or so, the farmer may farm several thousand. Thus, while making only a small profit per hectare, the overall accumulative profit on a farm-unit basis may actually be higher.

SCALE ECONOMIES AND LOCATIONAL STRATEGIES

The general principle of profit is that the average costs per unit of output decrease as the quantity of output or volume increases. Average costs tend to decrease with an increase in scale of operation until a critical point is reached, a point where diseconomies of scale occur. Diseconomies are the opposite of economies of scale. As output increases beyond a certain point, it would incur inefficiencies such that there will be penalties for growth. Even so, economists have become somewhat sceptical about the diseconomies of scale relationship inasmuch as a number of industries and activities appear to be unable to reach this critical point.

This relationship is shown graphically in Figure 4–3. Note that the average cost of producing one unit of output—for example, a bushel of wheat, a car, or a ping pong table—is much higher at quantity Z, where the cost is C, than at Y, where the cost per unit is B, and much higher than at quantity X, where the average cost per unit would be only A. However, when more than quantity X is produced, average costs per unit begin to rise rather than decrease. Obviously, a firm would have to have extremely good arguments for increasing the size of any operation beyond quantity X. Economists argue that the optimum scale size for a firm is where the marginal cost (the cost of producing an additional unit of output) curve intersects the average costs curve.

The general reasons that average costs decrease with increasing volume of output relate to four major cost components. Such components are applicable in different ways to all types of agriculture, to manufacturing, and to wholesale and retail sales. These basic ingredients are (1) specialization, (2) 'discount' purchases of raw materials, (3)

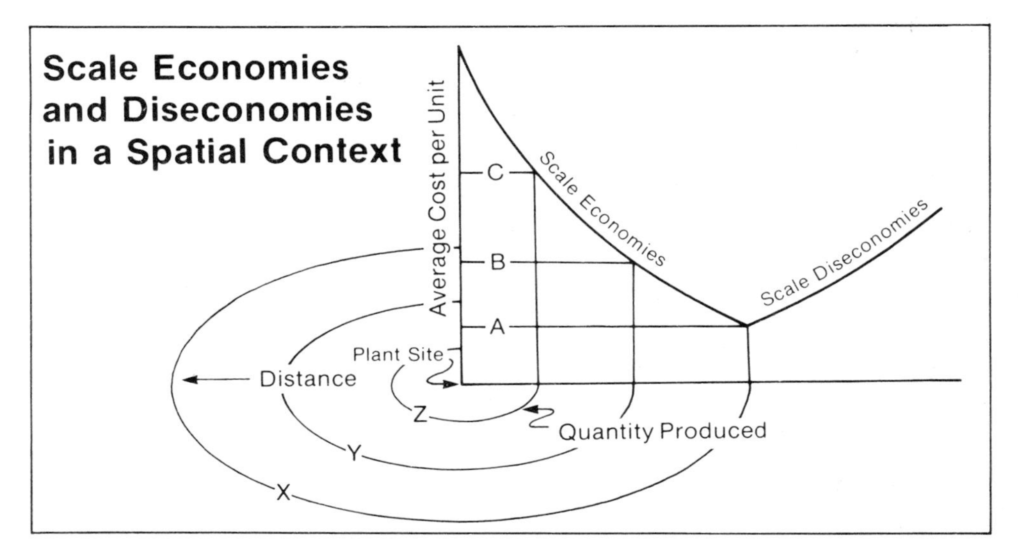

Figure 4–3 Scale economies and diseconomies in a spatial context. As quantity produced at the plant increases, average per unit costs decrease until quantity Z is reached. Beyond this production amount the average costs per unit increase as the result of scale diseconomies. With increasing production more territory is generally covered by the plant. However, increasing territorial extent, as the result of increased production, also necessitates higher transportation costs to the consumer.

vertical integration and bifurcation possibilities, and (4) possibilities for standardization.

Of these four, specialization is the most universal and perhaps the most important. For example, a small operation will have to use one person to do several different jobs. Hence that person does not become a real specialist in any one of these jobs, for the labour is broken down, often among several persons. In a large plant or on a large farm, however, each person can do a specialized job. On a large farm, there might be a manager, an equipment maintenance person, and various other persons doing specialized farm jobs. Labour efficiencies in the manufacturing of automobiles, where each person does a single job fast and well, are a well-known example of the kinds of efficiencies created from labour specialization.

Second, a large firm can get materials and services at a discount compared to prices paid by smaller concerns. A large operation often solicits bids for work to be done and then makes a choice: a small operation is often at the mercy of the sellers. Again, it is the old principle of volume, or quantity, being related to cost and hence to price. For example, a large farm operation would be able to purchase machinery, fuel, seed, and fertilizer, and get better financing and many items at less cost than a small operation. These are advantages which suggest why small farmers frequently band together in co-operatives which buy their supplies in bulk and sell their produce in larger lots. Large firms also use things in greater volume and do not have to maintain proportionally as great an inventory as small firms. This, of course, generally puts small operators at a disadvantage. The disappearance of many small farm and manufacturing operations is the result of just such competition based on this type of scale economy.

Third, large firms often achieve notorious success by owning and thereby controlling related operations. This is called vertical integration and is often achieved by mergers or amalgamations. In industry certain large firms, especially before antitrust legislation, controlled both forward and backward linked activities with regard to their operation. For example, an oil company might own not only the wells but the pipeline and tanker companies used to transport the petroleum to the refinery and to market, where it also owns retail service stations. There is still considerable vertical integration in the petroleum business, but in most countries full control of outlets is prevented by anti-monopoly legislation or by the maintenance of effective competition. Agriculture, like industry, provides examples of large-scale farm operations which are heavily involved in the processing and marketing of their products. And, unlike small enterprises, the large-scale operations provide the possibilities for taking full advantage of formerly unused or waste products. In a large-scale potato farm operation, for example, the cull potatoes might be processed into dehydrated potatoes, and the waste juices from this process might be used to feed cattle. From the cattle, a packing plant business is opened, and so it goes on. By achieving such size, a firm can capitalize on items that would be too small to bother with in a less extensive operation.

Vertical integration and 'spin-off' also often allow a large firm to bifurcate its activities spatially, thereby placing the particular components of the firm's activities in the most beneficial places. For example, power and light companies have long been spatially bifurcated in that they have their accounting and administrative offices in one location, usually downtown, and their service yards elsewhere. Likewise, large manufacturing operations often have their administrative offices in one place and their plants in another.

Fourth, standardization and possibilities for fuller utilization of the factors of production are also made more possible at increased scale. An industrial plant can begin to set up assembly line production with interchangeable parts; a large-scale farm operation can maintain tractors for each purpose and can therefore make more productive use of its machinery.

These relationships have several important geographical implications. First, it is clear that any given quantity of output has spatial implications. For example, the larger the output (volume of sales) of a store serving a surrounding area, the greater will be the trade area for that store (Figure 4–3). We have seen that as production increases, so does the territory required for that much output; as the territory increases, the costs to the firm, and hence the price to the consumer, must increase with distance from the firm. If we follow these arguments (in which it was assumed that price and quantity demanded are interrelated), then it follows that as scale increases, territory increases; and prices to the consumer increase with increasing distance from a plant. Note, however, that an increase in scale will have the effect of lowering the price at the plant but of increasing the transportation cost component to the consumer. This increase would result from the fact that fewer plants would be serving the market with increasing scale of production (Figure 4–4).

If we compare the economies of scale relationships in Figure 4–3 with transportation cost implications as shown in Figure 4–4, it is evident that scale considerations often loom paramount in determining the location of production. Note that a firm producing at production Y at a cost of B would only be competitive to a firm located some distance away and producing at production X at FOB cost A, in the V-shaped area shown in Figure 4–4 (FOB = Free On Board: a procedure whereby the price of the item at any given location is a function of the costs at place of production plus transportation costs to the consumer). Thus the higher cost firm would have only a very limited market as compared to a firm located farther away but operating at greater scale economy. If the two firms were located adjacent to each other, the smaller firm producing at cost B would be unable to compete. Therefore with respect to location of the firm, scale economies are often more important than transportation costs. Because of such economies a business may not be in the optimum location from the standpoint of accessibility to raw materials or to markets, but will be in a position to produce at high volume in order to achieve scale economies. Such a firm may prefer not to minimize such transportation costs but to minimize the cost of on-site production through scale economies.

In this regard scale economies often lead to geographical distortion. Thus perhaps a manufacturer can open a new, larger plant and derive competitive advantages over an

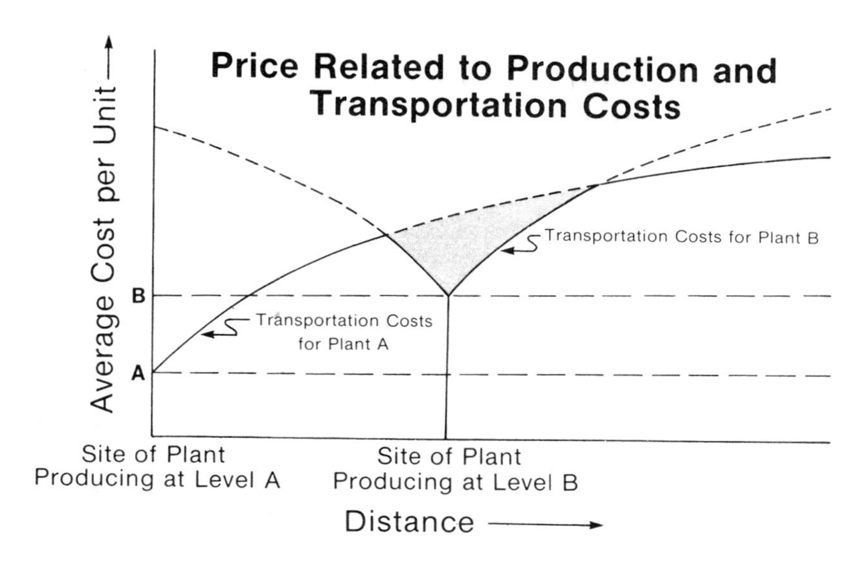

Figure 4–4 Relationship between scale of production and costs to the consumer at various distances from the plant. The lower the costs of production at the plant, the greater the competitive range of the plant for customers (compare Figure 4–3).

older, smaller plant more favourably situated. However, this is extremely difficult to accomplish except over the short run, inasmuch as the older plant may have the more favourable location and will also have possibilities for increasing volume and thereby achieving comparable scale economies.

Another geographical implication of the scale-economies phenomenon is its relationship to the number of producers in a given market. Obviously, if economies of scale can be achieved and if the total market demand is still less than, say, quantity X in Figure 4–3, then one firm could monopolize the entire production. Such a firm, if located so as to minimize total transportation costs as well, might make competition impossible. In fact, in a number of industries in the United Kingdom, such as chemicals, aluminium, and automobile production, only a few firms dominate the market. Clearly these are activities that can achieve tremendous scale economies.

On the other hand, diseconomies of scale may also occur, as demonstrated in the area of the curve to the right of quantity X in Figure 4–3, because extension of any of the four factors discussed above with regard to economies of scale results in an increase in average costs per unit. Transportation costs may also rise to a point where the savings of such costs will outweigh further increases in quantity from any one location. Should such diseconomies occur, then the obvious spatial implications are three: (1) more firms will enter the market, (2) any given firm will wisely decentralize its operation and set up branch plants, or (3) both will occur.

This phenomenon gives outlying or marginal areas an opportunity for production as they produce for a regional or local market. This trend toward decentralization and regional plants has become particularly prominent in the United States during the past decade, while in the United Kingdom it is a development which has been encouraged by government for an even longer period. New automobile assembly plants, regional offices of all kinds, and other regionally oriented manufacturing have become the rule rather than the exception. This leads to a less geographically concentrated distribution of employment, and hence population, than was formerly the case. Even so, there are many other activities that appear still to be achieving economies of scale and are moving from formerly local markets to strategic places like London, New York, and Paris, which are geared to national and even international markets.

EXTERNAL ECONOMIES AND ACTIVITY LOCATION

A factor encouraging concentration in a few areas is external economies. This simply means that where large concentrations of other activities are found, particularly those that are similar, many advantages accrue to a business enterprise. First, such firms receive better service with regard to obtaining a skilled labour force and find it easier to use all kinds of business aids, including legal, educational, financial, and municipal services. In many instances these are services such firms would have to provide for themselves if they were not in an area that afforded such agglomerative or external economies. More concentrated areas, particularly metropolitan areas, also contain an infrastructure of public facilities and private services that is not available in less fortunate areas. These are reasons why private business is loath to migrate to depressed industrial areas like the Appalachian area in the United States, or to relocate in the rural 'Development Areas' in the United Kingdom, even if central or regional government offers financial incentives.

Thus the large city-dominated regions seem to have the advantages in regard to external economies—for example, the industrial belt of the eastern United States, the 'Western Triangle' in North Sea Europe, and the petrochemical industrial complex among the Gulf states of the United States. In fact, most activities found in large industrial complexes benefit from such economies. Perhaps the classic example is the automobile industry in both the USA (Michigan) and England (West Midlands). The automobile assembly plants benefit by the proximity of all types of parts plants that manufacture special items and sell their products to the automobile manufacturers. Such a large service component nearby makes it difficult for the automobile assembly plants to move. After a time it is difficult to tell which is the dog and which is the tail; at any rate, it appears that the external economies' tail is wagging the dog in many industrial complexes. Where the manufacture or assembly of an item runs in spatial contradiction to the manufacture of parts, however, these two functions can be widely separated.

SOME TRANSPORTATION PRICING SYSTEMS

Several of the ways in which a firm may reduce its on-site costs have been demonstrated under the principle of economies of scale and external economies. But a firm can also adopt strategies to increase its price, thereby making greater profit by manipulating the transportation cost variable. Three procedures are widespread: (1) the FOB pricing system, (2) the postage-stamp pricing system, and (3) the basing-point pricing system.

The FOB system creates a spatial configuration that makes prices smaller for those nearer the source of production than for those farther away. According to the supply and demand principles discussed earlier, those people further away who are being charged higher prices will demand less of the item. Thus the density of coverage will decrease outward from the source. From a manufacturer's point of view, this may not be a very beneficial system inasmuch as it may not enable the plant to achieve the quantity of ouput, and hence the economies of scale, that it might under another price-distance system (Figure 4–5).

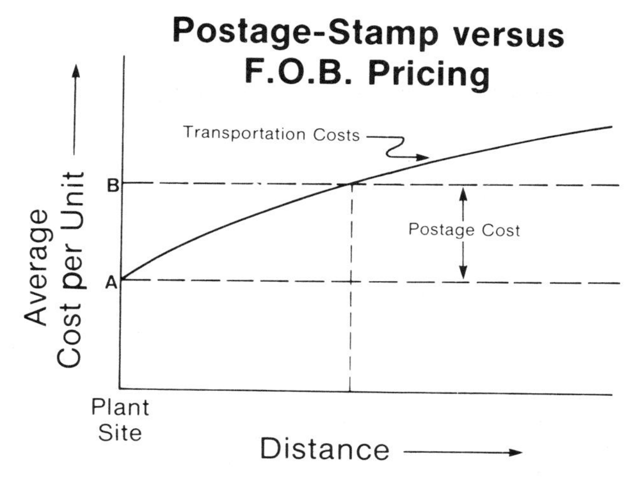

Figure 4–5 The postage-stamp pricing system versus an FOB system. Note that customers near the plant pay more than they would otherwise, whereas customers farther away from the plant pay less. This system is commonly used to boost volume of production and hence achieve scale economies.

Another pricing procedure is the postage-stamp rate system (Figure 4–5). Under this system all consumers are charged the same price regardless of location from the source of production. Thus those closest to the source pay more for the item, and those beyond a critical distance pay less for the item than under the first method. Here the producer location strategy would be to locate nearest to the highest number of potential consumers. In this way the producer could save the most on transportation costs and hence make a greater profit than if located in an area of sparse demand. (Incidentally, under the FOB-plus-transportation system, the plant would make equal profits from each consumer no matter where the consumer was located.) The postage-stamp method is heavily practised by national firms where national price advertising is important and where increased production volume gives scale economies that outweigh transportation cost considerations.

The third method, the basing-point system, is still practised in modified form by a number of large firms (Figure 4–6). Under this system various places are designated as

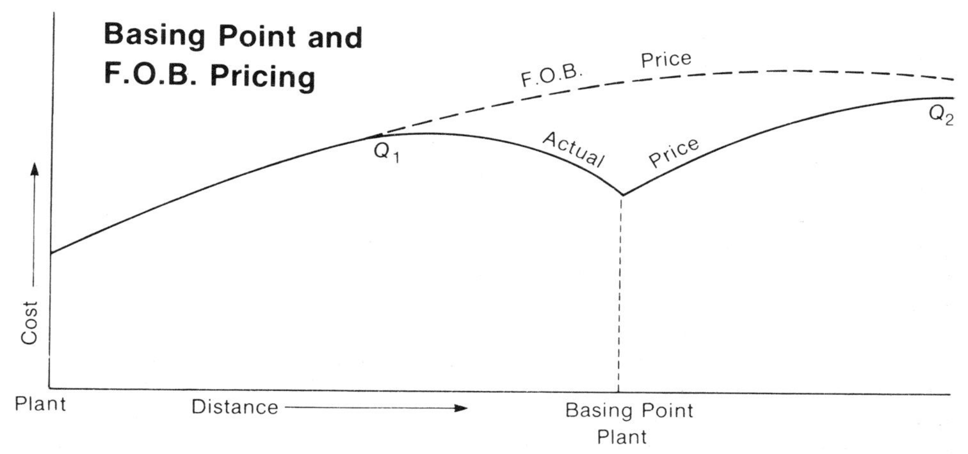

Figure 4–6 The basing-point system of pricing versus an FOB system. Customers near basing points (Q_1 to Q_2) pay less than under an FOB system. Such a situation, however, results in those persons to the left of the basing point to Q_1 paying more for the item (yet less than the FOB price) even though they are progressively closer to the plant.

basing points. Consumers near these points then buy the item as if it had come from this basing point. One result is that prices are much lower at these selected points than they are in the surrounding area. For many activities, therefore, lower prices at these points make for considerably lower production costs. One basing point system practised by the American steel industry, until it was declared unconstitutional, was Pittsburgh Plus. This meant that all steel purchased in the United States was priced as if it had been made in Pittsburgh, even though it often came from steel plants much closer. This generated considerable profits as transportation that was not being used was being paid for by the consumer. The US automobile industry bases its prices on transportation from Detroit, even though the cars are assembled in local plants. In this case, however, most of the component parts are indeed shipped from Michigan but at a rate lower than that charged for a completed automobile.

An interesting geographical aspect of such a pricing system is that transportation costs appear to drop just before they get to the basing point. Thus here is a case where transportation charges are less for further distances than for closer distances. It is difficult to gauge whether such a modified FOB system is beneficial or detrimental to the consumer. Those consumers who live near basing points undoubtedly benefit. Those who live outside of basing-point ranges, points Q on Figure 4–6, are perhaps just paying their fair share. Such a system, however, does favour those areas that are designated as basing points, as well as those areas near them.

SHORT-RUN MAXIMIZATION AND LOCATIONAL STRATEGIES

In the above discussion, simple profit maximization over the short run has been the guiding principle. The entrepreneur rather than the state has been in mind, with regional concerns relative to economic developments almost ignored.

It is quickly evident, however, that goals that may achieve maximum profits over the short run (say, several decades) may not be the same goals that would achieve maximum accumulated profits over the long run. This bias of the models presented is especially critical from a locational analysis standpoint. Locating a new factory, say, in one place may be the most profitable thing to do for the present, but not for the future. In fact, focusing on current profit maximization might allow opportunities for injurious competitive responses in other territories to occur and thereby may jeopardize the future profitability of the operation. A wise locational strategy might even be to operate at a loss for a few years in a new and expanding territory in order to ensure future competitive strength in that area.

It is difficult to find many activities that are based solely on short-run maximization procedures. For example, farmers might plant their total farms for the most profitable crops, but in so doing they might deplete their soil. Most crops require rotation to maintain soil fertility and avoid soil disease. This means planting a crop that is less profitable over the short run. Likewise, a timber company might make high profits by cutting all its available timber in one year, but then what will they do in the future? Of course, for most companies, taxes and other factors mitigate against making excessive short-run profits. Growers of beef animals could sell their entire herd and keep nothing back for animal propagation. Instead, they generally try to maintain a satisfactory profit, to operate at the least cost possible within a set of goals, and to improve their competitive position.

In other cases, profit maximization is further reduced in importance by other, perhaps noneconomic goals. Many case studies show that industrialists seek out the more amenity-laden areas for their plants, other things being somewhat equal. They place high value on an attractive community, good schools, public facilities, and especially on those areas that offer a good physical environment and recreational opportunities. Likewise, some farmers could probably obtain a greater profit by switching crops or location. But as potato farmers, or wheat or dairy farmers, they like the crops or stock they know and do not want to switch. They like the people and the area where they live and they are attached to the land and the region. It would take more than just the possibility of greater profit to get them to move.

Thus, before moving to consider other bases, we may conclude that although the economic bases are critical in the understanding of man and his activities, they must be used with caution. The 'other things being equal' restraint must be kept constantly in mind if we are to comprehend their significance. In fact, it was possibly the empirical knowledge of the earth that kept geographers so long from accepting and using more of the economist's concepts and tools. The principles discussed do have validity when used with proper qualifications; and when used properly, they are indispensable to geographical understanding.

BIBLIOGRAPHY

Chisholm, M. *Geography and Economics*. London: Bell, 1966.

Davies, R. L. *Marketing Geography*. Cambridge: Retailing and Planning Associates, 1976.

Losch, A. *The Economics of Location*. New Haven, Conn.: Yale University Press, 1954.

McCarty, H. H., and Lindberg, J. B. *A Preface to Economic Geography*. Englewood Cliffs, N.J.: Prentice-Hall, 1966.

McKee, D. L., Dean, R. D., and Leahy, W. H. *Regional Economics: Theory and Practice*. New York: Free Press, 1970.

McNee, R. B. 'The changing relationships of economics and economic geography', *Economic Geography*, July 1959, pp. 189–198.

Samuelson, P. A. *Economics: An Introductory Analysis*. New York: McGraw-Hill, 1967.

Smith, A. *The Wealth of Nations*. New York: Random House Edition, 1937.

5

Institutional Bases

This chapter focuses on the differing human characteristics of people as they are reflected within an institutional context. The diversity of religions, ethnic features, languages, and general attitudes demonstrates how greatly people vary in their social and cultural characteristics. Moreover, these diversities are often crystallized into institutional frameworks that can be examined with reference to their spatial impact on people and their distribution of economic activities.

It is through institutional frameworks, such as national governments, that socio-cultural values are translated into actual production, consumption, and occupancy patterns. People communally develop various governmental policies toward production and trade, have distinctive monetary systems, form national alliances, and use institutions in many other ways to determine economic policies that affect the placement of people. This chapter will not attempt to explain why the institutional characteristics of people vary from place to place. Instead, attention will be placed on exploring the effects of these institutional forces on the location of people and their activities.

It is evident that we see ourselves becoming progressively engaged in circular reasoning and explanations. On the one hand, the places where people are located are understood from the foundation of where their activities are best placed. Inasmuch as people are producers and constitute the labour force, they must be located near where they must work. Thus the discussion dealing with the locus of production-oriented activities provides considerable explanation for the location of people themselves. But in another context, people are the market for all production. Their location will strongly influence the range, scale and location of the productive activities and, of course, the services.

Finally, people, through their cultural eyes, decide on the nature of 'usable' land and resources; their attitudes and values are reflected in the nature and extent of their resource-related activities; and their decisions determine the kind of spatial response, if any, to be made. Thus people are the final judge and arbiter of what they will and will not do in any given place and in any given situation.

NATIONAL STATES AS INSTITUTIONS

The most important single institutional factor affecting the location and distribution of differing activities is the creation of national states or countries. Each country sets its own economic policy, decides on the priorities and allocation of its resources, and can, in short, set the stage for progress or decline in its territory. The importance of understanding the role of national states in the arrangement of things on the earth is apparent when we realize that there are almost 200 different independent nations. One-quarter of these nations, mostly in Africa, has emerged during the past several decades. Thus many nations are still novices in the governing of activities and in the setting of policy.

Nonetheless, many governments of the world are the prime decision-makers with regard to the spatial allocation of various activities. This is especially true in countries with Communist governments and in many newly emerging nations vitally concerned with economic development and the priorities of resource use. It is also much the case in dictatorships and in capitalist democracies alike. First, most governments actually own or control certain economic activities, such as railways, airways, postal services, banks, power plants, major factories, roads and highways and parks. (Communist governments control all economic activities.) Thus the extent and the location of such activities is strictly a government decision made within the value system of each national state. Another way in which government affects the location of things is through national taxation systems. Government channelling of resource use results in different distributions of man and his activities among various countries. Government policy and government works thus are often first-order factors in any such explanation.

For example, the government of Venezuela, a country rich in petroleum, until recently decided that it would offer favourable conditions for petroleum exploitation to foreign companies. By using this money properly, the Venezuelan government plans to build up a wider and more diversified range of activities. Oil is similarly used as a basis for economic development in Nigeria, Libya and in the various states in the Persian Gulf. In these countries, petroleum is deliberately used (at varying rates of depletion) in hope of achieving something better in the long run. In the United States there are strict controls of the amount of petroleum that can be taken from any given area or well. In the United Kingdom, the national government determines how the development of North Sea oil and gas is to proceed, and where its landward installations may be located. It has also to calculate the expected yield in oil company taxes to the Exchequer and how these very considerable revenues can be deployed to the greatest benefit.

In short, the decision as to when, how and where a particular resource in a country will be used now, in the future, or not at all is often made by the government, mainly through its export policy, regardless of the political form of that government. Thus more important to the economic geographer than the political nature of a country is the nature of its policy with regard to particular commodities and activities. An easy measure of such policy would be a country's trade relations with other countries as they might affect the distribution of people and activities. Such relations can perhaps be covered most simply by an examination of the formal political and trade alliances among countries of the world.

Regional Political Alliances Among Nations

In an area having similar cultural features—language, race, religion, and technological stage—and sharing common aspirations, the formal national policies of states will often reflect these common values. In the context of world regions, the national associations, expressed through various political, military, and trade agreements and alliances, are most revealing.

Six major formal political-military alliances are worthy of note: (1) the Warsaw Pact, which ties the Soviet Union politically and militarily to its satellites in Eastern Europe; (2) the Organization of African Unity (OAU), which links some 35 African nations having common interests in regard to development; (3) the Organization of American States (OAS), which contains some 20 Latin American nations and the United States in a collective arrangement to settle disputes and to render mutual military support against Communist interference; (4) the North Atlantic Treaty Organization (NATO), a defence alliance of some 13 European nations, the United States, and Canada; (5) its counterpart in Asia, the South East Asia Treaty Organization (SEATO), which includes some eight countries in that area; and (6) the Central Treaty Organization (CENTO), composed of Great Britain, Iran, Pakistan, and Turkey for purposes of mutual defence. These political associations, though primarily defence-oriented, clearly set the stage for furthering economic interests and forming world-wide economic blocs.

Two types of alliances are evident: (1) those that are primarily compacts among culturally similar countries for mutual defence (OAU, OAS, and the Warsaw Pact), and (2) those that involve patronage support from outside areas and cultures (SEATO, CENTO, and NATO).

National Trade Groups

The political-military associations are further brought into focus by an examination of formal trade associations of nations. Here the general attempt is to facilitate trade among certain selected nations, generally those having strong social and cultural similarities. Thus, paradoxically, trade, which presumably best occurs among different types of areas, is moulded through such arrangements to include primarily those countries that are culturally most alike. As might be expected, therefore, many have not been very successful, particularly those in South America, Africa, and Asia where most nations have more trade demands with nations in other areas than with their nearby neighbours. Nevertheless, these trade advantages undoubtedly encourage more trade among such similar cultural areas than would otherwise be the case. By the same token it discourages trade among areas that otherwise might logically appear to have need for interaction.

Historically the most territorially extensive trade alliance in the world has been the British Commonwealth of Nations. These 29 independent countries and about 40 dependencies covered almost one-quarter of the earth's land surface. Foremost among these nations were former colonies such as New Zealand, Australia, Canada, India, Pakistan, Kenya, Nigeria, and many other African countries. The commonwealth nations were given preferential treatment with Great Britain whereby for many commodities no tariffs were charged. With Great Britain's entry into the Common Market, some preferential treatment is still extended to some of her commonwealth countries' exports (for example, New Zealand dairy products in Europe). The success of this commonwealth system, which includes a tremendous diversity of physical, political, and economic systems, is demonstrated by the fact that about one-quarter of all non-Communist world trade occurs within this bloc—a bloc in which Great Britain was the kingpin. In recent decades, commonwealth countries have found it increasingly profitable to trade with other countries rather than with the 'motherland'. But historically, this system operated much to the advantage of the leader nation, which received industrial raw material and food from commonwealth members and traded them finished products in return. Latterly, however, as some of the member nations have become more industrialized—for instance, Canada and Australia—they responded less and less to this arrangement because they too prefer to import raw materials and export finished products. Or, as in the case of India and Australia, they have placed high import duties on manufactured products from other areas, including Great Britain, to encourage local manufacturing production.

Since the Treaty of Rome in 1957, six European nations—Italy, France, West Germany, and the Benelux countries—have banded together in order to form an economically integrated new community, the European Economic Community. The Common Market as it has become known has reached significant agreements on the flow of industrial materials and labour and is moving rapidly towards a fully integrated economic union, with common policies related to agriculture, industry, transportation, trade and monetary matters. In 1972 three other nations—the United Kingdom, Denmark, and Ireland—were admitted. Further new member-nations may strengthen the EEC (Figure 5–1).

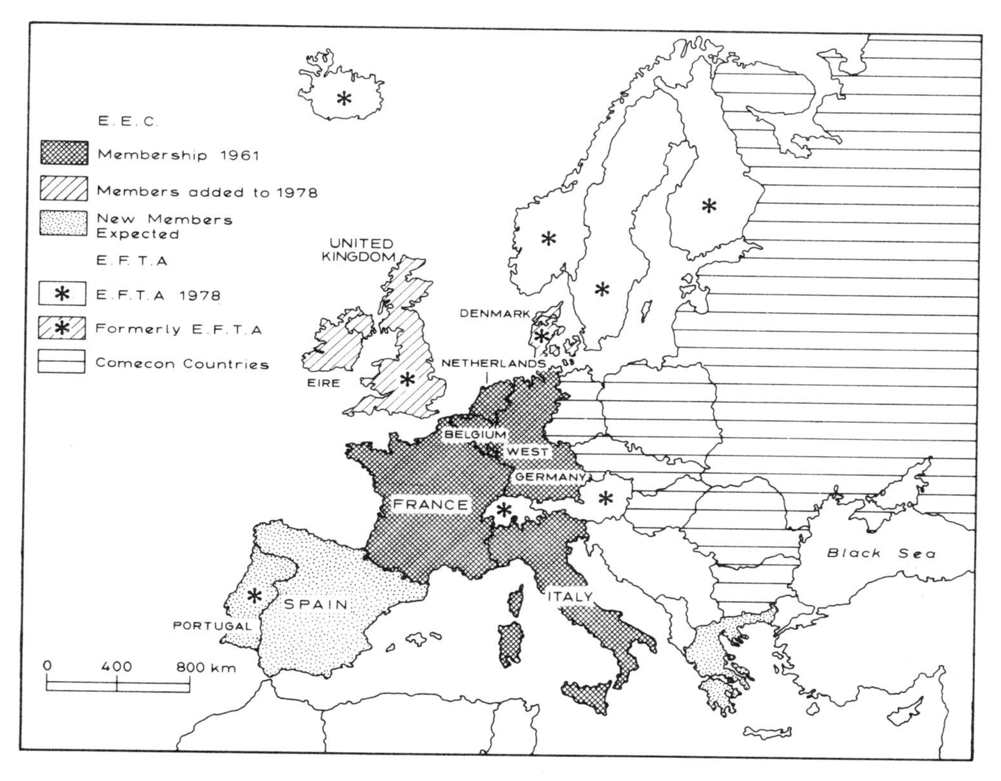

Figure 5–1 EEC countries and their relationships. The European Economic Community represents a most significant economic and political bloc in world affairs. At present, it is a nine-nation union which contains a population of over 260 millions as compared with just over 200 millions in the United States.

The ultimate hope of many Europeans is that the nations of the EEC will progress towards political union, perhaps leading to the emergence of a new 'United States of Europe'. But it may be noted that seven European States (Austria, Finland, Iceland, Norway, Portugal, Sweden and Switzerland) still remain outside the EEC as members of the European Free Trade Association. The aim of EFTA (more modest than the EEC) has been to eliminate tariffs on a wide range of industrial products and commodities between its members. The Soviet-bloc countries have also set up an Eastern European 'free trade territory' which includes the Soviet Union and eight of its East European Satellite nations, under the Council of Mutual Economic Assistance (COMECON). Neither of these trade groups is so important internationally as the EEC. Most of the grossed up financial and economic measures (like value of import and export dollars, gold reserves, industrial output) rank the present nine countries of the EEC superior to any single nation on earth. Only in overall gross national product does the United States exceed the European community.

In 1961 a trade arrangement similar to the Common Market, the Central American

Common Market (CACM), was formed in Central America. This included the countries of Costa Rica, El Salvador, Guatemala, Honduras, and Nicaragua. The purpose was to stimulate mutual trade through reduction and elimination of tariffs. However, as might be expected, most trade is outside this area.

The world's largest free trade association is the Latin American Free Trade Association (LAFTA), which includes some eleven major Latin American countries. However, since its inception in 1961, progress has been extremely slow. First, there is the problem of adequate transportation among the many countries (the Pan American highway is still not complete). Second, the products each has to offer have more demand outside of Central and South America than within it. Finally, there have been problems of governmental disorder and general disagreements on the extent to which tariffs should be lowered. Yet over a decade ago, an agreement was made by some 19 Latin American nations and the United States that a common market would be established. Slowly the LAFTA common market has changed production and trade patterns. For example, Chile is now buying cotton from Mexico rather than from the United States, Mexico now sells steel to several Latin American countries and Brazil continues to industrialize and to sell manufactured items throughout Latin America.

Commodity Agreements

Commodities that move heavily in international trade are usually controlled by commodity agreements among the affected nations. Whereas the trade agreements discussed above are regional in nature, these are topical. Wheat, coffee, and sugar are examples of internationally controlled commodities. Such controls are developed in order to stabilize prices, ensure continuity in production areas, and gain some commitments of various types from consuming areas. A multitude of crops are protected and given special treatment by reciprocal trade agreements between any two countries. Such favoured treatment is often characterized by guaranteed quotas, lowered or discontinued custom duties, and various forms of price support. In other instances, unilateral action is taken by a single country, as in the case of sugar, discussed below. The end result of such agreements is that certain areas are given preferential and protected status with regard to particular commodities, and other areas are discriminated against.

In 1948, more than 40 countries signed the International Wheat Agreement. These wheat-growing nations agreed to share the existing world market for wheat rather than for any one of them to glut the market during a highly productive year. Theoretically this agreement was to protect the world price of wheat. It has been only partially successful. The Soviet Union withdrew as early as 1949 because it thought its quota too low—it had a good wheat crop in 1949. Great Britain dropped out in 1953 because it thought the import price was pegged too high. The United States usually has stockpiled its carry-over (surplus) and has maintained two price systems, one for domestic consumption and one for world trade. Of course, those wheat-growing nations that did not sign the wheat agreement have had a field day, periodically selling at below the pegged world prices. Consequently, nations like the United States have adopted price support and various other plans to control domestic production. The end result is that the production and price of world wheat is highly distorted by such national self-interest and general commodity agreements.

Coffee is another crop that has gone through various periods of surplus and deficit and is likewise controlled, at least among certain countries. The stockpiling of coffee in Brazil and the initiating of artificially high prices has encouraged coffee production in other places, such as the Ivory Coast and Mexico. As a consequence, in 1963 in the United Nations an International Coffee Agreement was signed whereby exporting countries accepted export quotas and importing countries accepted a lower limit on coffee prices and further agreed to limit purchases from those nations not signing the coffee

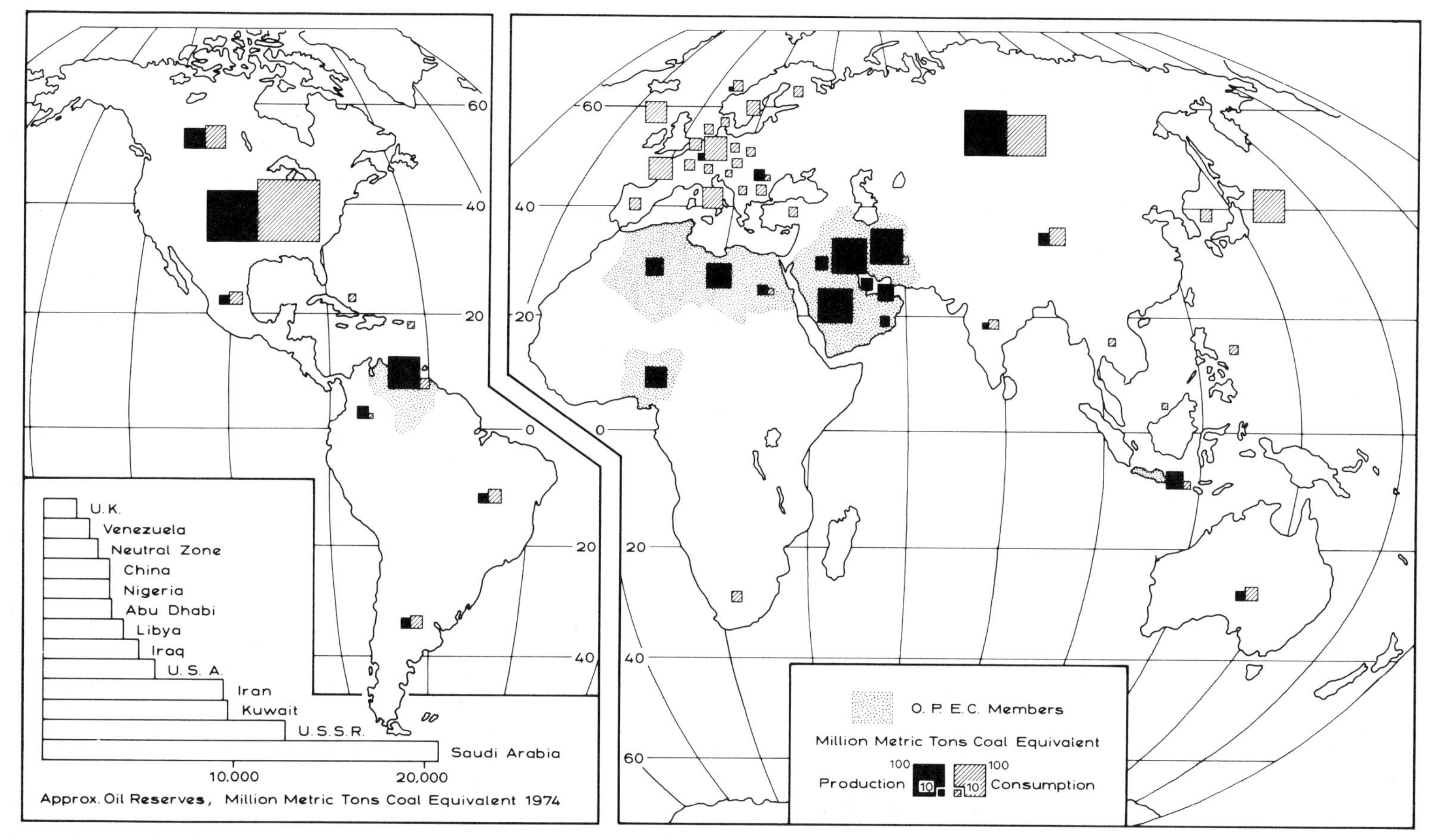

Figure 5–2 OPEC membership and the scale of oil wealth. This commodity cartel among nations, formed on the basis of a critical energy commodity, offers economic and political problems for many of the more urbanized and industrialized nations.

agreement. Controversy, however, continues. For some years it had been assumed that sufficient coffee was stockpiled in Brazil and other countries in a deliberate attempt to keep prices stable by controlling the available supply. Few were concerned when a freeze injured the coffee crop in 1976. However, the result was skyrocketing coffee prices, presumably because demand quickly exceeded supply. Many people contend that coffee, like petroleum and other commodities, is simply being artificially manipulated by such countries in order to create higher prices.

Sugar has been another item of long-standing controversy. It can be produced from a number of sources but comes primarily from sugar beets and sugar cane. The latter is normally a more efficient source, particularly as grown on large tropical-area plantations. But national interests have long protected beet sugar production in midlatitude nations like Great Britain and France, where production quotas or acreage subsidies have been applied to maintain farm incomes, ensure the continuity of home supplies, and act beneficially in an import-saving role. However, the problem is more complex in countries where sugar can be produced from both sources. Thus in the United States, where Hawaii and Louisiana produce cane sugar and California, Idaho, and Colorado produce beet sugar, the federal government has arbitrarily controlled the production and price of both types of sugar. In addition, it has set up import quotas in dealing with other countries.

A classic example of trade restrictions preventing general fulfilment of the logical transportation, physical, and economic features is the sugar embargo imposed on Cuba by the United States. Cuba is considered to have the comparative advantage over most other sugar producing areas of the earth. This is certainly the case with regard to shipment of sugar to the United States. In fact, before the USA's embargo, Cuba supplied over one-third of all sugar imports to the United States. The consequences of this action was that less efficient sources, both foreign and domestic, increased their production. Other things being equal, this would have had the effect of increasing prices drastically, but inasmuch as United States prices were already pegged artificially high, the increase was small. Almost every country, through tariff duties, taxes, subsidies, and other means, has exerted some controls over sugar pricing and hence over the location of sugar-producing areas. Sugar is not an isolated example, however; most crops moving heavily in international trade are controlled by some of these methods.

These cases demonstrate ways in which national policies can neutralize certain areas that are otherwise favourably endowed or provide advantages and encourage activities in areas that otherwise could not compete. Such policies are clearly injurious to the best use of the land from a physical and economic sense and constitute one of the major arguments presented by economists and geographers in favour of free trade. Nevertheless, from the standpoint of an individual nation in competition with the rest of the world, some of these policies make a great deal of sense.

Undoubtedly the most important commodity agreement among nations today is the Organization of Petroleum Exporting Countries (OPEC), founded in 1960 (Figure 5–2). It was initially formed by Iran, Venezuela, and Saudi Arabia but has grown to include Algeria, Ecuador, Gabon, Indonesia, Iraq, Kuwait, Libya, Nigeria, Qatar, and United Arab Emirates. Petroleum export dollars amount to 8 to 12 times the value of other export crops such as coffee, copper, sugar, rubber, or cotton. Income to the oil-producing nations from petroleum export now amounts to some 15 billion dollars annually and is expected to reach as much as 40 billion dollars annually by 1980. The Middle East alone accounts for 60 per cent of the world's known petroleum reserves. This organization ranks as the single most significant one among commodity cartels in the history of the world. In power and influence it may exceed the Common Market; indeed, without lasting agreement on oil-pricing being reached, these two powerful political-economic blocs appear to be on an economic collision course.

CULTURAL INFLUENCES OF INSTITUTIONAL ARRANGEMENTS

The foregoing discussion of various alliances and arrangements among countries requires augmentation. We can no longer be content to take all of these as givens; some general considerations as to 'why' must now be entertained. The affinities and disaffinities among nations are, of course, highly complex. To understand them fully, they must be placed within a historical and economic context, but much can be gathered simply through an examination of the general cultural features of various nations as these might affect their relationships. In particular, it is instructive to examine three great encouragements and/or encumbrances to co-operation among nations: (1) religion, (2) language, and (3) stage of technology.

Religion

Religion is one of the most pervasive factors affecting the nature of governmental policies and the general welfare, though often in a subtle fashion. First, religion affects people's ways of life and hence their values in regard to social, economic, and legal institutions and systems. The difficulties engendered by the caste system in India and by that country's custom of considering the cow a sacred animal are well known. Religious restrictions with regard to food occur in many groups: until recent times Catholics could not eat meat on Fridays, Jews and Moslems may not eat pork, Hindus may not eat beef. Some groups, such as Mormons and Catholics, place high value on large families and believe it is their religious duty to 'multiply and replenish the earth'. Thus seemingly simple solutions to some of the economic problems of the earth (for instance, instituting birth control, ending the plethora of unused cattle in India, reducing the consumption of fish, or raising the consumption of milk) may not be applicable in the face of religion-based customs.

Some argue that religion plays a less dominant role in the more economically advanced nations. But even in the United States where formal religion has been conscientiously separated from state government, the Protestant ethic is still strongly reflected in governmental policies. The tax-free status accorded to religious institutions is a case in point. The pervasive influence of religion can also be demonstrated with regard to the work pattern and holidays. Sunday is considered a 'day of rest'. Clearly a Christian work taboo, which still has great importance for productivity and the nature of the economy. Incidentally, the 'day of rest' custom is better suited to an urban-industrial economy than to an agricultural one, where uninterrupted periods of intensive planting and harvesting are required; yet it was more pervasive when the urban-industrial countries were predominantly rural.

The effect of religion is also evident on the landscape itself. Religious institutions own or use large amounts of territory. They directly control land for churches, schools, and various other religious or religion-based activities. In fact, in many communities the cathedrals, churches, mosques, and temples are perhaps the most distinctive structures. And a major land use in the Christian world (and in China, incidentally) is for cemeteries, clearly a need based on religion. Other cultures, such as the Hindu, Buddhist, and Japanese, commonly cremate rather than bury their dead. The number of associated activities required for such religious practices is considerable. In fact, one might reach the conclusion that only economically advanced nations can afford to put so much emphasis on economically nonproductive structures and activities without seriously overtaxing the economic system.

There are hundreds of different religions in the world, but eight major ones are Christianity, Islam, Hinduism, Confucianism, Buddhism, Shintoism, Taoism, and Judaism. On a world basis the first six of these constitute the main religious blocs. The 'Western' countries of northwestern Europe, the countries of North and South America, and the European outliers of Australia and New Zealand are Christian in general philosophy. So too are many other areas such as the Philippine Islands, Hawaii, and other Pacific islands—largely as a result of missionary work during the last century. The Middle East, North Africa, and Pakistan are primarily Islamic (Moslem), and the other Asiatic areas are strong in the tradition of Buddhism and its associated religions.

It is also important to note conflicts of a religious or quasireligious nature. Hindu versus Moslem in the Indian sub-continent, Arab versus Jew in the Middle East, Catholic versus Protestant in Northern Ireland, are all examples each of which has dire economic, as well as social and political implications. And we can add antireligious problems which affect many peoples: the anti-Christian policies adopted by many Communist nations, prohibitions on the preaching of Christianity in many Moslem states. In many countries, sex, creed, colour and religion have all been bound up together in the matter of job discrimination.

We may also recall that many religious groups set out to establish their own colonies in the New World, and in modern times even to found their own nations. In Israel, it may not be overstatement to suggest that the occupancy of land and the character of the settlements are determined by a religion-inspired policy. In some other countries, sectarian influences are so strong that it results in the widespread duplication of facilities, with more establishments than would be expected on the basis of population numbers alone. Thus in many more ways than most people realize, religion is reflected in politics and in national policies, which in turn affect locational decisions among and within countries.

Ethnic Groups

Another feature generally affecting the institutional arrangements among and within nations is what is popularly termed 'race'. In point of fact, biologically all people belong to one common pool of genes and form one mating circle. There is no scientific basis for distinguishing among the earth's people racially. All evidence points to one family and one origin. Nonetheless, there are vast outward variations among the population in kinds of skin colour, types of hair, facial features, and other characteristics.

The most generally recognized crude categorization today consists of three types: (a) the Caucasoid (white-skinned, and presumably Japhetic), (b) the Mongoloid (yellow-skinned), and (c) the Negroid (dark-skinned). There are many subtypes and many other classifications. Many of the major differences in world patterns of trade and general conflict in resource use appear to be among the Caucasoid, Mongoloid, and Negroid. Where these differences also tend to coincide nationally with religious and language differences, the implications for world activity patterns are direct and potentially violent.

Unfortunately, much of the trouble within countries is also of a racial nature, and where racial differences are strongly evident, this likewise leads to discernibly different spatial patterns among the groups. Again, the result is often a duplication, although certainly not an equal provision, of equal functions. Perhaps the most notorious case is that of South Africa, where the Caucasoids (Europeans) have forced the Negroids (Bantus) to live in special areas. Such an apartheid policy is ostensibly based on the assumption that the white-skinned population is currently so advanced relative to the Bantus that separation is necessary for the protection of the Bantus. South Africa is not alone in discriminating against peoples with ethnic backgrounds quite different from the ruling group (in the case of South Africa, a ruling minority). Most Caucasoid nations discriminate against Mongoloid and Negroid peoples. Certainly the immigration policies in Australia and the United States are racially biased. The Mongoloids also practise discrimination (China is a case in point). Moreover, it is unquestionably true that national states have been formed primarily on the basis of ethnic traits. The division of Hispaniola into two nations, 'black' Haiti on the west and relatively white Dominican Republic on the east, is a good example.

Finally, we would be delinquent not to mention a major land-use feature in the United States that is based primarily on ethnic backgrounds, namely, the extensive Indian reservations. Whereas Australians early hunted the Tasmanian peoples for sport and thereby exterminated them, the United States placed its indigenous population in territorial prisons. Most other countries have not taken such drastic action, and considerable intermixing has occurred. Still, racial conflicts are visible on the landscapes of many areas of the world. The ghettos and 'Chinatowns' throughout much of the Western world are examples. Many African and Asian towns are also community-segregated, each quarter reflecting the tribal origins of its inhabitants.

Language

Language, religion and race constitute an important trinity in cultural distinction; its importance in isolating various peoples of the earth is evident. Lack of communication presents barriers not only to trade but to general understanding. Language differences in particular present a barrier to the transmission of technological innovations and to economic development and progress. Not uncommon are the examples of different languages among upland and lowland peoples in the same general area, and a consequent disparity in their economic development. Generally, those groups able to communicate with the wider group are the most prosperous and the most economically developed. Groups that find themselves isolated linguistically as well as geographically are often the most economically backward. (These peoples are discussed in Chapter 7.)

A people whose language is spoken only in a very limited area can communicate with the outside world or with other small groups nearby through what is termed lingua franca, a simplified, generally understood trading language used over a wide area by people of varying native tongues. The need for lingua francas becomes evident when we realize that there are hundreds of different languages, belonging to more than 20 major language families, spoken on earth. Furthermore, even people whose languages belong to the same family cannot understand one another.

As with religion and ethnic characteristics, when languages differ within and between countries, difficulties and conflict often result. The obvious exception to this is Switzerland, which has consistently been one of the most peaceful of nations although German is spoken in the northeast, Italian in the south, French in the west, and Romansch in the far southeast.

That language affects the location of economic activities can be appreciated by comparing the relations of the United States with Canada and Mexico. US relations with Spanish-speaking Mexico have never been very well established, with American firms reluctant to invest in the country. The proximity of the two countries would suggest that a higher level of social and economic interaction would take place if no language barrier existed. Canada, like Mexico, is coterminus with the United States, and is highly industrialized and prosperous. US firms have large investments in Canada, and most products move quite freely across the border, which is one of the freest in the world. Thus the United States and Canada are closely tied. Yet within Canada, a nation which has two official languages, English and French, we can also see how language differences can affect economic life. The English-speaking (Anglophone) majority (17 millions, 71 per cent) is spread over ten provinces, while the French-speaking (Francophone) populace (5.5 millions, 29 per cent) is mainly located in Quebec, where French-speakers outnumber English-speakers by six to one. But even in Quebec, the language of trade and commerce is English, a circumstance which has discriminated against the progress of French-only persons. Their political rallying-point became the Parti Quebecois which, coming to power in 1975, set about legislation which strongly redressed the balance in favour of the Francophone community, particularly in the important areas of education and business. As a result, many firms in Montreal with nation-wide connections or US links have relocated themselves in Toronto or elsewhere in Anglophone Canada. There is no official estimate of how many of the million anglophones have left Quebec, but it may run into tens of thousands.

Within Europe, there are many instances where minority language groups within the nation-state claim that they are discriminated against by the more powerful majority group in economic as in other ways: the Flemings in Belgium, Basques in Spain, Ukrainians in Russia are historic examples. Indeed, similar claims are made within the British Isles, where Scottish and Irish Gaelic-speaking and Welsh-speaking communities of the 'Celtic fringe' rally around language, culture and homeland to press for more recognition within the framework of government-sponsored regional development programmes.

Stage of Technology

In practice, the stage of technology is often interrelated with the religious, ethnic, and language patterns just discussed. With regard to current and future economic activities, stage of technology is undoubtedly the critical item. Here again it is like a two-edged sword: on the one hand, we will take it as given in order to gain some understanding of why people and activities are distributed as they are on the earth. On the other hand, this entire text is geared to provide some further explanation as to why there is such a differential in the level of technology and economic well-being among the various parts of the world. Hence we are discussing stages of technology here only in a preliminary context that will prove valuable in the analysis of certain activities to be discussed later.

Many measures might be used to separate the levels of technology and stage of economic development among areas and nations: per capita consumption of goods, services, and energy, per capita production of goods and services (gross national product), and, of course, per capita income. Other measures might include productivity per worker and occupational characteristics. Regionalization of technology levels depends on the number of regions desired. For example, under a twofold division, one would need to distinguish only between the have and the have-not nations, between the poor and the rich, or between the developed and the underdeveloped. However, such a twofold division does an injustice to the wide differences among the so-called underdeveloped areas—some are intensive agricultural systems, whereas the economy of others is based on the most rudimentary forms of hunting and gathering. Thus under a threefold division the countries with 'primitive' economies are differentiated from underdeveloped and overpopulated countries (Figure 5–3).

First, let us examine the level of development among nations on the basis of the primary, secondary, tertiary, and quaternary breakdown discussed earlier. Remember, however, that this is on a national basis and must necessarily ignore anomalies within any given country. Therefore the information for those countries with large territorial boundaries might be considered the most unreliable in this regard. About three groups of countries emerge:

1. Countries in which most of the labour force (over 60 per cent) is engaged in primary activities, with few workers in the secondary, tertiary, or quaternary sectors. In this category are found China, India, and the countries concentrated in Southeast Asia.

2. Countries in which there is a mixture of all the employment activities. In this category are found the USSR and many smaller countries.

3. Countries, generally considered advanced, in which there is high employment in the tertiary and quaternary sectors, fairly high employment in the secondary sector, and very little employment in the primary sector. Such countries include the United Kingdom, the United States, Canada, Australia, New Zealand, and certain other countries in Western Europe (Figure 5–3).

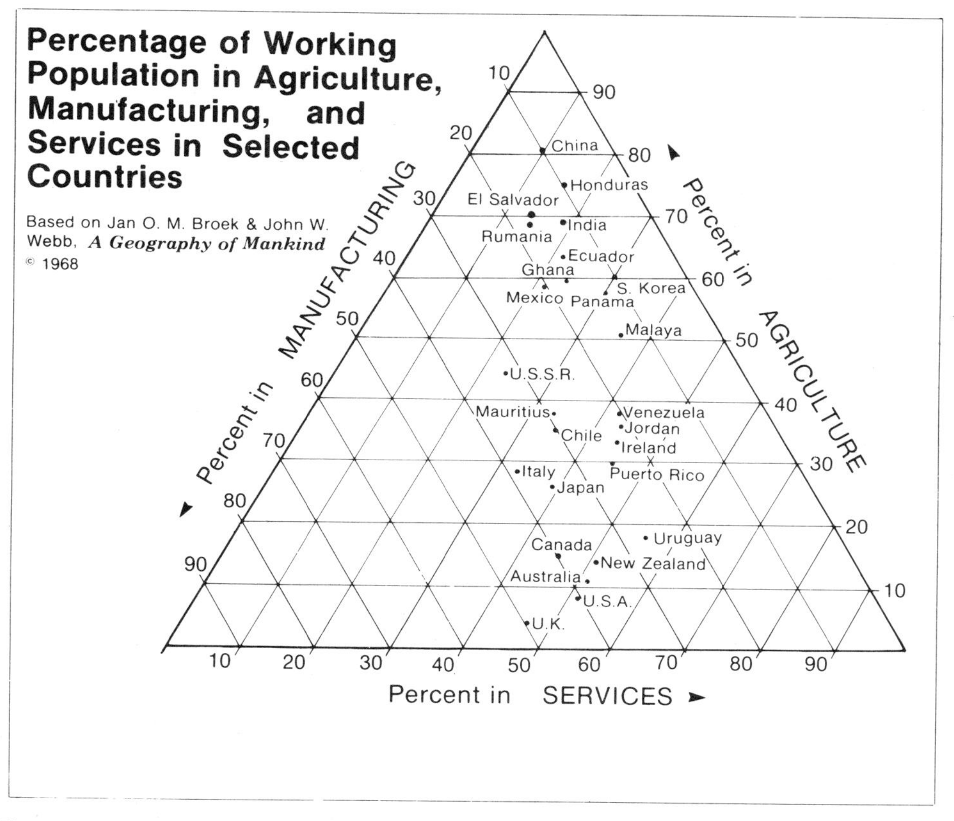

Figure 5–3 Employment proportions in selected countries. Example: the position of the dot for the UK indicates 46% in services (including transportation and utilities), 50% in manufacturing (including mining and construction) and 4% in agriculture (including fishing and forestry). (Based on Broeck and Webb. *A Geography of Mankind*, 1968. By permission of McGraw-Hill Book Company.)

One approach to the differentiation among nations as based on a multiplicity of criteria is that by Broek and Webb (Jan O. M. Broek and John W. Webb, *A Geography of Mankind* (New York: McGraw-Hill, 1968), pp. 326–327.) In their study a fourfold breakdown is presented: (1) advanced, (2) intermediate, (3) underdeveloped, upper group, and (4) underdeveloped, lower group. As might be anticipated, the nations in the advanced group include the United States, Canada, most of Western Europe, Australia, and New Zealand. These are the most urbanized of the world's nations aside from Japan. Japan is listed as an 'intermediate' developed nation along with the USSR, South Arica, Argentina, Chile, Uruguay, Venezuela, Panama, Cuba, Spain, Portugal, and Israel. The 'underdeveloped' countries in the upper group include the remainder of Central and South America, Southern Rhodesia, most of North Africa, most of the Middle East, Pakistan, India, China, Korea, Malaya, and the Philippines.

Such a static classification, although useful for general orientation purposes at any one time, places some very different countries in the same category. For example, Japan is considered an 'intermediate' country by the criteria used but is otherwise quite different from USSR, South Africa, and Argentina. In many respects Japan is highly developed. Likewise, India and China may have living standards somewhat similar to those of the Arabs of northern Africa but are otherwise so culturally and occupationally different that we would not generally group them together.

World 'Cultural' Blocs

If we consider the combined influence of religion, race and language and stage of technology, some notion of the differences in attitudes, values, and economic characteristics of various peoples can be placed in a spatial context. Certainly, these are not the only factors making up culture in the anthropological use of the term, but these factors do provide a fair approximation of how one might visualize the world in terms of the human bases.

From these bases, six broad types of cultural territories can be readily distinguished: (1) Western-Christian, (2) Communist, (3) Islamic-Arab, (4) Indian, (5) Southeast and Maritime Asian, and (6) tribal and other (Figure 5–4). Each of these could be further subdivided, but in general such culturally based regions are also fairly closely correlated with general stage of economic development and general ethnic, religious, and linguistic characteristics. Thus the Western-Christian portions are generally industrialized, whereas tribal and other areas are characterized by primary economic activities. Nevertheless, there is mixing in a number of places, but only the predominant culture type is shown.

In all but the Southeast and Maritime Asian and the tribal cultural regions, such similarly classified areas can be thought of as having strong economic and political ties. Thus these areas often coincide with trade patterns and certainly with the formation of trade agreements. Where trade crosses the first-order cultural realms, it is only because of strong complementarity among areas. Such interaction, however, is often tense at best. Some heavy world flows of this type include petroleum from the Islamic Middle East to Western Europe, tropical products from Africa to Western countries, and the shipment of Canadian wheat to the People's Republic of China. Even the reciprocal trade between Japan and the United Kingdom, two highly industrialized nations, is not without certain culturally based difficulties. Perhaps because of such cultural differences, the general rule is for most trade to occur within these blocs rather than between them. Nonetheless, the most urbanized and industrialized areas require the most material and have greater interaction with other dissimilar places than those at the less developed part of the scale.

Finally, there is no doubt that the national states are regionalizing and fortifying themselves into large political, cultural, military, and economic alliances. The forces of national groupings are seen here only in embryo stage: the future world units will be even more dominant and evident. Certainly the trend and the intention are clear. For example, many African nations are trying to co-operate for mutual political and economic benefit, and the Latin American countries and the United States are attempting to develop an integrated territorial and industrial complex. Western Europe is working toward what one author has called 'The Logic of Unity' (see Bibliography): in 1979 all the Common Market countries, including Great Britain, will send their elected 'Euro MPs' to the European Parliament in Strasbourg and Luxembourg. These trends, coupled with the already highly integrated Soviet bloc, will be major factors in determining the geographical reality in the future.

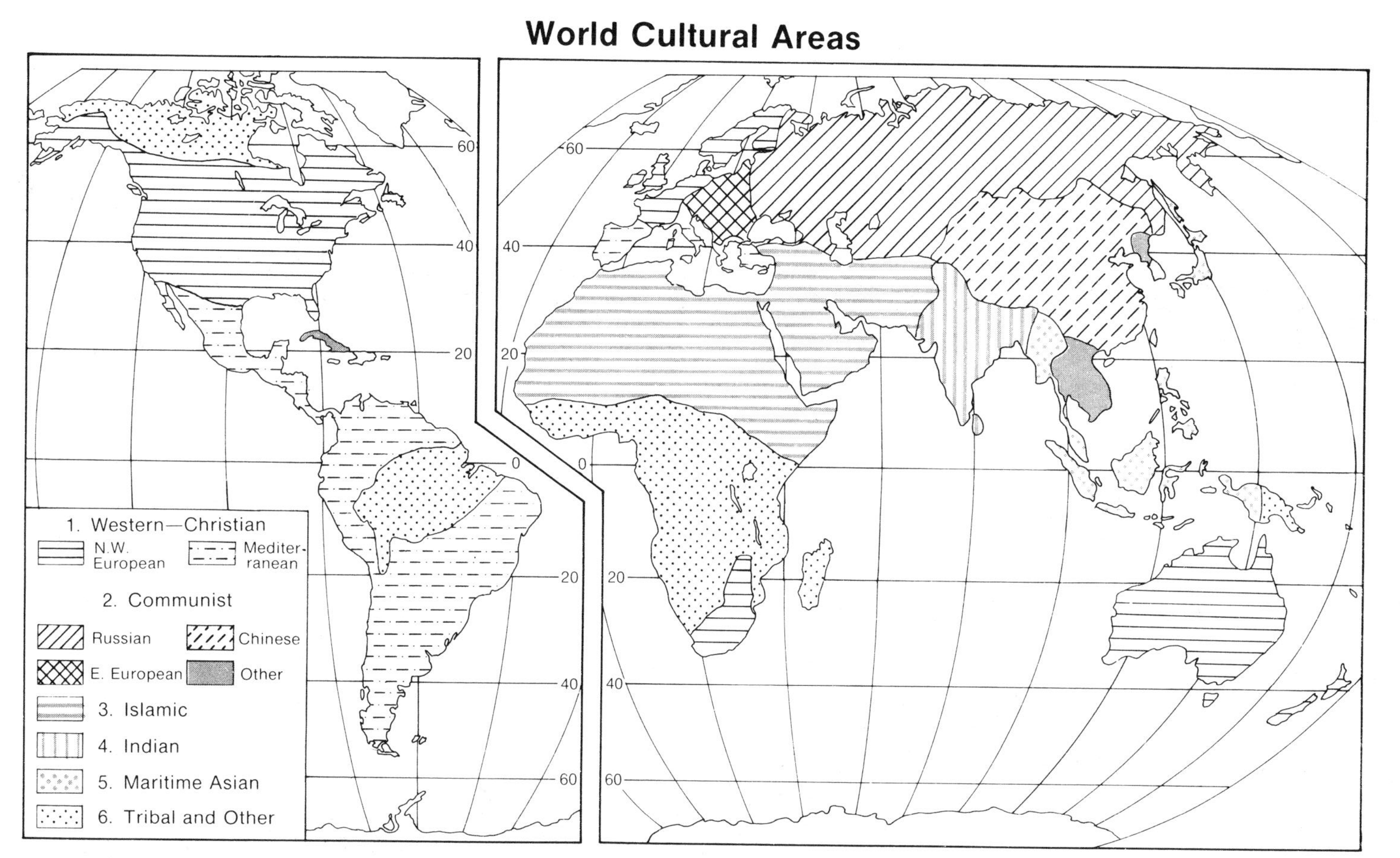

Figure 5–4 World cultural areas based on religion, language, and racial characteristics (generalized).

BIBLIOGRAPHY

Alexander, J. W. 'International trade: selected types of world regions', *Economic Geography*, April 1960, pp. 95–115.
Benson, P. H. *Religion in Contemporary Culture: A Study of Religion Through Social Science*. New York: Harper & Row, 1960.
Brunn, S. D. 'The political geography of ministates', *Geographical Review*, April 1972, pp. 275–278.
Downs, A. *An Economic Theory of Democracy*. New York: Harper & Row, 1957.
Grotewold, A. 'The growth of industrial core areas and patterns of world trade', *Annals of the Association of American Geographers*, June 1971, pp. 361–370.
Isaac, E. 'Religion, landscape, and space', *Landscape*, Winter 1960, p. 14.
Isaac, E. 'The influence of religion on the spread of citrus', *Science*, January 2? 1959, pp. 179–186
James, P. E. *One World Divided: A Geographer Looks at the Modern World*. Waltham, Mass.: Blaisdell, 1964.
Johnston, R. J. 'Governmental influences in the human geography of "developed countries",' *Geography*, January 1979, pp. 1–11.
Meinig, D. W. 'The Mormon culture region: strategies and patterns in the geography of the American West, 1847–1965', *Annals of the Association of American Geographers*, June 1965, pp. 191–220.
Minshull, G. N. *The New Europe. An Economic Geography of the EEC*. Sevenoaks, Kent: Hodder & Stoughton, 1978.
Parker, G. *The Logic of Unity*. London: Longman, 2nd ed., 1975.
Prescott, J. R. V. *The Political Geography of the Oceans*. New York: John Wiley & Sons, 1975.
Shanks, M. and Lambert, J. *The Common Market Today—And Tomorrow*. New York: Praeger, 1962.
Shortridge, J. R. 'Patterns of religion in the United States', *Geographical Review*, October 1976, pp. 420–434.
Smith, H. R., and Hart, J. F. 'The American tariff map', *Geographical Review*, July 1955, pp. 327–346.
Soja, E. W. *The Political Organization of Space. Commission on College Geography*, no. 8. Washington, D.C.: Association of American Geographers, 1971.
Sopher, D. *Geography of Religions*. Englewood Cliffs, N.J.: Prentice-Hall, 1967.
Tawney, R. H. *Religion and the Rise of Capitalism*. New York: Harcourt Brace & World, 1952.
Towle, L. W. *International Trade and Commercial Policy*. New York: Harper and Brothers, 1956.
Wenk, E. *The Politics of the Ocean*. Seattle: University of Washington Press, 1972.

6

Transportation Bases

The effects of transportation on the location of economic activities will be a recurrent theme throughout this book. Nevertheless, it is important that any general principles affecting movement be presented before analysis of any specific activity pattern is attempted. The principles of spatial interaction and of transportation costs supply a vital framework for assessing any given area or activity. Indeed, it might be said that anyone who fully understands the reasons underlying the earth's network of routes and their flow patterns has also gained much understanding of the economic geography of the world.

TRANSPORTATION AND LOCATIONAL CHOICE: A PARADOX

The general effect of transportation improvement is to allow more locational choice for any given economic activity. In a primitive economy, for example, with very limited transportation, people can only choose from resources close at hand, poor though those resources may be, to meet their major life requirements. As transportation improves, these people can fan out and begin to select from resources scattered over a much wider area. They may be able to use several of these new-found materials to satisfy their needs. And because the new resources may be much richer in quality, these people will be more efficient than formerly.

As transportation develops still further, expanding their range of selection, there is an even greater likelihood that such people will find several different high-quality resources. Their economy thus becomes more diversified in terms of material used but more specialized in terms of the division of labour and in terms of selection of areas for use. Finally, with improved access to extended areas, they may come into contact with other people and be able to interchange some of the high-quality resources and products. In this way trade is established and the economy advances from the subsistence level to the commercial level. With trade, the area may become even more highly specialized. It will also become greatly dependent on outside areas for its existence.

Transportation improvement results in areal differentiation. Instead of areas becoming more alike in terms of production, they become more distinct. Small peculiarities among regions are augmented as each region develops that activity in which it has a slight advantage. Therefore improvements in transportation tend to make even more apparent the natural differences over the earth, indeed, to accentuate them. That such is the case is becoming daily more evident in the production pattern of the world.

But transportation is also quite paradoxical in other ways. As the world becomes more accessible, materials can be brought in from longer distances. Thus, paradoxically, improvements in transportation lead to greater consumption of transportation. Although transportation is a cost in the acquisition process, it is usually true that the greater the use of transportation in a given area, the more advanced are the people in that area.

Advanced economies, which today select resources almost on a world scale, have become highly reliant on distant materials. Like the United States, the EEC countries, though collectively rich in natural resources, are such great consumers that they cannot supply their full needs from domestic sources. In many instances, they are heavy importers because nondomestic resources are higher in quality or more accessible (near water, for example) than local resources.

When such materials decline in availability as a result of wars, trade controls, or the like, technological skill is used to manufacture substitutes. Synthetics and artificial products of all types have been developed to remove dependency on resources no longer readily available. South Africa, for example, is rich in coal, but has almost no oil. Worried that overseas oil may not be always available to meet its needs, it has pioneered the production of 'syncrude', oil made by a process that treats coal with solvents. Such substitutes in turn cut down the necessity of transporting the original materials. Yet synthetics may also result in trade. There is constant play-off between activities (creating substitutes, for example) designed to minimize the use of transportation, and people's incessant desire to improve transportation and hence use more of it.

THREE PRINCIPLES OF SPATIAL INTERACTION

Materials will not be transferred from one part of the earth to another unless their increased worth at their destination justifies the costs of transporting them. The basic reason for most things having different values in different places is that the distribution of the earth's resources is unequal; consequently a surplus of a particular item is found in some areas while elsewhere that item exists in short supply or not at all. By connecting areas, transportation helps to remove the inequity of resource distribution. Because materials usually vary in value from one part of the earth to another, transportation is said to 'add value' to them. They are therefore always moved toward the place where they are worth more.

In more specific terms, materials will not move over the surface of the earth unless (1) there is a supply, or surplus, of an item in one place and a demand for that specific item in another, a condition hereinafter labelled complementarity, and (2) the item can 'reasonably' be moved from supply area S to demand area D. This latter prerequisite will hereinafter be labelled transferability. In addition to these two primary requirements, supply area S must represent the closest, or best, opportunity for which demand area D can receive such an item; in other words, there must be no intervening area that offers a better opportunity for trade of this item with area D than does area S. Thus it can be assumed that from the vantage point of area D, area S has a comparative advantage over other areas. (This principle was discussed in Chapter 4.) Implicit in these interactions, or flows, is also the assumption that appropriate methods of payment, or exchange, are available so that area D is able to pay for the item from area S.

Complementarity as a necessary prerequisite for interchange and physical flow of

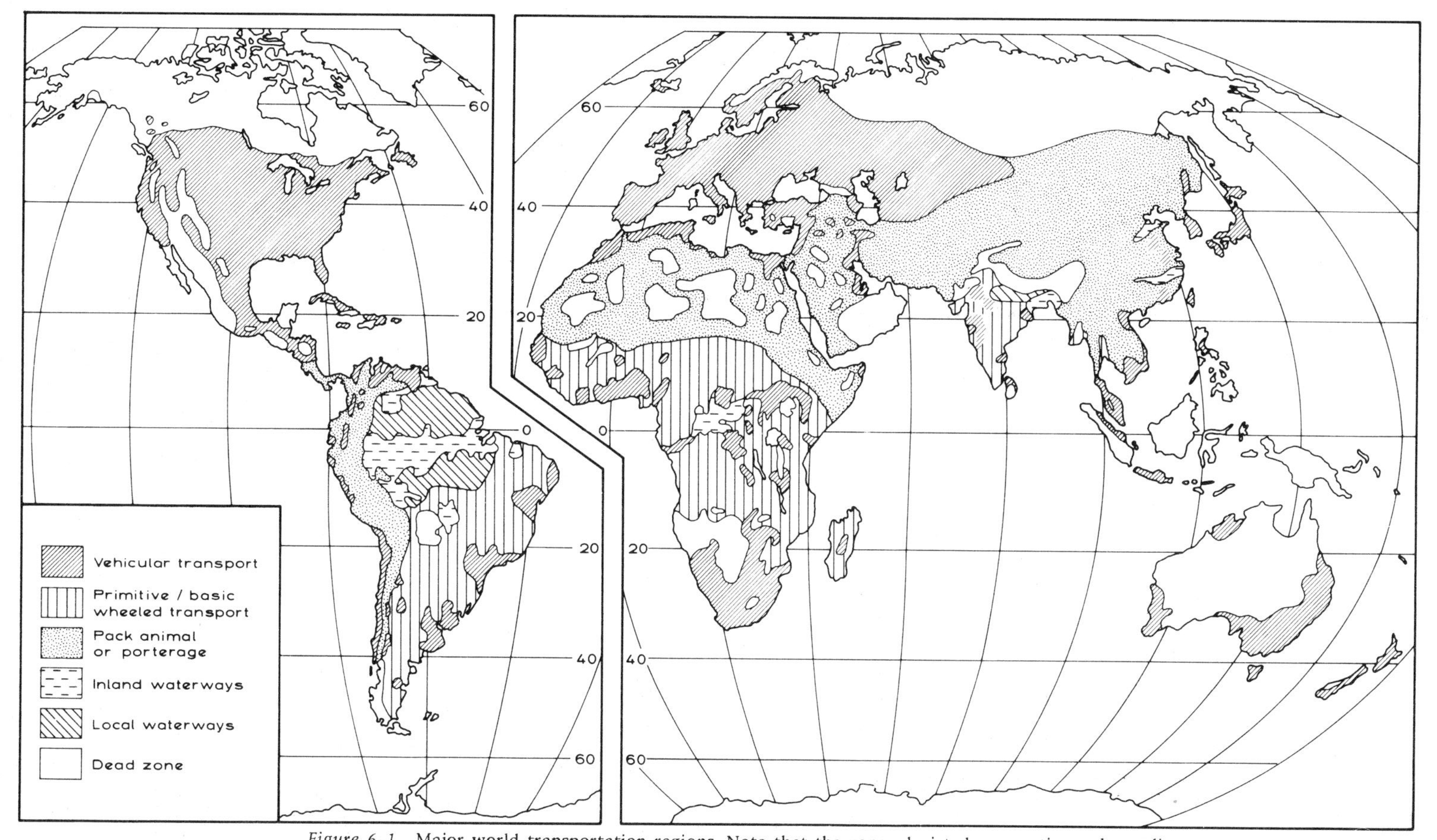

Figure 6–1 Major world transportation regions. Note that the zones depicted are continuously modified as the road vehicle makes further incursions, especially within the developing countries of South America, Africa and Southeast Asia.

commodities, products, and goods merely means that the demand area and the supply area must complement each other specifically with regard to the material to be transported. Just to have a surplus of food in one area and a demand for food in another tells us very little about the possibilities of interaction, because people demand specific things such as soft wheat, corned beef, mini-cars, quality wines and bib overalls. People are particular about their demands. An understanding of spatial interaction from the vantage point of complementarity involves a consideration of specific items of demand and supply.

Transferability, like complementarity, means exactly what the word suggests: how transferable are things over the earth? Some things are immobile. Some items are not moved because the high cost of moving them cancels out of the initial demand for them. For example, if you wanted 10 kilos of apples and were willing to pay £5.00 for them, and if they were available in another area for £4.50, specific complementarity would exist. But if transportation costs amounted to £0.70, your demand would no longer operate because the price to you would be £5.20—£0.20 more than you would be willing to pay. Your demand is not operable at £5.20 and cannot be met at £5.00. In this way the cost of transportation prevents many items from being moved over the earth. You either must go without apples or substitute something else in their place.

Such cancellations of demand create a large number of substitutions. As a general rule, when transferability problems prevent fulfilment of the specific complementarity, a substitution of another item for that initially demanded usually occurs. Moreover, substituting an item also often involves substituting one area for another—one area intervenes for another. This is another factor that discourages movement. Often the demand is short-lived and immediate. If the demand time is limited, a mode of transportation may be used that is faster and thus more expensive than others available.

There may be time restraints which affect where, when and how much of a given item can be supplied. Some products are more sensitive to distance than others, and seasonality plays a very strong role in the movement of agricultural commodities. The reader may care to note the changing price, origin and duration of supply of many common fruits or vegetables in local store or supermarket, with a view to relating the findings to transferability and intervening opportunity. A further interesting case might be the daily circulation of the local newspaper. There being 'nothing staler than yesterday's news', the paper's information value is lost only hours after leaving the presses. How does this value (demand) decline over distance from its sources of supply? And what happens where it meets a rival product?

The three fundamental bases underlying spatial interaction can be 'made operational' by the use of a gravity model. The gravity model gains its name because the original formulation was used by Newton with respect to the gravitational relationships among planetary bodies. Based on empirical observation, it is an inductive type of model in which 'mass' and 'distance' are related by regularities in movement. In the nineteenth century it was applied to the study of human migration and in modern times it has been employed in many kinds of social and economic survey.

In its simplest possible form, the gravity model can be shown as follows: $I = P/dn$ where I refers to potential interaction, P represents the size of a place, d represents distance and where the exponent n represents the expected deterioration of interaction with distance. Generally this exponent is 2, thereby squaring the distance measure. Thus P substitutes for the above notions of complementarity, d for transferability, and the exponent n represents intervening opportunity. As an example, assume that area A is 10 kilometres from a place, P, of 10 000 population. If so, A would have an interaction index of 100 with P, as compared with another area, B, only 5 kilometres distant, which would have an index value of 400. So, area B's interaction potential with regard to place P would be four times greater than area A's. The general interaction between two places, say, P1 and P2, can be determined according to the derived formula $P1-P2/dn$.

In our example, Newton's 'mass' has become the total population of a given place and 'distance' is measured in kilometres. We incorporate what may be said to be a common-sense idea that a greater amount of interaction will take place between large cities (say, London and Liverpool) than between large city and small town (Liverpool and Lerwick) or between small town and village (Lerwick and Lulworth). Also, that whatever the size of the places interacting, increasing distance separation will normally operate as a restraint on the scale of movement between them.

But these relationships must, once more, be recognized as 'other things being equal'. In practice, it is recognized that the propensity for interaction may be conditioned by wide differences from place to place in income, tastes and preferences. A single measure for 'mass' may be too crude and the separating 'distance' may be more realistically measured in time or cost rather than in geographical kilometres. Thus, in their more refined state, gravity models must be tailor-made for specific interactions (for example, rail travel, luxury goods, telephone calls).

FACTORS AFFECTING TRANSPORTATION COSTS

The major factor affecting the transferability of materials over the earth is the cost of shipping a particular item a given distance. Although factors like time and perishability may be present, they can often also be treated as transportation cost factors. Perishable items may require special transport facilities such as refrigerated ships, trucks, and railroad cars, or speedy shipment that can be supplied only by air. Thus the main requisite for understanding interaction, given complementarity, is the cost of transportation. The following are the major factors affecting the costs of transportation:

1. Length of haul
2. Area characteristics between complementary places
3. Mode of transportation
4. Nature of item shipped
5. Volume of interaction
6. Backhaul possibilities
7. Extent of competition
8. Government regulation and control

Length of Haul

In general, the greater the distance between any two places, the greater will be the total cost of transportation for any given item. Thus it costs more to ship Finland's timber to London than to Copenhagen. This cost relationship is an indirect measure of proximity, or accessibility, of various places on the earth relative to each other.

Perhaps the most interesting aspect of length of haul and transportation cost is that the cost per unit decreases with increasing distance. For most transportation modes, the greater the distance shipped, the lower the rate per unit. Actually the cost of transportation increases, but at a decreasing rate, with distance from point of origin. This occurs primarily because loading and unloading costs (terminal costs) are constant regardless of length of haul. Thus short trips are relatively more costly per unit kilometre as compared to longer trips. However, this varies greatly among modes. The practical result of this exponential cost-distance relationship is often a series of steplike rates, with each horizontal plateau becoming greater in distance and each vertical rise becoming less with increasing distance (Figure 6–2).

This pricing and distance factor causes much longer flows of material to occur than would be initially supposed. First, the more distant areas are charged much less per unit than their distances would suggest. For example, a place four times as far away as another place may pay only about twice as much per unit kilometre. Second, distant areas are often generally blanketed with the same rate. As might be expected, the breaks in transportation charges often make such places desirable as transfer points.

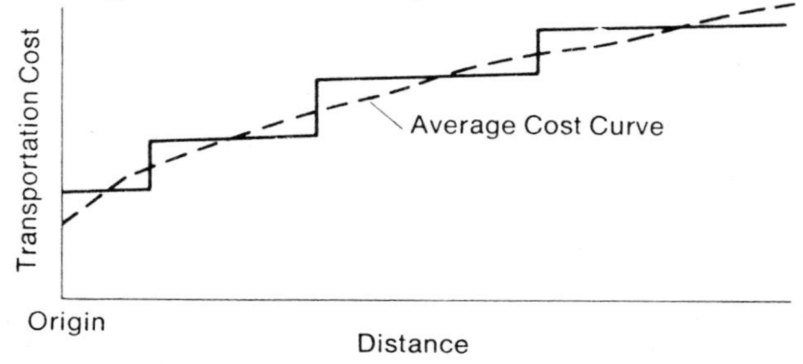

Figure 6–2 The cost of transportation as affected by distance shipped. Transportation costs increase with distance shipped but at a decreasing rate. Thus longer distance shipments cost less per unit kilometre than shorter distance shipments. The general increase is characterized by steplike changes in cost rather than by a smooth curve.

Another rather common method of favouring more distant areas is the 'postage stamp' rate procedure, that is, charging the same price per item no matter where it is purchased or sent. A first-class letter, for example, will go anywhere in the United Kingdom for the same cost to the sender. Postage stamp rates are also used by many firms. Nevertheless, under such a policy people near the source of the product pay more and those further away pay less than the true costs of shipment.

The distinction between actual costs, which are normally exponential, and rates charged should be noted, because the two rarely coincide in the real world. But in terms of economic location, both may prove important for complete understanding.

Area Characteristics Between Complementary Places

A first major modification of the distance-cost factor is the nature and composition of the territory between any two places. For example, areas characterized by water, land, mountains, plains, and rivers all have differing transportation costs. If it is necessary to change modes of transportation between places, costs will rise because of transfers of cargo at the break-of-bulk points, like ports, airports or road/rail freight depots.

On the other hand, land transportation is generally five or six times more costly than water transportation. For example, North Sea Europe and the East Coast of the United States are closer in terms of transportation costs than are New York and Omaha, Nebraska, though not in terms of time nor, perhaps, in actual costs when tariffs and various international trade controls are taken into account.

The result of the land-water surface cost discrepancy is to cause a refractive route phenomenon between such areas. Refraction occurs when light passes through two different media having different densities—air and water, for example. The denser medium (analogous to higher transportation cost area) causes the light to bend to meet the less dense material in a shorter distance than would otherwise be the case if light travelled in a straight line (Figure 6–3). This is precisely what happens given different transportation costs over various surfaces, for example, between land and water. The land transportation route is generally much shorter than it would be by a direct line route between A and B. If water transportation costs were zero, or so small relative to land transportation as to be of no locational consideration, the land route would take the shortest route to the sea (BXA) rather than a direct route between A and B (BYA). In Figure 6–3 it is assumed that there are two ports on the mainland: W and Z. Therefore neither route BYA nor BXA, nor any combination between them, is possible. Instead, shipment must go either by way of BWA or BZA. Clearly, the BWA route is less costly than BZA inasmuch as it uses the least land transportation. Yet it is the longest route in terms of total distance.

The goal of cost-minimizing at water-land breaks as described above may not be achieved in practice for a number of reasons. Ports cannot be put just anywhere along a coast, for example, as the locations of the English and French cross-Channel ports will show. And in other real-world cases, the short-sea facility may be dominated by the pre-existing deep-sea trades of the port. In short, other locational influences may be more demanding. But both for the builders of the transport facilities (road constructors, railway companies, shipping lines) and for the users of the facilities (road carriers, freight shippers, passengers), the principle of minimizing costs remains.

Cost differences over land also vary greatly because of many different surface characteristics. Where different territorial cost surfaces are traversed a 'bending' and shortening of costs, but a lengthening of distance, generally occurs, much the same as with water and land surfaces. Crossing plains is therefore less costly than traversing hilly or mountainous areas. This is easily demonstrated in North America where 'slightly higher rates' are charged for shipments west of the Rocky Mountains. (Terrain here, however, is not the only factor involved. Volume of traffic and backhaul problems are other factors.)

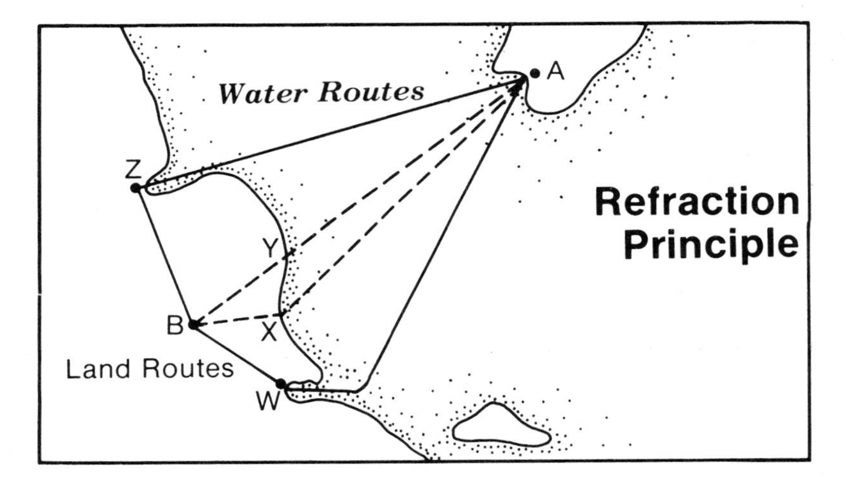

Figure 6–3 Variation in transportation costs over land and water surfaces. The cheapest route (AWB) is the one that maximizes the use of water mileage and minimizes the land mileage. The location of port possibilities may dictate a more costly route than the one that is theoretically the cheapest.

In order to be operationally understood, land surface characteristics must be very carefully examined. A rather rolling terrain may require considerable cut-and-fill for railway routes and motorways, numerous small bridges may be required to cross small streams, and low-lying and very flat land may require that roadbeds be raised. In very cold climates permafrost poses still another problem in construction and in many areas vegetation presents clearance and control problems.

The effect of terrain also varies greatly in importance among transportation modes. Air traffic is least affected, although terrain plays a cost role in the restriction of opportunity for airport locations, flight patterns as the result of mountain ranges, and by various weather conditions associated with terrain in certain climatic types. Most surface transportation involves greater costs in hilly areas because of differential costs of route construction and over-the-road operating costs. Yet pipelines can operate, albeit with more pumping, at grades of 40 degrees and above; trucks too can handle relatively steep grades but at high cost. On the other hand, any grade over 5 degrees is considered excessive for rail transportation. This explains in large part the circuitry of rail routes in hilly areas as they seek out low passes and river grades.

Thus it is evident that some parts of the earth have much greater cost advantages with regard to interaction than have others. Because of complementarity among areas, however, many routes are constructed in high-cost territory. In fact, a great deal of the transportation movement in the world is 'cross-grain'; that is, transportation routes run counter to the easiest directions of movement. Territorial character, while affecting costs and causing variations in routes, does not determine the nature and extent of transportation routes. It is the flows, the reflection of the principles of spatial interaction, that determine such route location.

Mode of Transportation

It does little good to discuss the costs of transportation with regard to distance and territorial characteristics without specifying the mode involved. Some modes have an advantage for short distances (for example, truck), whereas others have the advantage for longer distance (for example, rail). Nevertheless, some ranking of the costs of these modes can be made in terms of tonne-kilometre costs, that is, the cost of transporting one tonne of material 1 kilometre. In general, tonne-kilometre costs of various modes increase in the following order: (1) ships and barges, (2) pipelines, (3) trains, (4) trucks, (5) aircraft, and (6) human-powered carriers. If the highest cost transportation, that of a human porter, is considered as 100, then aircraft would cost about 25 per cent as much per tonne-kilometre, trucks 6 or more per cent, trains 1 to 3 per cent, and ships about 1 per cent as much. Actual costs will vary from region to region and over time.

Such cost data, however, does not settle the question of which mode is best. We may refer ahead to Figure 6–8 to illustrate the importance of speed. This figure shows the many alternative ways by which the industrial goods of Western Europe may be transported to the commodity-hungry oil states of the Middle East. Manufactured articles produced in the English Midlands, for example, and destined for the Persian Gulf can travel by road or rail across Europe (with cross-Channel ferry in each case), or by water via the Mediterranean Sea. Typical shipments may take only half as long by road or rail as by ship because of the slower speed of ships (about 30 km per hour versus 60–80 km per hour for trucks or trains) and the longer route required of ships—in this case, through the Suez Canal. Trucking may have the edge on railroading not because trucks move so much faster than trains but because no single railway connects Western Europe and the Middle East. Trucks may be held up at successive customs posts at national frontiers, but by train several freight depot changes are necessary which cause considerable delays. Trucking costs may be twice as high as railroad costs, but an overall saving may be made if several days are saved in transit. Choice between the three modes may well become a question of how much the time saved is worth, which in turn may depend on the value of the goods or the contract delivery dates set between sender and receiver. In fact, if the goods are of high worth to the customer and time is of the essence, the trip can be made by air in less than twenty-four hours (total elapsed time: ground–air–ground).

Another determinant of mode advantages is the amount of material that needs to be moved. Each mode has different carrying capacities. An oil supertanker can carry 200 000 tonnes or more. A typical general-freight cargo ship can carry 20 000 tonnes. (Incidentally, the supertanker has speeds of only about 16 to 17 knots whereas the general cargo ship has speeds of over 20 knots. But the 20 000 tonnes carried by the general cargo ship requires a 72 000 horsepower engine as compared with a 32 000 horsepower engine in the supertanker. These are large factors that clearly affect costs.) The most powerful freight train on fairly flat terrain can carry almost

10 000 tonnes (two diesel locomotives). By comparison, the typical large truck carries only 20–25 tonnes. Thus, for large shipments ship and train are clearly superior to truck and air.

Another key to differential mode costs is the ratio of the horsepower to freight hauled. A single 6000 horsepower diesel locomotive with a crew of only four, can pull some 200 freight cars or 5600 tonnes of freight. It would take 200 trucks, in all using some 40 000 horsepower and at least 200 drivers to carry the same freight. The horsepower tonnage ratio for a train is about 1.0 (6000 horsepower and 5600 tonnes) compared to 4.0 for trucks (100 horsepower and 25 tonnes). The hypothetical cost advantage is therefore at least 4 to 1. However, the railroad must build and maintain its own track (an additional cost), whereas routes are only partially paid for by trucks. The supertanker's horsepower-to-tonnage ratio is 0.16, and the fastest general-container ships have a horsepower-to-tonnage ratio of about 3.0. But the private automobile is by far the most inefficient mode inasmuch as over 200 horsepower is often used to move less than one-tenth of a tonne—a ratio of 2000 to 1.

Nevertheless, it is not difficult to understand why, despite their greatly differing costs, all these modes are used and needed. As the graph relationships shown in Figure 6–4 reveal, each has special advantages relative to the other. Some have a special advantage with regard to carrying particular commodities. Sometimes the specialities of each are used by combining several modes in one shipment. Combination most commonly takes place at break-of-bulk areas, where materials are transferred from one mode to another. Of course, such loading, unloading, and reloading also involve considerable costs, and it is only on fairly long trips, or where there is little alternative but to transfer, that such breaks occur. However, containerization (see below) has made the transfer of general freight much cheaper and easier.

With many raw materials and bulk commodities, the question 'which mode is best, or even cheapest?', can be answered only with regard to a particular territory and shipment. For example, barges and tankers are less costly per tonne-kilometre for shipping petroleum than are pipelines. But pipelines are little affected by grade and often avoid circuitous routes and port location restrictions imposed on ocean and river shipping. Thus the shorter distance needed by pipelines may override the slight tonne-kilometre cost advantage of ocean shipping. On the other hand, constant pumping and full flow are necessary for pipeline operation. Only the large fields and demand areas can support such a system. Until the 1960s the United States was the only country with major 'large inch' pipelines (built primarily because German submarines posed a threat to tanker shipments from the Gulf Coast to the East Coast during World War II). Now Western Europe, the Middle East, and the USSR also have major oil pipelines. In terms of flow, a 24-inch pipeline (about the average size) will deliver about 45 000 tonnes per day. Yet one supertanker will deliver more than five times as much.

In most areas of the world the supertanker is the most efficient carrier of petroleum, even where pipelines might run quite directly between supply and demand areas. Most often, however, both methods must be employed. Supertankers arriving in Western Europe via the Cape route from the Middle East can only enter a small number of great ports (because of the depth of water they draw and also because of their lack of manœuvrability). Thus the premier oil ports like Rotterdam and Marseille (Fos) are not only invested with oil refineries, but are also the source of long-distance pipelines which supply crude oil to many more oil refineries located closer to consumers in the interior of France and Germany. In the United Kingdom, most oil refineries are salt-water based, but many (like Grangemouth on the Forth, Llandarcy on the Bristol Channel and Stanlow near the Mersey) suffer from inadequate depth alongside. Their reception difficulties are overcome by short-distance pipelines which reach deep water where the large tankers unload. It might be noted that British east coast refineries are now served by underwater pipelines from the major North Sea oilfields (Figure 8–2), whereas the smaller North Sea discoveries are tapped directly by tankers fed by

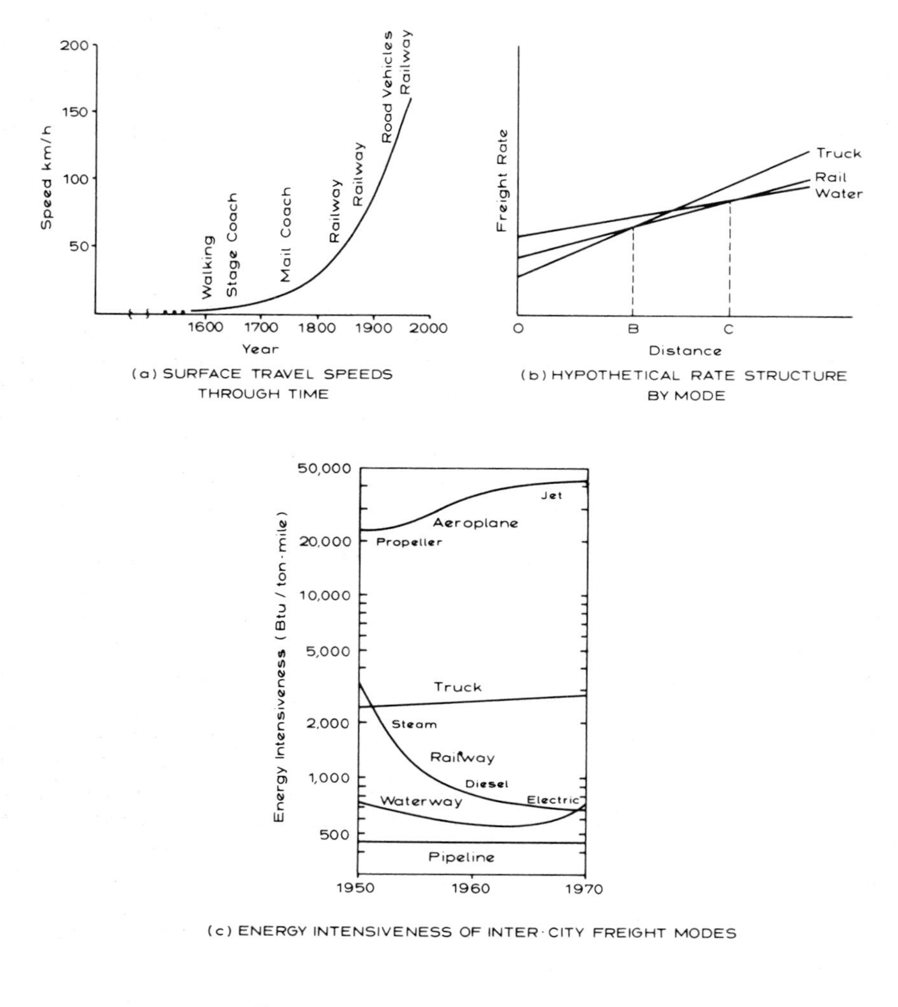

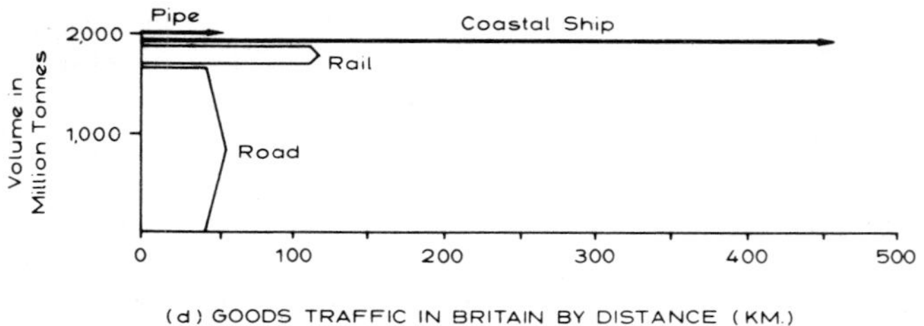

Figure 6–4 Relationships between the modes of transportation. Choice of mode for a given purpose depends on a multitude of factors, including speed, distance, cost. Increasingly, the fossil-fuel dependence of transportation also draws attention to energy consumed.

'collars' over the well-head. In this case, the mode of transportation is dictated by scale of operations.

In other cases, however, there may be little choice. This was so with one of the world's most recent oil-transportation decisions. The Alaska pipeline was chosen for moving petroleum from the Alaska North Slope rather than opting for supertanker for physical environmental reasons. The ice-bound waters in winter north of Bering Strait do not allow a meaningful supertanker option. The 800-mile pipeline, completed in 1977, moves crude oil to Valdez in the Gulf of Alaska from which point the petroleum is taken by supertanker to Japan and the United States, The Valdez port provides the nearest all-weather port for supertankers.

Nature of Items Being Shipped

The general principle with regard to most types of item shipped is to 'charge what the traffic will bear'. Therefore the higher the value of the material being shipped, the higher the transportation charge. In this regard the rate charged for products is generally higher than for commodities. This principle, like several of the others, tends to favour long-distance flows of raw material; thus items like coal, wheat, iron ore, and petroleum are shipped over much larger distances than might initially be supposed. On the other hand, many high-value items can 'stand the cost of transportation' better than can low-value items. Thus products of high value are moved over vast areas of the earth.

Another major principle relating to type of material carried is the difficulty involved in its shipment, particularly with regard to loading and unloading problems. Commodities that can be moved in large bulk by mechanized handling and special carriers and facilities can be transported at relatively low cost. In fact, the use of such facilities on the Great Lakes for the movement of wheat, iron ore, and coal has resulted in some of the lowest per tonne-kilometre costs in the world.

By the same token, highly perishable and breakable items, which require special care in shipment and handling (refrigeration, for example), are properly charged higher rates than otherwise. These perishable items include fruits, especially bananas, and vegetables. Some items are so perishable and bulky (milk, for example) that they must often be produced close to their points of consumption. In other cases, however, when the perishable products are of high value (flowers, for example), they can stand the cost of high-speed transportation.

Special refrigerator railway carriages and trucks were also developed to increase the distance range of perishable products but only at added cost. Most often, however, perishability difficulties are overcome near production areas by changing the form of the product through packing, canning, and processing. In other cases the produce is picked green and calculated to arrive at the market ready for immediate consumption. This is done, for example, with bananas and apples.

Volume of Traffic

The lower the volume of traffic between any two places, the higher the rate charged and generally the higher the actual shipment cost. This occurs because, for all modes except air and pipeline (pipelines must carry full loads), it is not much more expensive to carry a full load than part of a load. In fact, less than carload (LTC) shipments are commonly higher per unit cost than the general commodity rate. Full carloads are made up from LTC shipments by freight-forwarding companies. These companies charge the customer at the LTC rate, but ship at carload rates. In turn they provide the customer with other services, such as door-to-door pickup and delivery.

Higher volumes for many carriers also allow efficiencies in transportation. Trains can be longer and 'unitized'; that is, all wagons carry the same item and go to the same destination. 'Merry-go-round' coal trains, which travel only between coalmine and power station, are a good example. The best evidence of this point in the British Isles is the development of the 'Freightliner' network over which fixed-schedule freight trains are worked daily from inland freight depots to major cities and ports (Figure 6–5).

By any mode of transportation, where single items only are involved and volume is high, great efficiencies can often be obtained in the use of specialized loading/unloading equipment. Purpose-built carrier types, such as oil tankers, ore barges, grain ships and many lorry types are also obligatory for regular volume shipments. Major railway systems also match vehicle with commodity: customized wagons include piggy-back, automobile rack, refrigerator, insulated, cushioned underframe (for fragile shipments) and various types of gondola, hopper, tank and flat cars. For the carriage of general freight, however, the most important innovation of recent decades is the standardized, reusable container which is freely transferable between transportation modes.

Unitization in Principle and in Practice

Containers are now in widespread use throughout the western-industrial world. We may refer to the 'container revolution in transportation', really a part of the wider switch in general cargo handling to unit loads. The broader term 'unitization' includes the use of flats, pallets, and containers. Flats and pallets are used for the compact stacking, storage and movement of items like cut timber or crated vegetables. Containers, open or closed, are standard-size, robust boxes which can be employed for a wide range of dry goods, processed commodities and manufactured articles.

In concept, the 'lift-on, lift-off' container is almost as old as the railways, which used various types only for such things as furniture or hardware in single truck loads which could be transferred to horse-wagons. In their modern form containers were pioneered by the US army and then adopted by shipping companies in the regional cargo trades of the USA and Australia. On land, British Rail embraced the container principle in the development of its 'Freightliner' services (Figure 6–5 (a) and (b)). It was quickly joined in many parts of the world by rival or interlocking container systems, the hallmark of which is a 'through method of handling' which may involve two or more modes of transportation. The basic unit in such systems is the standard ISO (International Standards Organization) container 8ft × 8ft end dimensions, with lengths in multiples of 10ft up to 40ft). The back-up equipment includes a whole range of high-capacity 'lift-on, lift-off' equipment, including straddle carriers and gantry cranes; lorries, railway wagons, ships and even aircraft are specially built for the slotted-in box.

The practical implications of 'containerization' are many and varied. Here, we focus on aspects which may cause us to modify some of the principles already expounded. First, it is the box which travels, rather than a multiplicity of goods separately conveyed at rates which differ according to their value and difficulty in handling. 'What the traffic will bear' may now be applied to the container, whatever its contents. These contents may be of high or low value, but provided that they are not overweight, or demand heating or refrigeration, they travel at common rates. These rates will still increase with distance, but all standard container movements may benefit from volume handling, which suggests considerable economies of scale to the operator. Some of the savings made may be passed to the shipper, who may also take advantage from light packing (lower cost), safety from pilferage (lower insurance) and reduced warehousing (the weatherproof container replaces special storage needs in transit).

By successive leaps, we might argue that if the unit costs of moving goods by container make savings over older methods, the producer of these goods may gain suffici-

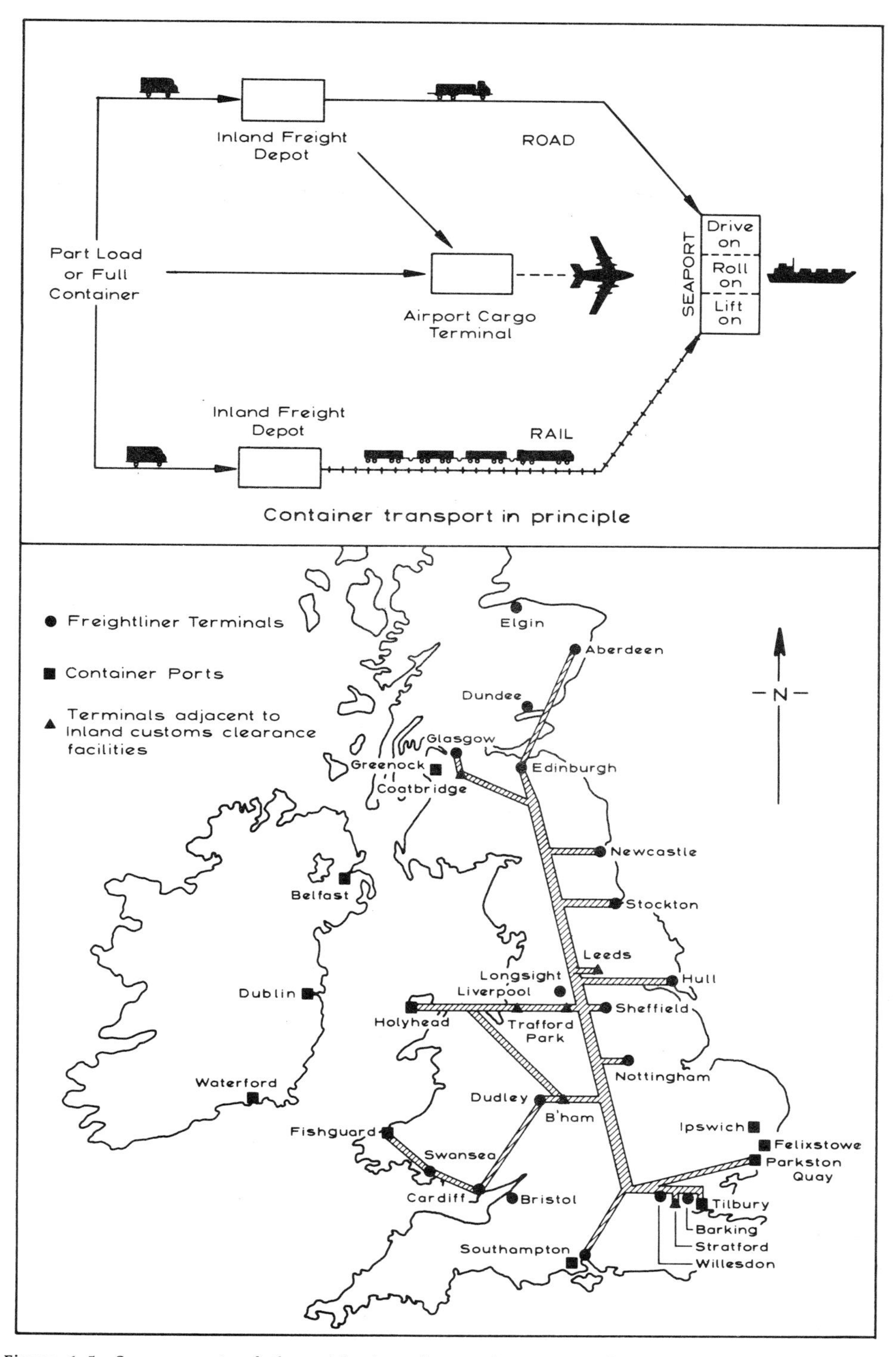

Figure 6–5 Some aspects of the unitization of general cargo. (a) Container transport in principle. (b) B.R. Freightliner Network and UK container ports. The standard freight container has become the basic unit in systems of handling, storage and transfer which affect all transportation modes (lightweight versions by air). Freightliner is the world's biggest overland container haulier, operating around 180 trains (equivalent to 5000 20 ft containers) every day.

ently for him to reach wider markets by reason of a cheaper product. In many cases this has happened between the industrialized countries; the extended interchange of such things as luxury, foreign-prepared foodstuffs, and automobile parts, together with a whole range of consumer durables like typewriters and refrigerators, has been aided by the use of container transportation. But if we return to the area characteristics between complementary places, we are reminded of the volume and value of trading imbalances which affect many countries, not least of all the underdeveloped ones which produce mainly raw materials and unprocessed commodities. Containers dispatched with, say, aero-engines or oil-drilling equipment from the United Kingdom to West Africa, may not be backhauled (see below) with ground-nuts or tropical timber which are conveyed more cheaply by other means.

The most positive effect of containerization is that it has modified the nature of seaports. As containers have been adopted more and more, fewer and fewer port cities have been able to compete in equipping themselves for container operations. The machinery is costly, a vast holding space is required, and dockside ship-to-shore access may only be achieved by clearing away existing warehouses. The rapid, almost wholly-mechanized throughput of containers may turn a ship round in a day, closing many berths and making dockers redundant.

Lastly, at a remodelled seaport, not only are activities changed, they are reduced. The containers are packed either at source or at inland depots where they are customs-sealed for overseas destinations (Figure 6–5 (a) and (b)). In the USA, such depots may be very far from the seaport; for example, Spokane and Butte are important 'international port' holding areas for the Pacific coast port of Seattle.

Extent of Competition

Generally, the greater the amount of competition among modes, the lower will be the rate charged between any two places. The tendency to 'meet or beat competition' is so strong that cut-throat competition sometimes occurs among carriers. So severe was this practice among railroads in the United States during their early development that many were forced into bankruptcy. American railroads have been particularly vulnerable to such competitive responses. In more recent times, to stave off the growing competition from road (freight) and from airlines (passengers), they have cut costs down to actual operating costs by temporarily neglecting roadbed and rolling stock maintenance. Labour costs have to be met whether trains are hauled or not, so that it is actually beneficial to use them rather than to leave them idle.

In the United States, regulatory agencies like the Interstate Commerce Commission no longer allow carriers to engage in wars of competition. Even so, rates are usually lower on routes that have several alternative modes available, for example, along the Mississippi River valley, where barge, truck, and railroad compete. Of course, carried to extreme, such a protectionist policy (setting rates to protect the high-cost producer) has the effect of subsidizing inefficient transportation modes. Thus it is conceivable that where transportation rates should be the lowest because of competition they might actually be kept higher by artificial means. Fortunately, the advantages of various modes with regard to factors other than cost create fewer competitive cost responses than would otherwise be the case.

Without government regulation, on the other hand, a route having monopolistic freight rights to an area might gouge that area by charging the maximum costs it could bear. Even with regulations, places served by single modes or single carriers usually have higher rates than other more comprehensively served places. However, low volume as well as other factors are also usually involved in most such high-rate decisions.

It may be noted that in international transportation, regulatory controls are often much weaker than within the nation-state. The shipping companies may form up together in 'conferences' to divide the trade on a given route between them, or, as in the case of large containerized shipping lines, they may form consortia with the shipping lines of other nations for the same purpose. If rate-fixing is related to such arrangements, companies not so aligned may be forced to take the 'crumbs' of the trade. The entry of the USSR's merchant fleets into the deep-sea carrying trades of the world, usually at lower rates, has begun to upset these cartel-like arrangements.

Most international airlines belong to IATA (International Air Transport Association) and have acted in unison on international passenger fares pegging them at a high rate per kilometre on high-density routes like the North Atlantic. More recently, however, this cartel has been broken by the entry of non-IATA airlines offering stand-by or 'bus-stop' services at lower rates. One effect of the new competition is that many IATA airlines are forced to raise rates, or even abandon their services, on the low-density routes which they formerly subsidized from the higher priced, high density services.

Backhaul Possibilities

Generally, the smaller the possibility for a return shipment, or backhaul, the higher will be the rate charged. This is a reasonable practice inasmuch as returning rail wagons or lorries empty costs nearly as much as returning them with a full load. Many bulk carriers at sea are especially handicapped since they must load unremunerative ballast before proceeding to the next port.

The result of backhaul demands are such that if A and B are 50 km apart and the flow is only to B, then the shipping cost will perhaps be the equivalent of 75 or 100 km rather than 50. The rate per tonne-km might be based on about a 30 km shipment. What normally occurs is that the rate is set considerably lower from B to A and considerably higher than could otherwise be justified from A to B. The general attitude toward reducing the B to A rate is presumably a reflection of the standard dictum that any freight is better than no freight.

Within highly-industrialized areas in Western Europe, the backload problem may be minimized by 'call-up' systems which inform the haulier of return freight, just as taxis are redirected by radio-telephone. Similarly, within the British rail network computerized systems are used to find the whereabouts of each freight truck, and container companies operate in the same way. Yet almost all trade routes have unbalanced flows. Raw materials move to factories and finished products move out in return. Inasmuch as there is almost always some weight loss in the manufacturing process, the flow is invariably greater toward manufacturing areas than away from them. In the United States, major rail freight hauls are towards the industrial belt of the Northeast. Thus there is commonly a box-car shortage in the West and South. This is made more apparent because of the seasonal nature of agricultural production in these areas. Perhaps the most acute problem of the latter kind is experienced in the west-central provinces of Canada, where the Prairie wheat harvest must be moved out by rail both over the Rockies and to lakehead ports before the winter freeze-up.

It should be noted that the principle of backhaul runs counter to the principle of volume. In the case of volume, the heaviest flows receive the cheaper rates. In the case of one-way flows, these high volumes may nonetheless have high rates because of the paucity of backhaul possibilities. Indeed, the low-volume backhaul may have extremely low rates.

Regulation and Control

The effect of regulation and control on the transfer costs of commodities and products is enormous. First regulatory agencies often set rates among carriers within countries for the reasons already described. Second, various restrictions are often placed on various modes with regard to routes and general network configuration. This is especially true with regard to the interstate highway system in the United States and the motorway network in the United Kingdom. In each case, central government has borne the lion's share of costs, has made almost all the decisions regarding the extent of the network and what places the routes will or will not serve. Route pattern in such cases is largely dictated by the centrally-perceived needs of the nation, which include prestige, defence, the existence of large towns and cities, and, of course, by politics. Thus in the USA, compared to the railroad network, the highway system appears to be overbuilt in the South and underbuilt in the industrial belt. In the UK, where the national motorway network reproduces the modernized Inter-City railway network, it also reinforces the economically-stronger central areas at the expense of the peripheral ones.

Nowhere is regulation and control more evident in affecting costs than in international trade. Such controls are of two types: those that restrict trade by raising the costs of transportation and those that facilitate trade by offering special advantage to certain areas through trade blocks of various kinds. The effect of most regulation, however, is to raise the costs of transportation by tariffs and other import duties. Tariffs set for revenue-raising purposes, although numerous, are calculated to have little effect on flows.

However, the general effects of tariffs on transportation costs can be considerable, as shown in Figure 6–6. Note that international trade tariffs add costs to what already often amounts to a break-of-bulk condition, inasmuch as most international trade moves through ports from which transfers to land transportation have to be made. Thus tariffs are an additional charge at break-of-bulk points and therefore a restraint on world trade.

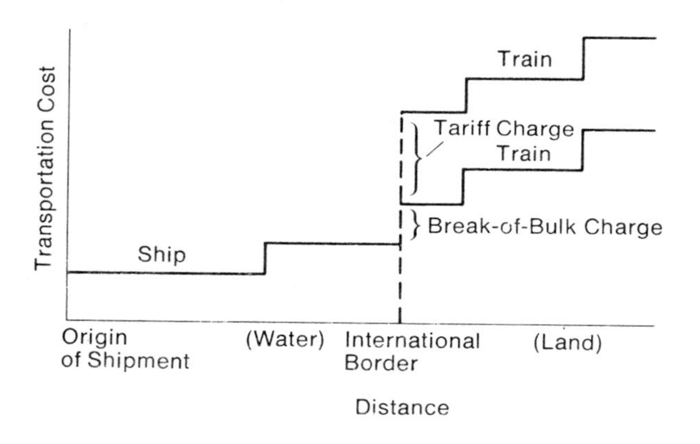

Figure 6–6 Effect of tariff on transportation costs. A tariff charge has the effect of increasing transportation costs. Such charges commonly occur at international borders. If a transportation mode change is involved, this further increases the 'shipping costs' at this point. Even if no tariff is charged at a break-of-bulk point, a new step-system of transportation cost is developed.

Regulation and control also have an effect on land transportation between countries (Figure 6–7). Only certain border crossings (X) are permitted. (Physically, too, there are often no places to cross these borders other than the designated ones.) Thus the length of haul must often be much increased to meet these border-crossing requirements. Given a tariff charge in addition to the route distortion, a major increase in transportation costs can result.

Tariffs are generally imposed to protect local economic activities and sometimes to favour one area of the world over another. For example, the UK has no import duties on tea but heavy import duties on coffee. This, of course, is a favour to Britain's former colonies. By contrast, the United States has no tariff on coffee but heavy quotas on cane sugar imports. This latter quota system favours local sugar beet production, which is noncompetitive with sugar cane production in other areas of the world. Similarly, the United States has almost prohibitive policies with regard to Argentine beef—another commodity in competition with similar activities within the United States. Although most countries talk in favour of free trade and do lower some import duties, heavy import duties are the rule, and outright embargoes are not uncommon. (Nontariff distortions of trade through trade agreements, quotas, and the like, which also strongly affect movements among countries, were discussed in Chapter 5.)

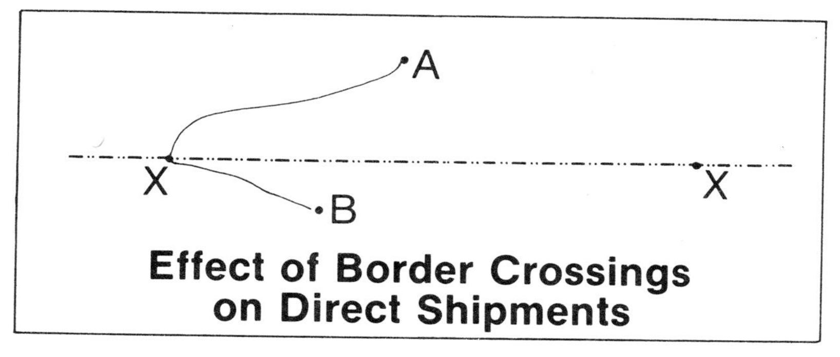

Figure 6–7 Effect of border crossings on direct shipments. Assuming a shipment from place A to B and an international border with crossings only at places X, the direct shipment route between A and B is negated by the necessity to cross at a specific border crossing. This has the effect of lengthening the distance shipped, thereby increasing the cost of shipment over what otherwise would occur.

In cases where competition is injurious to a country, embargoes, quotas, and especially high tariffs are imposed. The agriculture policy of the EEC, for example, is highly protectionist towards its member states, to the detriment of agricultural producers in distant areas outside the EEC. And within the United Kingdom, great concern has been expressed over the importation of Japanese motor vehicles, Far Eastern textiles and a wide range of manufactured goods that sell better than domestic products. Wherever tariffs, quotas and other restrictions are applied, there are increased costs of transportation and general impediments to movement. As such, they diminish or curtail many potentially heavy flows among areas.

TRADE ROUTES AND SPATIAL INTERACTION

Mindful of the general principles of spatial interaction discussed above, we can begin to appreciate the manner in which the regions of the earth are interrelated and tied together. The routes provided by the various modes of transportation are a direct mirror of the extent of interaction over the earth as affected by costs of transportation (and hence the possibilities for movement) and by organizational and institutional arrangements between countries. The pattern of routes is also a direct confirmation of actual complementarities among areas.

First of all, trade routes and their flows reflect the uneven distribution of population over the earth, and differing habits and customs. This is particularly significant in the large, overpopulated areas of Southeast Asia. Foodstuffs must be imported by many of these areas to bolster the local agricultural economy. Rice and wheat from the United States and Canada have been major commodities shipped to these places.

The routes, their direction, volume and flow-character are also indicative of the inequalities of resources distributed over the earth. Some areas are endowed with rich deposits of materials for industry, whereas other areas have few of the resources so critical to attaining an advanced industrial economy. Consequently, such materials are moved from the 'haves' to the 'have-nots'. Paradoxically, the have-nots in this case are often the most industrially advanced nations. Conversely, some of the nations quite richly endowed with materials of industry are also the least advanced industrially. Thus the flows of these materials are preponderantly from the underdeveloped nations in Southeast Asia, Africa, and South and Central America to the industrially advanced nations of Western Europe, the United States, and, increasingly, Japan. It should not be inferred, however, that the industrially advanced nations have no such resources; they just do not have enough.

Such flows also reflect the climatic inequities over the earth; notably those between the tropical and semitropical climatic areas, on the one hand, and the midlatitude areas of the industrially advanced countries, on the other. Inasmuch as many agricultural commodities can be produced efficiently only in certain latitudes, climatic areas favourable for producing particular crops have a distinct advantage. Some areas have good potential with regard to industry but have poor food resources: Great Britain follows a long-standing procedure of importing raw materials for industry, as well as foodstuffs, and exporting its manufactured products in return. In fact, this is a typical pattern for most of the economically advanced nations. This cycle increases transportation to a much greater extent than might be expected because of long-distance flows of both raw materials and finished products.

Another feature mirrored by trade routes and flows is the degree of economic development. The more economically developed nations constitute the nucleus of world trade. Of particular importance are Western Europe and the United States. Except for the Communist-bloc countries, most countries of the world are heavily hinged to this dual industrial complex on either side of the North Atlantic Ocean.

People in the more industrially advanced countries have much higher per capita incomes than do those in less developed countries. They also have rather diversified tastes in food and material. Thus they demand food from around the world. By contrast, people in the less developed countries of the world carry on a much more limited trade. Most of their trade consists of raw-material and commodity exports to the richer nations, in return for which they receive small flows of high-value products as well as food for their burgeoning populations. The United States, for example, is the world's foremost rice exporter. Canada is the world leader in wheat exports.

In this regard the greater the degree of economic development in a country, the greater the trade. This is confirmed by the fact that the greatest trade route in the world, in terms of value of items shipped, is the North Atlantic route between Western Europe and the United States. The route between highly industrialized Japan and Western countries, particularly the United States, is also heavily used. In this case, Japan imports many of its raw materials from the United States (for example, timber, scrap iron) and exports finished products such as automobiles, photographic equipment, and other high-priced and high-quality items in return.

There are many trade flows today in which many items of the same kind are interchanged among the industrially advanced countries. For example, there is a heavy flow of automobiles both to and from Western Europe, the United States, and Japan. Similarly, there are reciprocal flows of textiles, foodstuffs, and many manufactured items. One example which at the present time does not follow this pattern is the trade flow between the Middle Eastern Oil States and Western Europe. In this case, a single commodity—oil—moves one way and a considerable range of manufactured goods flows in the other. But the value of the oil greatly exceeds that of the goods, so that the oil states build up large credits in the European countries. This situation is a great spur to the West European countries to sell more goods in the Middle East. This example not only demonstrates one of the world's newer reciprocal trade flows in which there is a considerable imbalance, but also reveals how quite different modes of transportation must be employed to conduct it. The oil is conveyed by pipelines and supertankers and over quite different routes than the general cargo ships, trains and 'juggernaut' lorries which deliver manufactured goods (Figure 6–8).

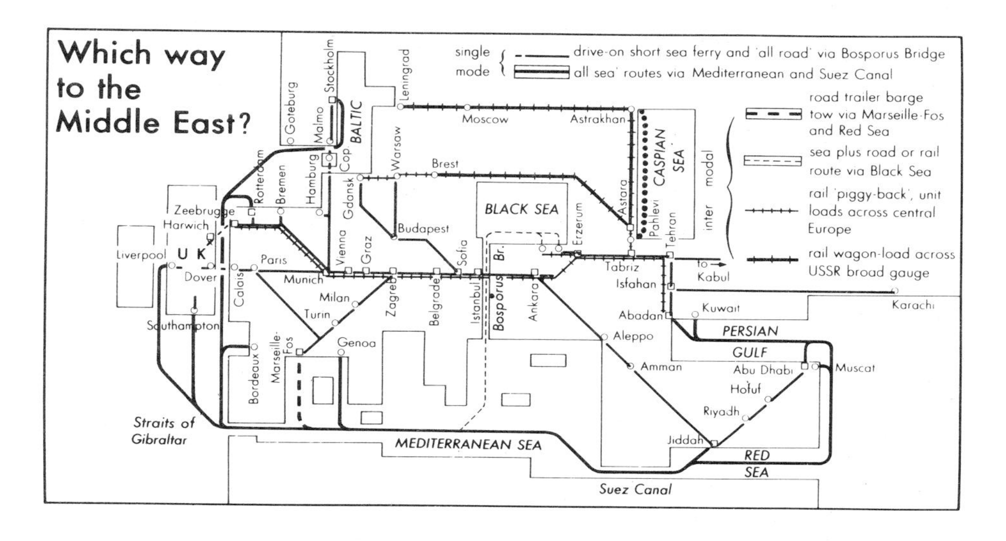

Figure 6–8 Which way to the Middle East? All modes of freight transfer are greatly in competition in the attempt to link 'North Sea Europe' with the growing markets in the oil-rich Middle East. Modal choice may depend on value of the commodity, weight, amount, delivery dates, and many other factors. The standard container makes mixed modes an attractive alternative to the 'all sea' mode. Not to scale. (From Williams, A. F. 'Transport strategy for a world without oil'. *The Geographical Magazine*, September 1978. By permission of the editor.)

THE IMPORTANCE OF TRANSPORTATION IN SUMMARY

Transportation both sets the conditions for potential interaction and mirrors actual spatial movements. Before movement can occur between any two areas, there must be specific complementarity, the possibility of effecting transfers, and various institutional arrangements. If these conditions are fulfilled, then a flow may occur, provided no better area can be substituted. Thus the physical flows and route patterns over the earth summarize and affirm such theoretical possibilities.

Many factors affect the transferability of things over the earth. Among them are degrees of perishability, fragility, and bulkiness. These and other factors are most easily seen in terms of general costs of transportation, which reflect considerations such as mode, distance to be shipped, mode-change needs, volume of movement, backhaul possibilities, the extent of competition among modes, the nature of the item shipped, and various controls and regulations. In combination, these considerations favour some areas and disadvantage others. Thus transportation-cost parameters are largely responsible for a great multitude of spatial inequalities.

In another sense, transportation trade routes simply mirror the earth's differences in resources. Some areas have great iron ore deposits but no coal; some have great forests but no farmland; some have one climate and some another. The physical connection of these diverse areas and potential interchange among them is summed up in transportation. Transportation, by enabling resources to be moved to more profitable areas, makes possible the use of otherwise useless resources. Paradoxically, by so doing, it also allows an area to specialize in its greater surplus—a surplus that otherwise would be of little use.

By shifting things among areas, transportation is largely the cause of territorial specialization. With poor transportation, areas are often unable to use some of their resources because of the need for a critical complement, such as coal for iron ore. When an area must obtain everything locally, a relatively inefficient system generally results. Thus isolated areas often have subsistence-type economies and must depend largely on themselves simply to survive. In contrast, an area highly connected to other areas can often enjoy a great diversity in life style by concentrating on the export of one or a few items in which it has a comparative advantage over other areas. This principle of territorial specialization is the fundamental essence of trade and, by definition, of transportation—truly a two-edged sword.

BIBLIOGRAPHY

Alexander, J. W., Brown, E. S. and Dahlberg, R. E. 'Freight rates: selected aspects of uniform and nodal regions', *Economic Geography*, January 1958, pp. 1–18.

Bird, J. *Seaports and Seaport Terminals*. London: Hutchinson, 1971.

Clark, C. 'Transport: the maker and breaker of cities', *Town Planning Review*, 1958, p. 239.

Couper, A. D. *The Geography of Sea Transport*. London: Hutchinson, 1972.

Fullerton, B. *The Development of British Transport Networks*. London: Oxford University Press, 1975.

Hay, A. *Transport for the Space Economy: A Geographical Study*. London: Macmillan, 1973.

Hoover, E. M. *The Location of Economic Activity*. New York: McGraw-Hill, 1948.

Olsson, G. *Distance and Human Interaction: A Review and Bibliography*. Bibliography Series no. 2. Philadelphia: Regional Science Research Institute, 1965.

Rath, E. *Container Systems*. New York: Wiley, 1973.

Robinson, H., and Bamford, C. G. *Geography of Transport*. Plymouth: Macdonald & Evans, 1978.

Sealy, K. R. *The Geography of Air Transport*. London: Hutchinson, 1972.

Smith, D. M. *Industrial Location—An Economic Geographical Analysis*. New York: John Wiley & Sons, 1971.

Symons, L., and White, C., eds. *Russian Transport, An Historical and Geographical Survey*. London: Bell, 1975.

Taaffe, R. N. 'Transportation and regional specialization: the example of Soviet Central Asia', *Annals of the Association of American Geographers*, March 1962, pp. 80–98.

Taaffe, E. J., and Gauthier, H. L. *Geography of Transportation*. Englewood Cliffs, N.J.: Prentice-Hall, 1973.

Ullman, E. L. *American Commodity Flow*. Seattle: University of Washington Press, 1957.

Ullman, E. L. *Geography as Spatial Interaction*. Ed. Ronald R. Boyce. Seattle: University of Washington Press, 1978.

Wallace, I. 'Containerization at Canadian ports', *Annals of the Association of American Geographers*, September 1975, pp. 433–448.

Wallace, W. H. 'Railroad traffic densities and patterns', *Annals of the Association of American Geographers*, December 1958, pp. 352–74.

Warntz, W. 'Transatlantic flights and pressure patterns', *Geographical Review*, April 1961, pp. 187–212.

Williams, A. F., ed. 'Transport in Britain' Series, *The Geographical Magazine*, London, March–Sept., 1978.

Wolfe, R. I. *Transportation and Politics*. Princeton, N. J.: Van Nostrand, 1963.

Part Three

THE LOCATION OF ECONOMIC ACTIVITIES

'. . . *the need to know economic geography—to know the location of things relevant to production and livelihood—is greater than ever before. Increasing affluence gives us greater and greater mobility, but we cannot use this new freedom intelligently unless we understand the locational pattern of the earth and how and why it is changing . . . We can search for locational* principles, *or the logic behind the myriad details of particular locational patterns. The growth of economic geography as a scholarly field—as opposed to grade-school geography, which has traditionally stressed the memorization of locational facts . . . is closely related to this search for logic and order in the observable patterns.*'

McNee, R. B. *A Primer on Economic Geography*. N.Y. and Toronto, Random House, Inc., 1971.

7

Primitive and Traditional Economies

This chapter focuses upon some of the important characteristics of the noncommercial economies. By 'noncommercial' we mean those in which there is little trade or exchange, and as such they are the exceptions to the rule. If the various economic activities were to be discussed on the basis of the number of people involved, the primitive economies would hardly receive mention. Studying them is important, however, in differentiating commercial systems from others and consequently, in understanding the locational patterns of man and his activities in the advanced economies.

Most of the world's people are engaged in serving other people (tertiary and quaternary pursuits) and in manufacturing, in short, in producing things that can be traded or exchanged. Moreover, most people of the earth are specialists in that they do not directly provide their own food, shelter, clothing, tools, or other material items. If the noncommercial subsistence economies lack these features, there is nevertheless a considerable variety in their ways of life. A distinction is thus made between the primitive, noncommercial and subsistence economies of herding, gathering and rudimentary agriculture—hereafter labelled *primitive economies*—and the intensive, rice-dominated economies – hereafter called *traditional economies*.

PRIMITIVE PEOPLES AND ISOLATION

In some respects the more primitive economies mirror in present-day terms the plight of the entire world population in earlier times. Physically and economically isolated from the mainstream of world development, the people living in primitive economies represent a residual population. They are often left in foreboding environments that provide few opportunities for a technologically poor people and thus offer little latitude for experiment and no grace for failure. In almost every respect these people live on the margins of the good earth. Those in the high latitudes and desert areas are fortunate if able to eke out a subsistence living from the meagre table that nature has provided.

In comparison with the productivity in foodstuffs in the midlatitudes, these peoples live in the harsh lands. Even many tropical areas might be so considered as pertains to agriculture. Problems of disease, pestilence, and other related problems render these lands harsh for effective development. Moreover, the agricultural know-how as developed in midlatitude lands is less than effective in these areas. To date, productivity has been increased in these lands through the expansion of farmed areas, rather than by increased yields as in the economically advanced areas like Western Europe and the United States (where, indeed, the trend is for increased yields to be obtained on less land rather than more).

Explanations of why these primitive peoples are found in such unfortunate locations vary greatly. One popular explanation is framed in contrasts. On the one hand, the general cultural attainments and economic progress characteristic of advanced peoples owe much to their locale, which offers an optimum environmental challenge and stimulus. On the other hand, the primitive peoples have been treated unfairly in terms of physical environment. If the challenge is too harsh, these people can do little but remain relegated to a hand-to-mouth existence, as are many primitive peoples today. If the challenge is too easy, as perhaps it is in some tropical areas where food is readily available, then people again fare poorly with regard to economic and technological progress. Instead, people focus on superstitions, taboos, ceremonies, and general shamanism (whereby tribal priest or medicine man controls the group via unseen gods, demons and ancestral spirits).

A second theory suggests that the ancestors of today's primitive peoples were the most adventuresome of the world's prehistoric (paleolithic) peoples. These were the people who ventured farthest from the heartlands of population. Thus the primitive peoples of today may have been the great frontiersmen of yesterday pushed or squeezed farther and farther outward until they found themselves more and more on the margins of the good earth and trapped into a stagnant existence.

The more verifiable theory, however, is that of acculturation. Evidence demonstrates that the more contact a group has with other peoples, the greater will be its technological and general economic progress. Thus the peoples of the western world, heavily engaged in exploration and trade, learned many things of value through the resulting contacts. The noncommercial peoples are, by definition, those who have long been isolated from contact with other peoples of the earth.

One might wonder why primitive peoples do not move to other, better lands. Perhaps explanation lies in the complex interplay between physical environment, cultural traits, stage of technology and the claims to such better lands already staked out by more powerful groups. Certainly, the whole early history of mankind can be rooted in the theme of almost constant migrations, during the course of which considerable economic progress was made. But most migrations were forced, and the survival of more primitive groups has frequently depended not only on how well they could adapt to harsh, uncompromising environments, but also on their isolation. Learning how to survive, often by developing the most cunning skills in primitive hunting, fishing or herding, they are nevertheless unaware that they are occupying the least productive parts of the earth. The lands on their periphery are already occupied by other peoples, whose occupation stands as a major barrier to further migration.

The major deterrent to migration, we may summarize, is cultural, and the greatest problem primitive peoples face in the modern world is their weakness in the face of cultural contact with advanced societies, for contact among peoples of great cultural and technological diversity tends to devastate the smaller and technologically inferior group. Perhaps the most well-publicized evidence of this comes from the Amazon area of Brazil. Here, some twenty-seven Amerindian tribal groups are threatened by the construction of the trans-Amazon highway (the Trans-Amazonica) and the deprivations wrought by the new colonists and by land speculators who are clearing the forests to establish huge cattle ranches. At the beginning of the twentieth century the forests sheltered about 2 million Indians. Now they number fewer than 100 000. If these Indians join the settlers, they fall quickly before their diseases or succumb to the most corrupting influences of their way of life.

Perhaps the only hope for the continuing survival of such peoples is through even more isolation from the urban-industrial nations. But as has been amply argued, survival at any price is not acceptable. Contact has been made and forces are already at work.

SOME SMALL-SCALE LOCATIONAL PRINCIPLES

The primitive peoples are fairly easy to understand from a small-scale locational standpoint; they are found at or very near the source of their food, which is a dictating element in their locational arrangement. In simple gathering societies, food is consumed where it is taken, or in fishing-oriented societies the village is near the body of water supplying the fish. In herding societies the nomads move their camps with their herds in search of new pastures. In primitive agricultural systems the villages are found close to the fields. Perhaps the only locally challenging aspect of noncommercial economies is the choice the people make among alternative food sources.

These people are almost totally involved in obtaining sustenance for their daily existence. There are few specialists; all are involved to some degree in providing the basic necessities. Thus primitive economies tend to be highly co-operative and communal; the individual works not for his own benefit or that of his immediate family, but for the clan as a whole. These noncommercial economies tend to favour homogeneity among individuals more than do the commercial economies.

Another characteristic of primitive economies is the small number of people generally found in any given camp, village, or gathering. The general rule is for assembly in small bands, mainly a reflection of the limited nature of food in many areas, where the land is unable to support large numbers of people in a single cluster, given the nature of the economy. Although such people are nominally part of some national state, they are little affected by political boundaries. Most of them know little about plant or animal domestication, soil characteristics, or specialization and trade. Since nature dictates the location and locational responses of these peoples, they are almost entirely at her mercy.

Yet peoples living in primitive economies are clever and resourceful in ways that commercial people today are not. They use nature's resources, whether animal or vegetable, to a fuller extent than do other people. The Eskimos' traditional use of all parts of available animals for food, fuel, clothing, tools, and shelter is well-known. By contrast, we undoubtedly throw away more than these people use. Equally inspiring is the balance with nature that many primitive peoples, like the early American Indian, have achieved without scientific knowledge. They neither upset their resource balance nor overuse or overkill their food supply. Again in contrast, people today, with their tremendous misuse of many resources and eradication of many wild animals, show little conservation skill. Primitive peoples, on the other hand, can little afford to abuse nature. The penalty for such abuse is often death.

Primitive people carefully protect the resource that is the mainstay of survival. If they are herders of animals, they avoid killing them but concentrate on using renewable features. If they are paddy rice farmers, the preservation of the soil is paramount for survival, and all other resources, including people themselves, are bent toward this goal.

Nonetheless, scientists now believe that primitive people in the past have modified much of the natural vegetative and animal landscape. They have done this through ignorance, by burning, by migration paths, and in other ways. Many of these changes have not been beneficial and have resulted in major territorial migrations. Yet, ironically, such forced migrations may have proven beneficial in that they discovered much better areas and became sedentary.

ABSENCE OF CITIES: A KEY TO UNDERSTANDING

By understanding why primitive economies have no cities, we understand much about their nature and also gain further insights into the nature of commercial economies. In this regard there are at least five conditions for cities: (1) an agricultural or food surplus; (2) a means of transporting such surplus to a city and a method of distributing it to people there who are not involved in food production; (3) a technology beyond the neolithic, so that housing, water, and other necessities can be provided; (4) a culture amenable to accepting and protecting the city; and, most important, (5) a purpose for the city, or a function for it to perform. Cities cannot have any economic base they wish or be located just anywhere. Economies having no cities are, therefore, lacking in these basic features.

The satisfaction of these requirements leads in turn to other characteristics associated with cities, for example, literacy, law, an exchange medium, and a division of labour beyond mere family job allocation. These latter features are attributes of the term civilization, which has the same Latin root as the term city.

The difficulty of obtaining a dependable and permanent food surplus that could be given to 'nonproductive' people is undoubtedly the first barrier to the building of cities in primitive economies. Inasmuch as a city is also permanently placed, such surplus must be available to that place—an impossibility for a migratory economy. It is estimated that even with the development of agriculture, under very favourable conditions in ancient times, it took about 100 farmers to provide enough food surplus to support one person in the city. Today, only about one-twentieth of a farmer's effort in the United States is needed to feed one city person. Given the fact that only the most 'advanced' primitive economies have agriculture, it is obviously impossible for them to achieve the first prerequisite for cities.

Equally difficult to obtain is a means of acquiring such surplus for city needs. The earliest method of acquisition may have been religion-based taxation. Some surplus was also extracted by force, but military force requires specialization of labour—a feature generally lacking in primitive economies. They do have a strong religious framework, but it is based on each person providing directly for his or her own needs. The system that has endured longest and is most used today is taxation by the state, which was perfected by the Romans.

Once acquired, food must be physically transported to the city. In the very early cities on the Euphrates, for example, food was moved to the city through an elaborate canal system not unlike that in Southeast Asia today. Yet, most primitive economies are land-oriented. If fishing is practised, the rivers are not used for irrigation or for the growing of crops nearby. But perhaps most significant, people in a primitive economy, as will be demonstrated, have no real need for cities. Until such a need is evident, cities will not emerge, nor will these economies emerge from the primitive stage.

One of the main arguments of the moment with respect to primitive peoples and underdeveloped nations is the question of industrialization that provides employment in the cities versus more efficient agriculture. Until recently, most emphasis has been placed on achieving the 'take-off' stage in industrialization. Certainly, industry and cities are intermeshed in Western society. However, others have pointed out that simultaneous with industrialization has been the revolution in agriculture; people were freed from jobs on the farm at the same time that jobs were being provided in the cities. If people are prematurely attracted from the farm to the city, as appears to be the case in India for example, then tremendous unemployment occurs in the cities and agricultural productivity declines. Consequently, experts today are arguing more and more that improvements in agricultural productivity must occur before serious developments in industrialization. This argument is particularly pertinent inasmuch as many of these peoples are already suffering from problems of undernutrition and malnutrition.

TYPES OF NONCOMMERCIAL ECONOMIES

The truly primitive economies include, in ascending order of economic advancement: (1) primitive gathering—characterized by hunting, fishing, and the gathering of materials close at hand and supplied by nature, (2) primitive cultivation—characterized by the practice of a rudimentary, domestication type of agriculture in an attempt to improve nature's offerings, and (3) primitive herding—various nomadic economies in which selected native animals are domesticated and aided in their survival. All three types are migratory and nomadic over the long run and are limited to a level of subsistence barely above the minimum for survival. In order to supplement their main food activity, all engage at times in simple gathering of things that nature provides.

As primitive, subsistence-type activities progress to highly specialized activities, more and more people are generally found in each group. For example, while there are only a few thousand primitive gatherers and about 100 million primitive agriculturists, there are almost 1 billion persons engaged in paddy-type, intensive subsistence agriculture. Finally, there are almost 2 billion persons directly involved in commercial-industrial economies. About one-third of the world's population resides solely in urban environments. Thus it is evident that in terms of the amount of world population, the primitive

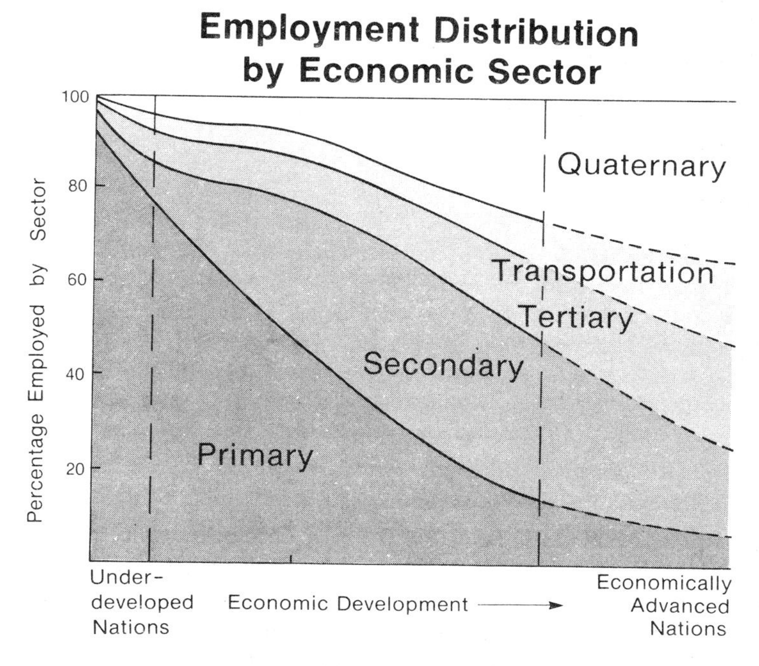

Figure 7–1 Relationship between stage of economic development and percentage of labour force employed by economic sector. Note that there is a tendency over time for the primary sector to contain a smaller proportion of the labour force, whereas the quaternary sector takes an increasing share. Based on the US from about 1850 to present.

economies are seemingly insignificant. In terms of the amount of area utilized, however, the less specialized activities are comparable to the more advanced industrial economies.

We may further sharpen our comparisons by classifying economies according to the proportions of the population directly engaged in the various economic sectors (see Chapter 2). In primitive economies, almost the entire population is directly engaged in primary activities (Figure 7–1). Practically absent as full-time occupations are such sectors as manufacturing, trade, and personal services. In other words, there is little specialization of labour among activities except for male–female distinctions in jobs. (Incidentally, in the hunting and herding economies, males secure the food, but in the primitive agriculture economies, females have this job.) In contrast to primitive economies, advanced commercial economies usually have less than 10 per cent of their population engaged in the primary sector. The largest occupational category is the quaternary or service category.

The diminishing percentage of workers in the primary sector presents another paradox. Those economies that are more advanced are also those that tend to be more productive in the primary sectors, particularly agriculture. However, because of technological developments, the proportion of persons in the primary sector diminishes. Advancement therefore appears to be reflected in the ability of a people to produce more primary commodities with fewer people.

The items that the primitive economies lack constitute the critical differentiation between noncommercial subsistence economies and commercial economies. The nature and purpose of such things as cities, specialized agricultural areas, and industry can thus be better understood by examination of the reason that some peoples of the earth are not able to obtain these.

Primitive Gathering Economies

The primitive gathering, hunting, and fishing economies are becoming more and more of a rarity on earth. At most, they account for only a few tens of thousands of persons, who are limited to small groupings in the vast areas where they are found. These are the peoples who are restricted largely to the more difficult environmental areas: (1) the far northern latitudes, (2) the central tropics of the Amazon and Congo basins, (3) the harsh margins of certain desert areas such as the Kalahari (home of the Bushmen) and the Australian deserts, and (4) isolated parts of certain Pacific islands, namely, New Guinea and various Melanesian islands. In many of these types of areas they are not the dominant people, nor are theirs the dominant economies. As such, these people occupy only small pockets of the earth and engage in the most rudimentary economic practices.

It is often said that such people, although limited in number, occupy vast areas of the earth at very low densities. In fact, such primitive peoples are huddled together in small bands over widely scattered areas. The vast areas on maps generally labelled 'occupied' by these primitive economies—the northern high latitudes, many parts of the tropics, and the fringes of vast deserts in Africa and Asia—are not really so occupied inasmuch as they contain only a few small bands of people wandering about a vast, empty place. Since there are large seasonal movements, most areas are void of population during most of the year. In this regard these peoples represent true residual populations now occupying areas long ignored by the remainder of the earth's population. On the other hand, as civilization grows, it does encroach on these areas and peoples, thus making such types of economies even smaller and less significant on a world basis.

Detailed anthropological studies of peoples in primitive economies have focused on their social structure, their customs in regard to courtship, marriage, birth and death, and their religious practices. As residual populations, these peoples have provided anthropologists with much valuable data and many important insights about prehistoric societies.

In the far northern environments hunting and fishing are the mainstay of economic survival, but the predominant economic activity in these environments is primitive herding—a more advanced system. In the tropical areas gathering of a fairly abundant assortment of foodstuffs, supplemented by hunting, is the main activity, but the predominant economic activity in the tropics is the more sophisticated slash-and-burn type of migratory agriculture. On desert margins, food gathering has been replaced by the herding of animals from place to place in order to increase animal yield. The primitive gatherers (whether of plants, animals, or fish) represent those peoples who have made the least progress, even within a primitive framework.

From a locational standpoint, however, the gathering economies are fairly easy to understand inasmuch as they are so closely tied to the natural food resources. These economies seek fulfilment of people's simplest and most basic needs: food, shelter, and tools. In this regard people living in such economies are the true scavengers of the earth. They merely assess the bounty of nature lying before them and work out the locational strategies that will best allow them to partake of this bounty.

Figure 7–2 illustrates such a locational strategy in the case of Forest Pygmies, who appear to sustain their needs by a series of 'least effort to gather' movements around a central camp. In the equatorial forest such movements may be local, but in the far

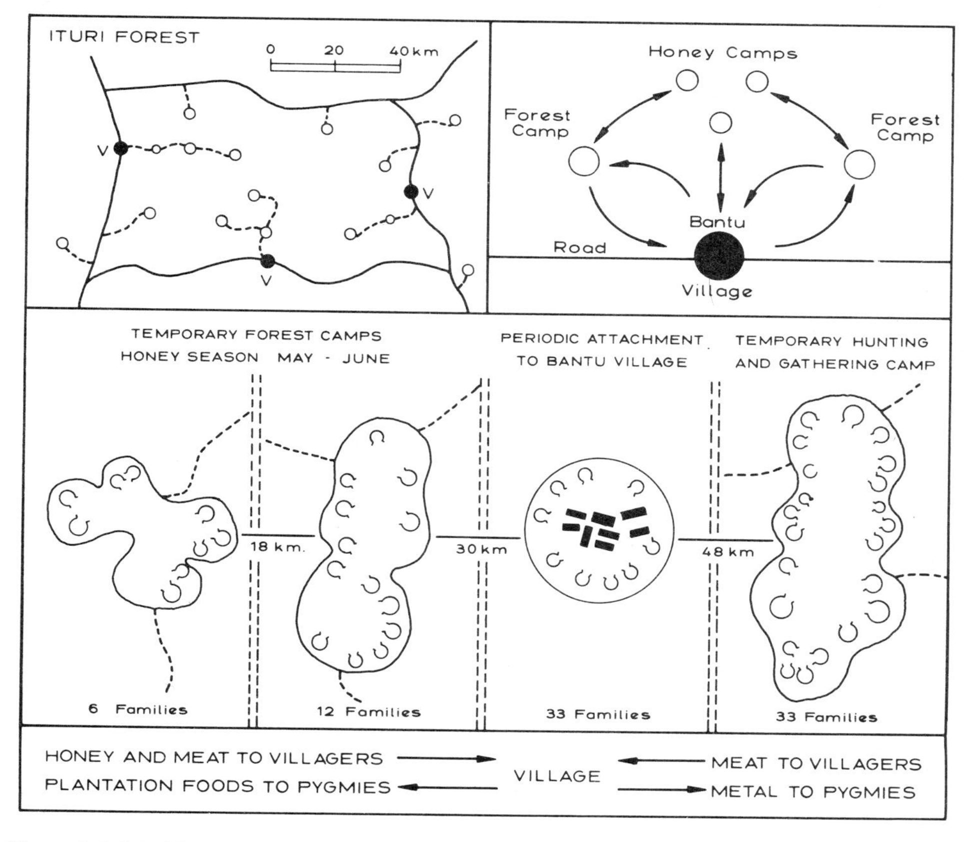

Figure 7–2 Primitive camps and cyclic gathering. Pygmy 'clan' movements around the Ituri Forest, N.E. Zaire. Nurtured by rain forest fruits, roots and meat, such groups may be independent of the settled (village) peoples or subservient to them. (Based on Turnbull, C. M. 'The Lesson of the Pygmies'. *Scientific American*, January 1963.)

northern tundra areas they may be over great distances. In northern Canada animals migrate from the forest lands in the south to the tundra areas in the north during the several months of summer. Fish are most easily caught in winter. This means that to enjoy these resources, people must move with the resources at various times of the year. Second, no one area provides a year-round food resource. Therefore, people here are at the mercy of these animal paths. Their locational strategy is, therefore, to locate in those places affording them maximum interception of the range patterns and trails of animals. They do not domesticate or herd these animals; they merely take, in a sensible and renewable way, the animal resources at hand. Their only domesticated animal is the dog, which is useful for hunting, a good scavenger, and helpful for transportation and protection.

Such peoples are diminishing in number as a result of their plight when they meet more advanced peoples. For example, the Alaska Eskimos have not been able to maintain their old way of life; they have become more sedentary, and using guns and the motorized sled, they have overkilled much of the animal resource. More and more Eskimos are finding themselves at the mercy of both civilization and nature.

It has been argued that because of their culture, such gathering peoples reject many opportunities available to them. However, given the foreboding environments in which these people are found, it is difficult to find much that they overlook. The general framework appears to be based on a highly rational survival instinct. In very meagre areas, such as the desert regions of Africa and Australia, these people are eating just about everything humanly possible—for example, lizards, frogs, snakes, ants, all manner of roots, eggs, and game. Perhaps the surprising feature is the large number of things that are not considered taboo.

Primitive Herding Economies

Primitive herding is carried on in two of earth's major physical environments: (1) the far northern areas of Europe and Asia, and (2) the margins of many Old World deserts. The best-known people in the first area are the Lapps, who herd reindeer (or at least provide protection for them) and who attempt to increase the reindeer yield over that which might occur naturally. The best-known people in the second area are the desert nomads, such as the Bedouins in North Africa. The people in the second area are in many ways more advanced than those in the first and are many times more numerous. In fact, the nomads of Saudi Arabia, Afghanistan, Mongolia, and parts of Northern Africa number some 10 million persons.

Both herding economies, however, attempt to increase animal reproduction by protecting the herd from enemies, leading them to new forage areas, and otherwise attempting to work in co-operation with the animals. Both economies are also based on an extremely sparse and only seasonally adequate vegetation for animals. In both cases seasonal migration is involved.

Food is extremely scanty in both environments, so a considerable overall area, estimated at more than 40 hectares, is needed for each animal. In most desert areas, 10 days is about the maximum length of stay before nearby food is depleted. Because of such sparse vegetation and the great loss of energy consumed in the seasonal migrations, only a very small percentage of the world's animals (certainly less than 5 per cent) is so herded. The herds are generally small, as are the groups who herd them.

Such environments also largely dictate the kinds of animals chosen for herding, which are those native to the area and most easily domesticated. In the dry areas animals that can survive on little water and that can forage on scanty and short vegetation are chosen for herding. These include the sheep, the goat, and the camel.

The distances to which such animals are followed or herded vary greatly in different areas. Figure 7–3 shows two examples, the Lapp-reindeer movements in northern

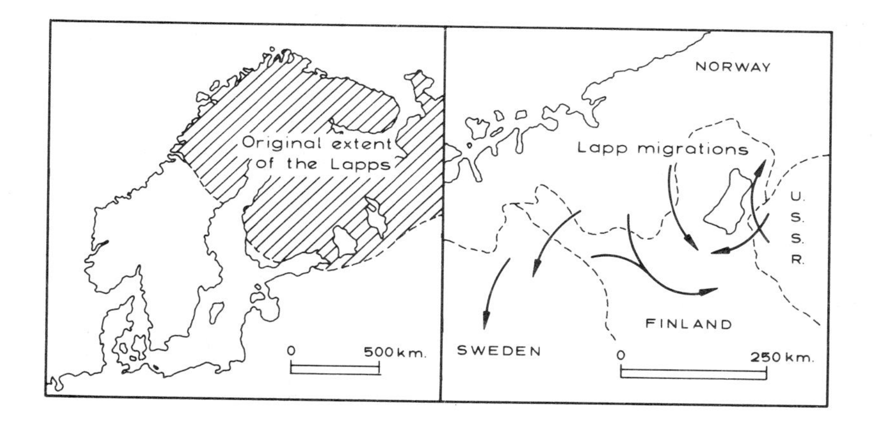

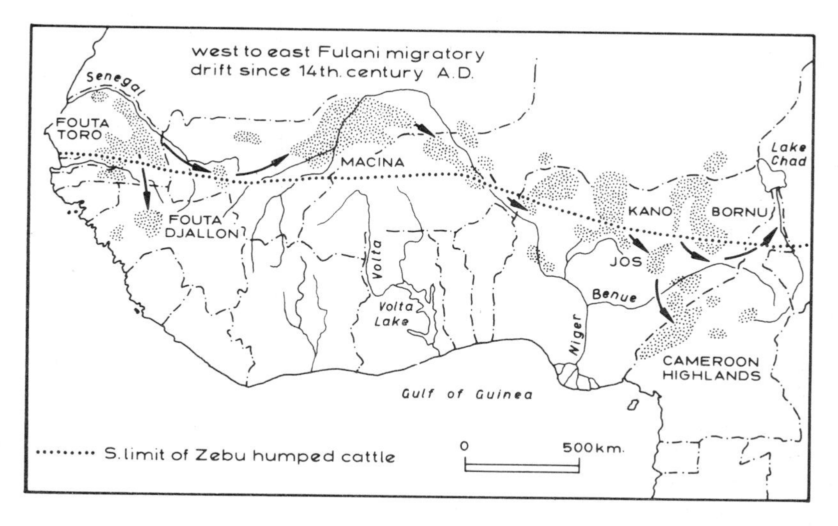

Figure 7–3 Two primitive herding economies: the Lapps and the Fulani. The nomadic Lapps may have introduced reindeer breeding in the twelfth century; in modern times their seasonal movements have been restricted by closure of political frontiers. The Fulani cattle tenders are now similarly restricted within the various countries of West Africa.

Scandinavia and the Fulani-cattle movements in West Africa. In Eurasia such distances may amount to hundreds of kilometres in a year. In desert areas where mountains are nearby, the horizontal distance travelled may be only 68 to 80 km, and the vertical distance a thousand metres or more. Obviously, if the mountains provide an excellent opportunity for the seasonal migration of animals over a small area, possibilities for permanent settlement occur. In fact, much agriculture in Europe and in the western part of the United States is based on seasonal migration of domesticated animals (transhumance).

An important characteristic of primitive herding peoples is the care with which they use their animals. In most cases animals are not killed except on special and ceremonial occasions. Instead, renewable resources are used, such as wool, hair, milk, and even blood, which (among the Bedouins and others) is tapped from the live animals. The diet of such people is supplemented by the gathering of available plants and nonherded animals. Unlike the primitive gatherers, however, the gathering done by herding peoples is somewhat incidental to their herding. They are therefore somewhat more specialized and thus more dependent on a single food source than are primitive gathering societies, which have a diversity of foods.

Because of their greater specialization, many herding peoples also engage in limited barter and trade. Theirs cannot be considered a strictly subsistence-level economy. In fact, the nomads of much of Asia are far less subsistence-oriented than are the rudimentary agriculturists—the next higher order of primitive economies.

Primitive Agricultural Economies

A surprising 200 million persons, a population equal to that of the United States, are involved in a primitive agricultural economy in tropical areas. Such primitive agriculture includes the shifting of fields and villages over time and a general method of clearing the natural vegetation (slash-and-burn). Crops are then grown on the basis of the simplest notions of plant and animal domestication. The surplus is so limited, however, that crops must be grown the year around, thus limiting this type of agriculture to tropical areas.

Very simply, the practice is as follows. Small bands of people gather in tribal villages from which they go forth to clear fields for planting by banding and burning. The ash-laden fields are then planted in the most rudimentary fashion by use of sticklike hoes. No field preparation other than this is involved. Soon the natural vegetation takes over again or the soil is depleted, and new fields must be found. Nonetheless, careful study of such field rotation systems has revealed a conscious and rather structured pattern of field rotation. Consequently, many of these people are sedentary over the short run and are able to carry on an exchange with nearby tribes. Those who have been most fortunate also have time for nonfood pursuits such as making pottery, carvings, and various religious and prestige items.

The primary difference between this primitive economy and others occurs in the division of labour. Women do practically all the field planting and field work in the primitive agricultural system, whereas men do most of the hunting, fishing, herding, and gathering in other primitive systems. The former system frees men for other pursuits during certain parts of the year and has resulted in the development of considerable cultism and ceremonialism, perhaps certain innovations as well.

One result of partially cleared fields is the ability to farm on extremely steep slopes. Thus the need to be highly selective in land is diminished. On the other hand, slope farming creates some spatial disparity in resources between land and stream so that spatial juxtaposition does not occur. Consequently, the primitive agriculturalist spends much time travelling from village to fields, village to stream, and village to various hunting and gathering places. The village location is, at best, a compromise among these uses.

Traditional Economies: Sedentary Subsistence in Midlatitudes

From primitive economies, we turn to traditional economies hallmarked by their ability to maintain themselves in a truly sedentary manner. The village way of life still dominates in many parts of the world, notably so in India and South-East Asia. These areas, dependent on rice, are considered separately below. Here, we confine attention to the arable-livestock bases of village life in midlatitude areas. In such areas, closest to the pervading influences of the urban-industrial society, there may be few communities still wholly dependent upon the local resources which surround them. Yet not so long ago millions of people were so dependent, living in areas like the north European Plain and northern China in the Old World, as well as in the early colonial areas of the New World.

Some important economic and spatial characteristics of the rural, sedentary way of life emerge if, as in earlier matters, we 'strip down' to fundamentals. For a settled way of life to be sustained in one place, without reliance on inputs from elsewhere, the necessities—like food, shelter, warmth and clothing—must be provided from internal resources. We may, therefore, restrict attention to the relationships between village location and its surrounding lands. It can be claimed fairly that these relationships were of paramount concern to any settler-group: none could survive without water, arable land, grazing space, building materials and fuel supplies.

We may follow Chisholm in calling these essentials 'the universal economic needs of an agricultural community' (Chisholm, M. *Rural Settlement and Land Use*. London: Hutchinson, 1962). All five are required by the settled village, but in unequal measure. This is because the disadvantages posed by distance are variable. Each need is therefore valued differently by the community. Water, required daily and difficult to store, is valued much more highly than fuel, which can be collected irregularly and stockpiled. Arable land is more time- and labour-consuming than grazing land; building materials, although bulky, are required only occasionally. Each basic need may be held, therefore, to bear different 'units of cost' to the community. If we relate basic needs in this way, the cost to the community of locating in a given place might be tallied by multiplying each 'unit of cost' by its distance separation from that place. The sum of such calculations for all five needs may then be compared with similar calculations in alternative places.

While such an analysis is a gross simplification of local circumstances which affect village locations in reality, the underlying principles can be supported by reference to many examples in the temperate lands of Europe, where local boundaries appear to have been defined by localized farming considerations. Figure 7–4 formalizes some of these examples. First, we depict in idealized spatial arrangement essential life-support features of the mediaeval 'three-field' subsistence community (Figure 7–4a). In its idealized state, this community is located to minimize efforts to produce its own grain, meat and milk. Its permanence depends most on its ability to keep its ploughlands in good heart by strict rotation (with fallow) about the site of the village, with less-demanding (but still necessary) pursuits at further distances away. It might be noted that to 'nest' a number of such communities in varied terrain, this idealized layout might give way to others. Thus the old, elongated or 'strip' parishes in southern and eastern England commanded the full range of local opportunities: a fair share for each of clay-vale pastures, light ploughland at the scarp-foot (where the village is located) and downland grazing above.

The same effective relationships between basic needs and the minimization of effort is suggested in the second idealized representation (Figure 7–4b). Here, we have in mind the nineteenth century crofting 'township' of the Highlands and Islands of Scotland where the management of the village lands was a communal affair and codified by regulations. The cultivated infield, for example, is backed by a fenced common pasture where each crofter is permitted to graze a limited number of sheep and cows. Again, this draws attention to the necessity for villagers to abide by strict rules which relate man and the land if they wish to maintain themselves in one place: both overgrazing and undergrazing will put the common pasture in jeopardy, reducing its value to the community as a whole. Note, too, that the infield is allocated in strips which, among other shared advantages, permits access for all to the shoreline; locally, fish may provide an important protein addition to a sparse diet.

New elements are introduced in the third model arrangements (Figure 7–4c). Well established by the nineteenth century, the sea-seeking village on the shores of the Atlantic was in its heyday in the early part of the twentieth, and it can still be exampled from Norway to Newfoundland. Based on fish as the premier local resource, it can be said to value its fishing grounds more highly than arable or grazing land. But even if such a community is partly or even mainly commercial (producing dried cod,

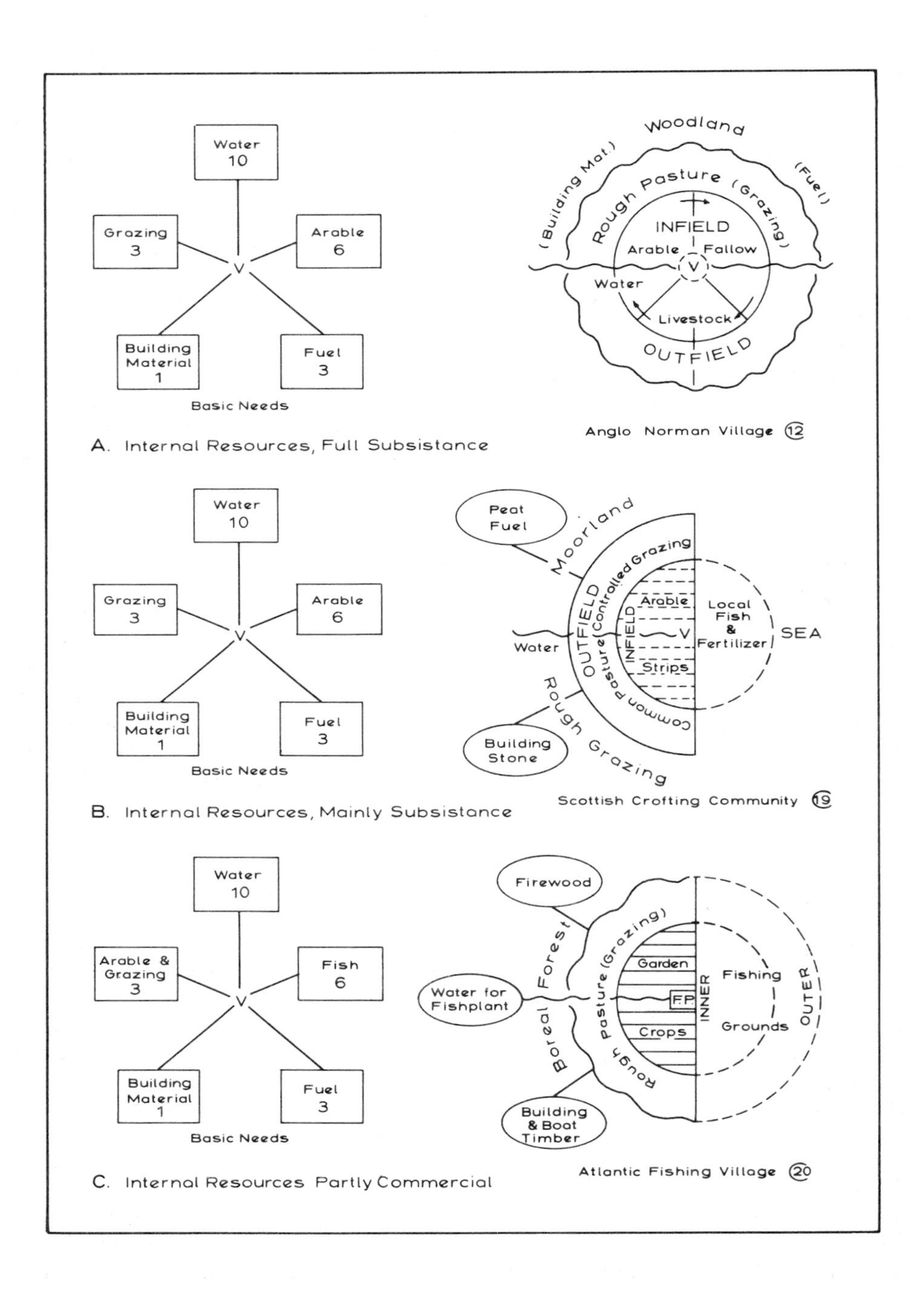

Figure 7–4 The settled village: basic needs (left) and the disposition of internal resources in idealized layout (right) in three different social systems. Note the graduation from 'full subsistence' (A) to 'partly commercial' (C). In the third case, cash inflow from fish decreases dependence on land-based local resources, but in all three systems important dietary supplements may come from hunting, trapping, or gathering.

easily stored, dispatched but two or three times each summer to distant markets), it is also locked in by cold and seasonally remote from incoming supplies. It is still dependent, therefore, on locally-produced potatoes, greenstuffs and meat. As in the previous examples, its backlands may yield the occasional rewards of gathering and hunting, but its struggle to survive may depend as much on the disposition of its landward resources as on its ability to catch fish.

The three model circumstances used above can only serve to demonstrate some of the most fundamental economic relationships between the settled community and its local resources. The 'basic local needs' analysis is most rewarding if we compare the real world locations, resources and activities of communities of like-kind in the same general area (especially so if the hypothetical units of cost can be replaced by data which can be represented in cash flows or energy budgets). From our cursory examination, however, some important principles do emerge. If we assume a situation in which, with basic economic aims given, we are seeking to establish the best possible sites for permanent habitation, then the exercise is one of 'least-cost location'. If, however, we assume that the basic economic aims of the community are not fixed, it is then an exercise in 'maximum profit location', for the reason that time and energies released by the choice of an optimum location can then be put to additional productive ends.

If neither of these two circumstances can be achieved in a real world in which so many other influences intervene, we can nevertheless see more clearly some of the influences which govern the sedentary way of life at its very roots. It may be, too, that we may also perceive something of its precariousness if reliance must be placed on local resources within a fixed area. Primitive and traditional sedentary communities were, and still are in the underdeveloped world, particularly vulnerable to crop failure and disease. The agrarian history of Europe is punctuated with natural disasters, from the Black Death in England to the great Potato Famine in Ireland. Such events seriously impaired the agricultural economy in many countries, just as do the periodic harvest failures in the village world of India today. Finally, it might also be observed that modestly-intensive land-using strategies which have been successful for generations may founder with the need to provide still more food supplies from given village lands for rising numbers of people. The consequences of this are best illustrated in relation to those countries with intensive paddy-type economies.

Traditional Economies: Intensive Agriculture in the Paddy System

The rice-dominated system of Southeast Asia accounts for about one-third of the world's population, is sedentary and occupies a considerable portion of the earth in an intensive way. The paddy-type agricultural economies have been referred to as primitive or rudimentary agricultural systems, as subsistence or noncommercial economies, as sawah (rice field) agriculture, as traditional economies, and as overpopulated and underdeveloped areas. In one sense this system is a little of all of these, but in another, it fits none of them perfectly. It is not the most intensive agricultural system in terms of yield, perhaps not even in terms of intensity of use per given unit of land (certainly the vegetable-garden agricultural system and the factorylike poultry, dairy, fish-farming, or hydroponics operations of advanced countries are far more intensive). Even the rice yields are higher in such commercial economies as Australia, Egypt, Japan, Italy, and the United States than they are in most of the paddy-type agricultural economies. Furthermore, this agricultural system is neither rudimentary nor simple. It is fairly complex in its interconnections between man, animals, crops, and the soil. It is, of course, highly labour and land intensive, but one might counterargue that much of Western production is capital intensive. Moreover, the paddy system is not strictly a subsistence one. Trade occurs and surpluses are sent to markets in exchange for other items. Finally, it is not strictly a rice economy inasmuch as other crops, as well as animals, are involved. In short, this is a particularly difficult activity to categorize. Perhaps the most fitting statement is that it is a high-density, sedentary, soil-based economy, lying between subsistence and commercial systems.

Perhaps something should be said about the general choice of crop grown. Rice is not the most productive crop in the world: manioc, potatoes, and several other crops produce more calories per acre. Second, it is not an easy crop to grow inasmuch as it requires considerable labour for planting, diking, and soil renewal. Third, it is one of the most water-demanding plants found: the major species used is actually characteristically found in a swamp environment. Even in the subtropical areas, irrigation is required to obtain reasonable yields.

Rice has other characteristics, however, that outweigh such deficits and make it an ideal crop for sedentary settlement. First, as a grass, it depletes the soil very little. With annual renewal through composting, manure, nightsoil, ash, and other renewing elements, such soil is little affected by use. Certainly the longevity of the sedentary system amply testifies to this. Second, the flooding of the flat land required for rice growing, necessary during much of the growing season, prevents the land erosion that so often occurs with many other kinds of crops. These two factors alone make for a continued food resource whereby sedentary settlements can be maintained. Third, rice keeps extremely well, as do most grains, so that it can easily be stored for winter use or held in reserve in case of crop failure. Finally, it is a crop with many food uses so that a reasonable diet, based on rice as a staple, can be developed. For example, it can be eaten whole kernel in various forms and used as flour for various purposes.

Rice is so deeply entrenched as the main staple in much of the Asian diet that it is grown even in many upland areas where flat land and humid conditions are absent. Ranking with some of the most impressive landscape features on earth, terraces have been built in mountainous areas. Here elaborate irrigation and soil-renewal procedures have been developed. Even so, these two forms are the exception to the general rule of paddy rice farming.

The nature of the paddy system is understood by noting the biotic inter-connection among people, soil, crops, and animals. These form a solidly linked system whereby soil replenishment is the key to survival. Rice and other crops are planted that diminish soil productivity very little if non-edible material is returned to the soil. Animals are used to eat various crops and household residuals, and manure is returned to the soil. In this regard, many of these animals are best classified as scavengers—for example, the chicken, pig, duck, and fish. These animals eat the scraps of various vegetative, human, and animal components. They also provide the farmer with a supplement to the rice diet. Consequently, paddy farms commonly have ponds in which various kinds of vegetable matter and garbage are placed to feed carp and ducks. People themselves are also tied to the biotic cycle. The most well-known practice is the carrying of nightsoil to the fields. Nothing beneficial to maintenance of the soil is overlooked. Fortunately, rice has another very admirable characteristic—it is largely disease proof. Very few crops could withstand nonrotated and heavily contaminated cycles such as this and remain viable.

One result of the close biotic tie between people and the soil is the effect the proximity of human residence has on field fertility. Inasmuch as the soil renewal material comes from the homesite of the paddy farmer and directly from human wastes, there is a tendency for the farmer to renew the close fields more heavily than those farther away. Vegetable gardens are found near villages and receive much of the human-generated fertilizer. This fall-off is particularly noticeable around many farm villages—soil fertility and the resulting rice yields tend to decrease as one proceeds from the village outward. This occurs despite the fact that every effort was made initially to place homes on the land least valuable for rice farming. Here proximity to settlement actually improves the soil rather than diminishes it, as is more commonly the case.

To the Western observer the major detriment of the system is its buildup of population. Given a system in which the land is fully occupied, migration or expansion is impossible, and primogeniture does not operate, the tendency is for an extended family system to develop so that more and more people must make do with less and less land. In fact, the average size farm in much of the paddy farm area is one acre or less. This means that there is a labour glut.

Being economically rational, the paddy farmer attempts to conserve those items in smallest supply—that is, land and capital—and to maximize use of that resource in greatest abundance, namely, labour. Consequently, people do most of the things done by draft animals or machines in commercial economies. Moreover, they tend their land so meticulously and in such detail that many of the jobs could not possibly be done by machines, even if they were available. As long as no other occupational outlets exist, there is no incentive to substitute machine labour for that of people. Consequently, people in such areas are caught in a vicious cycle. Children are needed to supply labour, yet the abundance of labour makes for little incentive to find substitutes for it.

One place in which the vicious cycle has been partially broken is Japan, where mechanized garden cultivators are rapidly coming into vogue. In Japan, where industrial jobs are plentiful and where other jobs exist in the cities, people are leaving the crowded farm areas for nonfarm work. In such cases, farm labour is becoming more scarce and mechanization can thus logically proceed. Of course, the use of mechanization also means more cash farming so that crops can be traded for items manufactured in cities.

BIBLIOGRAPHY

Brooks, E., Fuerst, R., Hemming, J. and Huxley, F. *Tribes of the Amazon Basin in Brazil*. London: Charles Knight, 1972.

Chakravarti, A. K. 'Green revolution in India', *Annals of the Association of American Geographers*, September 1973, pp. 319–330.

Chisholm, M. *Rural Settlement and Land Use*. London: Hutchinson, 1962.

Clark, C., and Haswell, M. R. *The Economics of Subsistence Agriculture*. London: Macmillan, 1964.

Eden, M. J. 'Last stand of the tropical forest', *The Geographical Magazine*, June 1975, pp. 578–582.

Hardoy, J. E. *Urbanization in Latin America: Approaches and Issues*. Garden City, N.Y.: Anchor Press/Doubleday, 1975.

Harris, D. R. 'New light on plant domestication and origins of agriculture: a review', *Geographical Review*, January 1967, pp. 90–107.

Jen, M. N. 'Agricultural landscape on Southeastern Asia', *Economic Geography*, April 1948, pp. 157–169.

LeClair, E. E., Jr, and Schneider, H. K. *Economic Anthropology: Readings in Theory and Analysis*. New York: Holt, Rinehart and Winston, 1968.

Lewthwaite, G. R. 'Environmentalism and determinism: a search for clarification', *Annals of the Association of American Geographers*, March 1966, pp. 1–23.

Peet, R. 'Inequality and poverty: a Marxist-geographic theory', *Annals of the Association of American Geographers*. December 1975, pp. 564–571.

Pullan, R. A. 'Burning impact on African Savannas'. *The Geographical Magazine*, April 1975, pp. 432–38.

Redfield, R. *The Primitive World and Its Transformation*. Ithaca, N.Y.: Cornell University Press, 1953.

Rostow, W. W. *The Stages of Economic Growth*. Cambridge, Mass.: Harvard University Press, 1960.

Seavoy, R. E. 'The shading cycle in shifting cultivation', *Annals of the Association of American Geographers*, December 1973, pp. 522–528.

Spencer, J. E., and Stewart, N. R. 'The nature of agricultural systems', *Annals of the Association of American Geographers*, December 1973, pp. 529–544.

Stewart, N. R., Belote, J., and Belote, L. 'Transhumance in the central Andes', *Annals of the Association of American Geographers*, September 1976, pp. 377–397.

8

Extractive Activities

Fishing, forestry, and mining constitute the extractive activities. In terms of numbers of people, all combined they are minuscule in comparison with agriculture, the other primary sector component. At most, the extractive activities employ only 2 or 3 per cent of the world's working population. Yet they are very dramatic activities and they can be very destructive. Indeed, a famous French geographer grouped all kinds of exploitation whose aim is to take raw materials from the earth—whether mineral, vegetable or animal—without restitution as 'Destructive Occupation':

> *'Among the various forms of destructive occupations some are of a normal and systematic character, while others, on the contrary, are carried out with an unrestrained intensity that makes them well deserve the German name Raubwirtschaft—'robber economy', or more simply, devastation. Destructive, or robber, economy is in a way a special form of wild-fruit picking, but it makes a far more violent attack upon nature. . . . It seems strange that this devastation should be the particular accompaniment of civilization, while so-called savages know it only in its less extreme forms.'* (Jean Brunhes, *Human Geography*, trans. Ernest F. Row. London: George G. Harrap, 1952, p. 147.)

Nonetheless, the extractive pursuits are fundamental to the maintenance of any commercial economy. In fact, these activities are the first step in a long ladder of processing, manufacturing, and distribution activities, which could otherwise be nonexistent without the extractive commodity inputs.

The extractive activities, although only a small segment of the primary sector, are treated separately for at least two reasons. First, they are distinctive in their production characteristics, which largely consist of harvesting the fruits of nature. People do little to augment this bounty at the source of supply—a situation quite unlike agriculture, in which people enlarge the natural offerings tremendously in places they choose.

A second and more important reason for separating agriculture from extractive activities is the difference in the degree of locational complexity between the two. Obviously, extractive activities must occur at the location of the resource: fish can be caught only where the fish are, mining can occur only where the minerals are, and timber must be cut where the forests are found. Thus one of the locational elements is given—the place where an extractive commodity is produced is at the location of the resource as fixed on the earth.

Here, we highlight some of the major variables affecting the extractive resources to be used, such as location, access, supply and substitution, before turning to the special characteristics of each activity—fishing, forestry, mining—in turn.

VARIABLES AFFECTING RESOURCE USE

The first major variable affecting the resources to be used is the location of the market. Other things being equal, those resources located closest to the market will be used first; if the demand is small, only those deposits close to the market will be utilized, but if the demand is large, then production may occur outward to great distances. Of course, this must be tempered by the distributional characteristics of the resource. If mineral deposits are plentiful, then even great demand can be satisfied locally. If the resource is parsimoniously distributed, then it will often be transported at great distance from markets. During the great growth periods of London, for example, sand and gravel deposits were available in great quantities from the valley of the River Thames, but coal had to be imported by sea from Northeast England.

A second variable is the quality of the deposit—its size, purity, and value. If the deposit is such that it can profitably be used, that is, it has met the accessibility conditions mentioned above, then this particular deposit will be considered in competition with others. The determination of deposit usability is subject to clarification based on quality. Conversely, extremely large deposits can cancel out some aspects of purity and value. Likewise, a high-value commodity, such as gold or diamonds, can be profitably worked in small deposits. Similarly, low quality ore can be made usable only by large-scale production, requiring a large deposit. Thus the factors that determine the quality of a deposit, and hence its usability, are somewhat substitutable one with another.

To complicate the problem further, accessibility to the market and quality of the deposit also often run counter to each other and are therefore often played off against each other. Hence a high-quality resource located some distance from the market may be as usable as a lower-quality resource located close to the market.

The manner in which these two variables might operate is demonstrated in Figure 8-1. Let us assume that there is only one market. Radiating from the market there are three distance-cost zones: 1, 2, and 3. Within this territory the qualities of particular deposits are graded from A through D on the basis of overall quality, excluding accessibility. Now assume that between the market and zone 1 deposit types A, B and C are profitable, that between zone 1 and 2 only deposits types A and B will do, and that beyond zone 2 a workable deposit must be of type A quality. Thus the two variables of accessibility and quality are intermeshed in order to create the actual pattern of resource use in a region. Poorer quality deposits are often used close to the market, zone 1, but are not usable further away, say, in zone 2 and and zone 3. Yet there is a quality limit required near the market also, as shown by the fact that deposit type D remains unused even when in zone 1. Likewise, in the second zone, zone 2, there are many deposits of equal quality to those in use in zone 1, but because of accessibility these are unused.

Thus it is evident that considerable locational choice remains. A map of forest, fishing, or mining use reflects varying qualities of deposits in conjunction with considerations of accessibility to the market, not to mention various controls and regulations so characteristic of these types of activities. Nor are the contaminating effects of time removed from the actual pattern. It is very evident that all similar-quality resources are not being equally utilized. A challenging geographical problem is to determine the

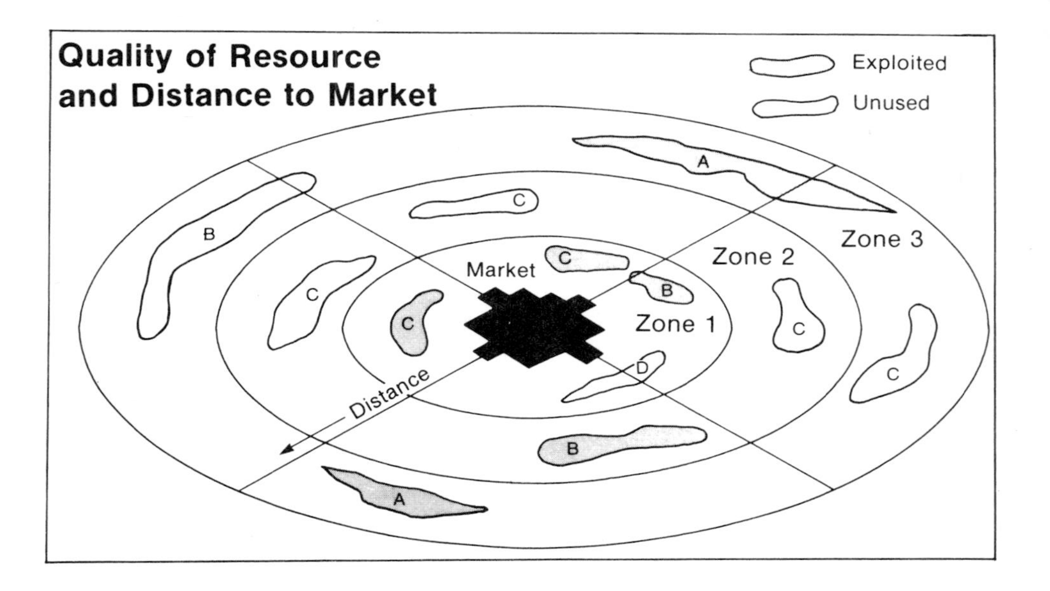

Figure 8–1 Quality of resource and distance from market. Note that the greater the distance from the market, the higher the quality required for resources to be viable.

critical points at which accessibility and quality features are being equally played off against each other. For example, what ore content and deposit size must be found to balance out a particular distance cost from the market?

As indicated in Chapter 4, there is a strong relationship between the scarcity of a resource and its price. And as price is one of the factors affecting the 'quality' of a resource and hence its use, this must be explored in greater detail. Given price or value differences, how are they likely to affect the pattern of resource use? As a general rule, the more widely distributed the resource, the lower its price and the greater the number of production sites. Consequently, the more widely distributed an extractive resource, the smaller will be the average distance it is shipped to market. Thus the lower the value, the shorter the haul, and paradoxically, the more important the consideration of accessibility becomes.

The fact that some commodities are produced close to market and are fairly widespread over the earth does not make them any less important. For example, sand and gravel operations rank among the most important types of mining by almost any measure: volume, total value, and employment. Because of their wide distribution, production occurs relatively near the demand source. Consequently, their value per unit of weight is also very low; they presumably cannot tolerate long-distance transportation costs. Yet if they were parsimoniously distributed over the earth, their value would certainly increase, as would their average distance of shipment to markets. On the other hand, gold and diamonds are restricted to very few areas and consequently have a very high value relative to transportation cost. Such resources can be sought out and utilized almost without regard to distance from markets. Thus the factor of accessibility must be carefully interpreted in terms of resource value, which in turn generally reflects the degree of spatial scarcity. Moreover, it must be appreciated that the price of any item will thus vary from place to place, depending on spatial proximity of that place to the resource. Most extractive resources are neither as cheap nor as ubiquitous as sand and gravel, nor as expensive and scarce as diamonds and gold. They fall somewhere in a highly compromised and highly complicated middle. For most of the extractive resources, the location of production involves a fairly complicated and rather subtle interplay among three variables: (1) proximity to market, (2) quality of the resource, and (3) price of the resource at the market place.

Together with these three factors, we must take into account the circumstance that extractive activities are also highly regulated and controlled in their nature and place of production. One reason for this is their potentially depletable nature and another stems from the fact that many extracted commodities move in international trade, or, as in the case of fishing, are obtained from international waters. Minerals in particular move heavily in international trade inasmuch as no one nation possesses the full range of mineral resources it requires; at least, most nations find it more profitable and wiser to import some minerals than to exhaust their own or to use lower-quality domestic resources at higher costs of production. For national defence purposes, for example, many minerals are carefully kept in reserve as long as they can be obtained elsewhere. Fish too are caught in internationally disputed areas. To protect this resource from depletion, international agreements and territorial water rights are common. Such characteristics add still another variable necessary to spatial understanding of production.

The location of the market is of further importance, especially if we consider any world resource reserves (exploitable but not used). The nature of mining and fishing is such that the existence of resources must be discovered. The exact nature and location of reserves is, therefore, not precisely known; indeed reserves increase drastically from time to time and from place to place. But the major effect of markets is shown by the fact that, logically enough, people try to find resources close to markets and relatively neglect the more distant potential resource areas. Those areas that are heavily industrialized and are using resources in abundance also appear to have the greatest number of usable extractive resources. But this may be an illusion inasmuch as the remainder of the world has been examined much less intensively.

DEPLETION CONSIDERATIONS

Minerals are clearly fugitive—once they are used they are exhausted insofar as a particular site is concerned. Because minerals are depletable, numerous questions arise as to which resources should be used where and when and in what quantity in any given country. Questions of short-run versus long-run needs, reserves for national defence, and ways to achieve maximum use are often critical. The national debates in Great Britain and in Norway related with the exploitation of North Sea oil (Figure 8–2) is a case in point.

The general rule has been that the highest quality resources are used first (that is, those with the highest ore content, or greatest size, or greatest ease of acquisition). Man generally works from the top downward on the resource pyramid thus formed. New methods to make lower-quality resources increasingly usable have been perfected with time, so that the question of what is a workable deposit changes. For example, a hundred years ago only 'pure' metallic copper was deemed minable. Several decades ago, copper oxides with more than 15 per cent copper were considered usable. Today through large-scale mining methods and improved processing techniques, copper sulphide ore deposits of one per cent copper or less are being worked.

A similar transition has occurred in iron ore. Several decades ago it was assumed that rich Mesabi iron ores would soon be exhausted and production would have to occur elsewhere. Ores having 50 per cent or more iron ore content did indeed become scarce, but few foresaw the technological development that would permit low-grade

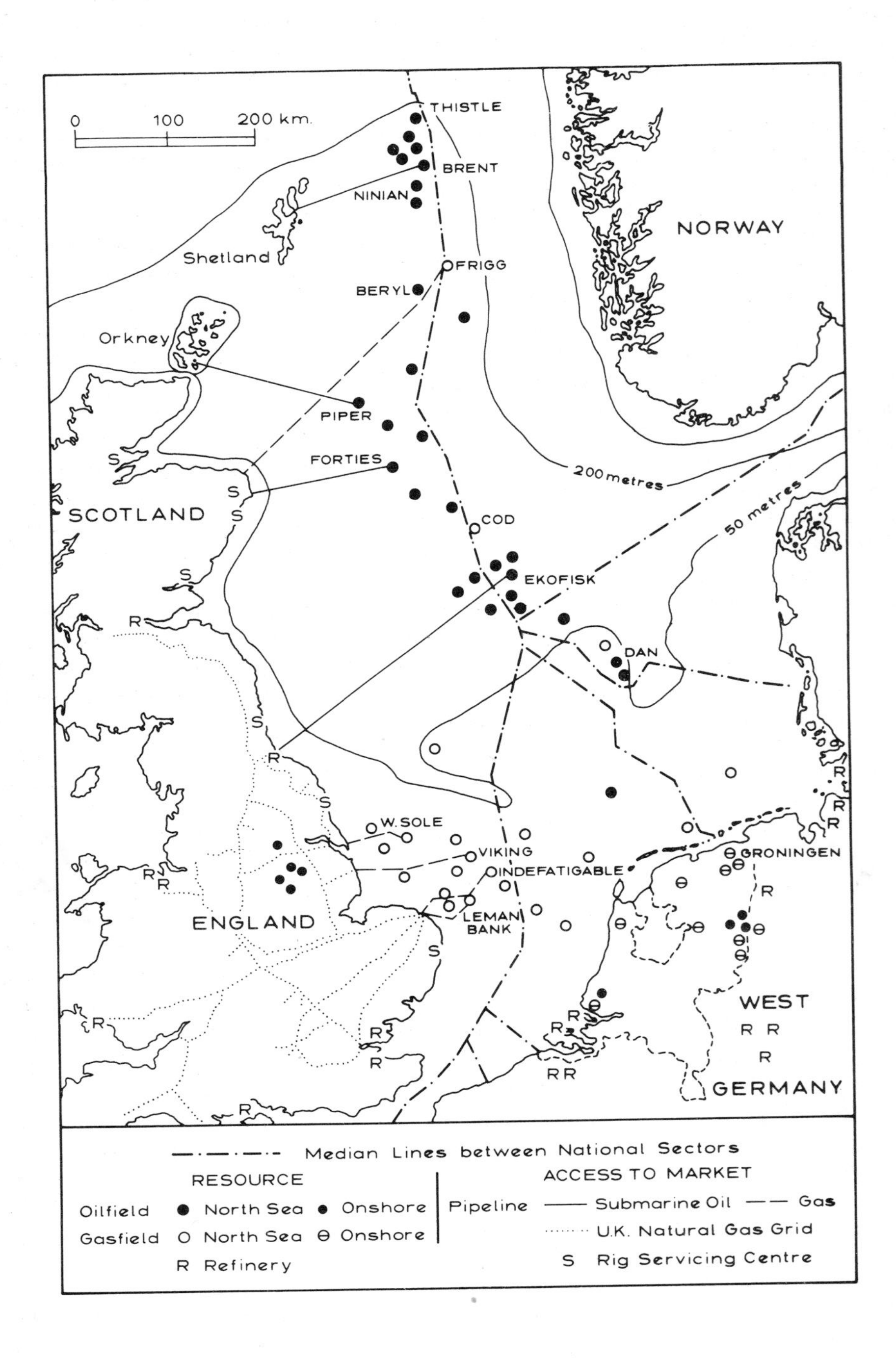

Figure 8–2 The discovery and exploitation of oil and natural gas in the North Sea. The carve-up of the North Sea was achieved by extending median lines seawards from territories of neighbouring states, the medians being equidistant between neighbours. Note deep waters which deter the laying of underwater pipelines to Norway.

taconite ores to be used. (Taconite, which contains about 30 per cent iron, is the mother rock from which the richer iron ore came.) This has been made possible through the process of beneficiation, by which otherwise worthless material can be efficiently removed. The point is that in deeming a resource usable, the general trend has been toward reduced deposit richness but increased size in the deposit. Consequently, many small mines formerly worked are no longer considered profitable; they do not contain enough total deposit to make the investment worthwhile. Geographically, this means that more and more production is coming from larger but fewer sources.

Forestry and fishing are also potentially depletable resources. Numerous examples are available to show where such resources have been destroyed: the formerly forested areas of Upper Michigan and Minnesota, the demise of the sperm whale in the North Pacific, the demise of the Arctic fur seals, the declining catch of Pacific salmon and North Sea herring. Although more attention is being paid to making such resources available on a sustained yield basis through careful control and management, the general pattern is one of depletion rather than continuation.

One major economic difficulty in the deliberately controlled or slowed-down use of such resources is the problem of short-run profits versus long-run public gains. The general desire of the typical entrepreneur is to get as much as possible of the resource now, while it is in demand and while he or she has access to it, rather than later when perhaps demand has shifted and he or she has lost out to a competitor. This is particularly the case in the fishing activities, where if one fisherman or nation does not get the resource, another fisherman or nation might.

Another important geographical implication of depletable resources is that activities connected with their extraction have little lasting impact locally. They employ few people, the commodities are usually shipped to distant markets so that few connected activities occur in the area and most important, such operations are often transient. About the only contribution made by earlier mining operations has been the ghost town, a phenomenon not only to be seen in the Western USA but also in Canada, Australia and elsewhere. Large-scale, factory ship fishing operations in the Atlantic have so over-fished many areas that local fishing villages close to the resource are abandoned. Even more dramatic is the detrimental effect on the natural environment from large-scale operations such as slate-quarrying (North Wales) and strip-mining for coal (Appalachians). Because of the negative aspects of the extractive activities resulting from the depletable-resource framework, they are of dubious value for purposes of long-term economic development. Yet these are the activities that are highly prominent in underdeveloped parts of the world—for example, petroleum from Venezuela and the Middle East, and iron ore from Labrador in Canada, and from Chile. These extractive activities provide the raw materials of industry and are hence on the periphery of the industrial and urban complexes characteristic of the advanced economies.

EXTRACTION CONSIDERATIONS

Because extractive activities generally represent depletable resources, their aim must be to maximize the resource take. In mining, the goal is to obtain all usable minerals in a particular deposit. In fishing and forestry, the goal is to harvest only the 'ripe' resource in a manner that ensures a sustained future supply. Full take, however, is sometimes impossible in mining because of the inability to extract minerals from low-grade ores that may be part of a worked deposit; in the case of underground mining, considerable ore must be left to support shafts or left in seams too small to work.

In the cases of fishing and forestry, the problem is often one of overuse rather than full use. The optimum extraction framework would, of course, be to harvest only those ready resources in such a manner as to ensure a sustained yield (Figure 8–3). In the case of fishing, only the mature fish would be caught, preferably after spawning. In net fishing the net size would be geared to allow smaller fish to escape. Recently there has been much controversy over the inadvertent killing of dolphin in the course of catching tuna. In the case of forestry, only trees that have reached their full growth would be cut. These are logical goals but are often difficult in practice; selectivity costs money and takes away from profits.

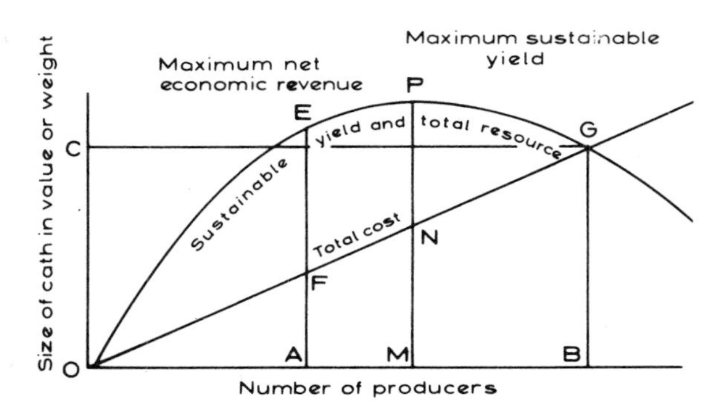

Figure 8–3 Optimum sustainable fish yield (after Coull; see Bibliography). Biological optimum in a fishery occurs at the point of maximum sustainable yield (P), the desirable goal of marine biologists and international legislators. The maximum economic yield (E), however, occurs when the distance between the graphs of cost and yield is greatest, and is always reached before the maximum sustainable yield. Why, in practice, is it impossible to tell with precision when the point of maximum sustainable yield has been reached?

In fishing, such goals are difficult to practise, not only because net and other techniques have not been fully developed but because of the nature of the resource. Fish are not directly owned by any one party until caught. Given a number of fishermen in common waters, the incentive is to catch all one can before someone else does. Thus without restraints on catch, the supply of many types of fish would soon be exhausted. Most fishing is therefore subject to elaborate international agreements with regard to national fishing water rights involving area, season and quantity. This is particularly the case with regard to whale fishing in the Antarctic, salmon fishing in the Pacific, and cod fishing in the North Atlantic. Many nations control the fish catch near their coasts by imposing 12- and even 50-mile territorial fishing limits. Despite all such attempts, many kinds of fish are rapidly being depleted as large-scale fishing techniques become more developed.

Again, in forestry the goal might be to 'farm' trees to obtain an optimum sustainable yield. But this is an ideal goal for several reasons. Timber has historically been looked upon as a nuisance; most settlers in North America, like the Saxons in England many centuries before, wanted farm land rather than forest land. Much of this frontier attitude still persists among many small landowners who can, after all, only sell their timber once in a lifetime. Compared with fish, which often have a three- to five-year cycle, timber growth in most areas takes almost 30 years, a length of time beyond the 'farming' time limits of all but the giant corporations. And even these have been reluctant to practise tree 'cropping' rather than tree 'mining'. Finally, it is often difficult to utilize only the mature trees because of the difficulty of getting such trees without injuring others. This can be accomplished only in a labour-intensive environment, such as in Finland for example, where large-scale machinery is less involved.

In much of the United States, particularly in the Pacific Northwest, the practice is to clear off cleanly all vegetation in a particular area (clear cutting) to replant the area with trees. Since they would all be the same age and be ripe at the same time, the area would be cleared once again at the next harvest time. The strategy here seems to be that of achieving both short- and long-run economies. Short-run economies are achieved because large-scale machinery can be used to harvest the trees indiscriminately; long-run economies are achieved because of standardization in the age of trees the next time around.

This sounds good in theory, but in practice it presents many problems. Replanting has often been delayed, the denuded landscape causes serious erosion problems in hilly terrain, and the area is ruined for many other uses, such as recreation, wildlife and water collection purposes.

The difficulty in achieving satisfactory extraction practices in forestry is perhaps best demonstrated by the existence of the national forest reserves in the United States, Germany, and many other industrial nations. In Great Britain, by far the greater proportion of forest activity is under the public sector management of the Forestry Commission which ensures the cyclic regeneration of the land under its care.

FULL UTILIZATION AND CONSIDERATIONS OF SUBSTITUTION

Even if forestry, fishing, and mining resources were extracted perfectly, according to the principles cited above, there would still be questions as to full use. A tree used only for timber has considerable waste in the form of bark, small branches, slab cuttings, and sawdust. Fish used only for human food involves considerable waste in terms of the removed portions that could be utilized for pet food, fertilizer, and various oils. Ores often contain more than one mineral.

The general trend has been for fuller utilization of these former waste products by turning them into by-products. In all major forest product regions, otherwise unusable wood pieces are used in the making of a wide range of compressed-board products; and even sawdust can be pressed into small logs for use in fireplaces. The very low copper ore deposits in Bingham Canyon, Utah, contain quantities of gold and silver which have supported the cost of removing the copper from waste material. Natural gas associated with oil wells has become a major fuel. Much waste, however, still occurs in areas where there is little demand for such by-products: the further from the market that production occurs, the less likely it is that such extracted material will be fully utilized.

This last relationship can also affect the production of the major commodity, be it timber, fish or mineral. Indeed, most extractive activities are plagued by substitution possibilities with other commodities better placed in relation to the market. Such substitution may occur both within and between production components. For example, stone may be substituted for wood in construction and coal for wood as a fuel. Indeed, energy may be obtained from coal, petroleum, natural gas, nuclear fuels, wood, and hydro and solar power. Likewise, in many cases metals may be substituted for wood, and vice versa; aluminium may be substituted for steel. Of course, there are many food substitutes for fish. Hence the demand for these activities may vary greatly from place to place as various substitutable items are in greater or lesser supply.

Thus each commodity is in competition with a number of others for the market. Most do not have a monopoly position (uranium, kyrolite, and certain rare minerals are exceptions). The production of these commodities fluctuates greatly as the demand mix changes. For example, until recently the use of petroleum has been on the rise relative to coal. Today coal is being developed as a substitute for petroleum, both as a fuel for heating and in terms of various chemical derivatives. The use of trees for pulp and paper is on the rise relative to their use as a fuel. Beef is on the rise relative to fish, although fish consumption is rapidly increasing in many underdeveloped countries. Such substitutability allows spatial patterns to shift from one extractive activity to the other, thus creating a complex pattern of use.

FURTHER NOTES ON FISHING AS AN ECONOMIC ACTIVITY

Simply stated, fishing must occur where the fish are. What are some of the locational conditions? How does accessibility and proximity to markets specifically affect fishing areas? Which areas will be used and which will remain unused? How can one type of fish be substituted for another, or for other food? These are some of the questions that arise because of the discrepancy between where particular kinds of fish are found and where fish are taken.

Location of fish relative to the market has always exerted a tremendous effect on where fish were caught. Because fish are highly perishable, they must be delivered quickly, whether to the market for fresh consumption or to canneries for preservation. Consequently, most fishing has occurred near the major markets, which have traditionally been in Western Europe and the eastern United States. Lately, Japan has also become a major market, and fishing has become most concentrated in the North Atlantic and the Pacific oceans (Figure 8–4). Fortuitous circumstances have also resulted in the great concentrations of fish in these midlatitude and northern ocean areas in the Northern Hemisphere. It seems that fish are mostly found, and certainly most easily caught, in shallow (that is, continental shelves and banks), cool, mixed waters, near forested continents.

There are more than 30 000 kinds of fish, and they are found from the equator to the poles and at most depths of the sea. But we focus here on those places where people have found fish to be most easily caught, most usable, and most accessible. That we are still in a hunting stage, rather than in any stage of aquaculture, is demonstrated by the fact that this potentially rich 70 per cent of the earth's surface to date provides only about 3 per cent of the earth's food.

In fact, a great number of fish are found in tropical waters; but unlike midlatitude fish, they rarely run in schools. Almost all fish caught commercially are found in schools or, like the shrimp, the oyster, and the sponge (which are technically not fish but which, for our purposes, will here be treated as such, as will the whale), in highly concentrated conditions.

There are also large numbers of fish in the Southern Hemisphere, but until recently, because of lack of accessibility to the markets, they were largely untouched. With the development of 'floating factories' and nationally owned fishing fleets such as those of Japan and the USSR, fishing at greater distances from the markets can be developed. One major fishing area is off the coasts of Peru and of Chile, where the mixing of currents favours the proliferation of plankton on which the fish feed. Peru has taken advantage of this resource to the extent that it is now a leading fishing nation in the world in terms of tonnage (see Figure 8–4). Major fish caught in this area include the anchovy, tuna, cod, and many others. A similar new area is just developing off the southwest coast of Africa in which anchovy, flounder, herring, red snapper, and sardines are caught in great abundance. Today almost one-quarter of the world's fish tonnage is gathered from the waters of the Southern Hemisphere.

Although employing only a small percentage of the world's population, fishing is in-

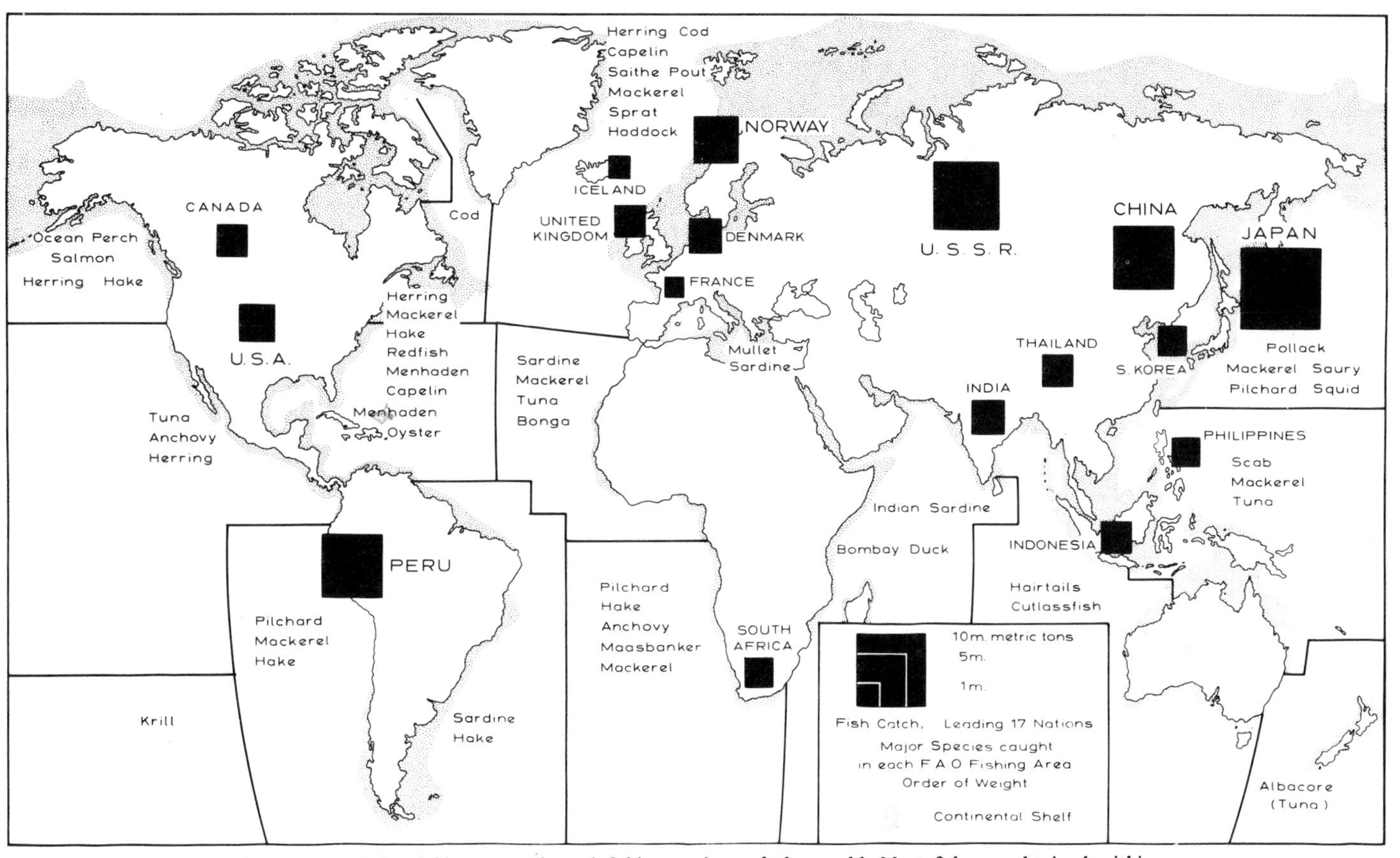

Figure 8–4 Major fishing grounds and fishing nations of the world. Most fish are obtained within 300 km of land. Of total world catch (about 70 million tonnes), about 50 per cent comes from the Pacific and 40 per cent from the North Atlantic. Most highly productive sea-areas are heavily over-fished.

creasingly becoming the domain of large corporations using large-scale operations. One special implication of these large-scale and technologically advanced fishing operations is to allow fish to be taken at greater distance from market areas. This condition must be greatly responsible for the rising importance of fishing grounds on the west coastal areas of South America and Africa. Another consequence is that the many small fishing villages and isolated canneries are disappearing. The general trend is to trade with large ports.

In order to understand the location of fishing, it is necessary to differentiate among the various kinds of fish. The first critical breakdown would surely be between freshwater and saltwater fish inasmuch as this has obvious geographical implications. Freshwater fish are relatively unimportant, however, and amount to only about 3 to 5 per cent of the world's commercial catch, and about 10 per cent of the world's fish consumption.

As might be expected, there are a number of ways fish can be classified. It might at first seem most logical to use their biological classification, but a better classification for geographical purposes is according to general location and movement habits. Here four types of fish might be distinguished. (1) *Pelagic*. This includes all fish that feed on the surface of the water and roam widely. Such ocean life includes mackerel and herring, menhaden and tuna. They are caught by seiners (nets). (2) *Demersal-mobile*. These are fish that live near or on the bottom of the ocean floor. Only those that prefer shallow bank waters would be commercially exploitable. Such fish include the cod, haddock, halibut, hake, sole, and flounder. As a general rule, these fish are caught by trawlers rather than seiners. (3) *Demersal-sedentary*. These fish live near the bottom and remain pretty much in the same general area throughout their lives. They include oysters, lobsters, shrimp, and sponges. In general the demersal fish cost much more than the pelagic because of the different costs in acquiring them. (4) *Anadromous*. These fish spawn in fresh water, usually rivers, but spend most of their lives in salt water. The most important fish in this category is the salmon. As might be expected, these are some of the easiest fish to catch because they run in schools and return to rivers to spawn. Consequently, many simple ways of catching them have been outlawed, and control is exercised over when and where they may and may not be caught.

Finally, a very useful classification is according to eating habits. The simplest breakdown is between those that eat plankton, which include the herring, menhaden, and mackerel, and those that eat other fish. The latter group includes the tuna and bluefish, which are usually caught by trolling with hooks rather than by snaring, as is the first group. The whale, on the other hand, is usually caught by an attack method, such as electric and explosive harpoons.

Most fish are caught for food, but several are used for other purposes and amount to sizable tonnages. These include the menhaden and hake. Menhaden are found in the Atlantic Ocean from the Grand Banks to Brazil but are concentrated in schools near the East Coast of the United States. Here the plankton eaters are easy prey for fishermen and are caught in large numbers; they are by far the largest group, in terms of tonnage, caught in the United States. These fish are used for a variety of purposes: as livestock feed; as fertilizer; as bait for mackerel, cod, and tuna fishing; for their oil; in the manufacture of oil, paints and varnishes; and even in the tempering of steel. The hake has also become popular as a bottom 'scrap' fish, where it is taken in large numbers by Russian and Japanese fleets for use in making fertilizers, fish flour, and in the manufacture of glue.

These fish demonstrate a current trend in fishing products. More and more, resources formerly considered useless are being used in large numbers for industrial purposes. In fact, industrial use now amounts to one-third of the world's tonnage, in contrast with only 10 per cent or so several decades ago. Undoubtedly many other uses for sea life will be found in the future.

Some fish types, however, are perilously close to extinction, mainly because of over-fishing. The Pacific salmon is an example. It must spawn in fresh water, so it has been particularly injured by land-based conditions. Industrial wastes have polluted many streams, dams have destroyed the runs of many salmon, and soil erosion has ruined lakes and streams that were formerly important spawning grounds. The present catch is only one-quarter of what it was several decades ago. The case of the Pacific Northwest salmon has become even more critical because the Russians and Japanese, with their large-scale floating cannery operations, have also found ways to catch the salmon successfully in the open ocean. To preserve the salmon, the United States has limited both the types of gear and the times at which the fish may be caught. But this has relegated the American salmon-fishing industry to primitive and relatively obsolete fishing equipment. Moreover, there are so many salmon fishermen that very few make much money from the operation.

This condition has led to a continuing controversy between the United States, Canada, and Japan over the salmon catch. The Japanese, by international treaty, agreed to restrict their catch to areas west of the 175th west longitude, supposedly restricting them to Asiatic salmon. But with the ability of the Japanese to catch American-based salmon on the open ocean, new controversies continue to arise. What the future holds for the salmon industry is therefore quite uncertain.

Whaling is a resource that would have been destroyed without international agreements. In 1946 the whaling nations of the time set up the International Whaling Commission, which tightly regulates almost every aspect of the industry in the Southern Hemisphere. Japan is the leading whaling nation, and about 60 000 whales are killed annually. Basically, the regulations of the commission restrict whaling to certain areas for individual countries and prohibit whaling in warm-ocean regions. They also forbid outright the killing of white whales and grey whales. For others, the regulations specify size, forbid the killing of calves or of cows nursing their young, limit the total amount of oil that can be taken, and restrict the operations of shore whaling stations to six months of the year. Nevertheless, certain types of whales, such as the blue whale, have been officially declared extinct.

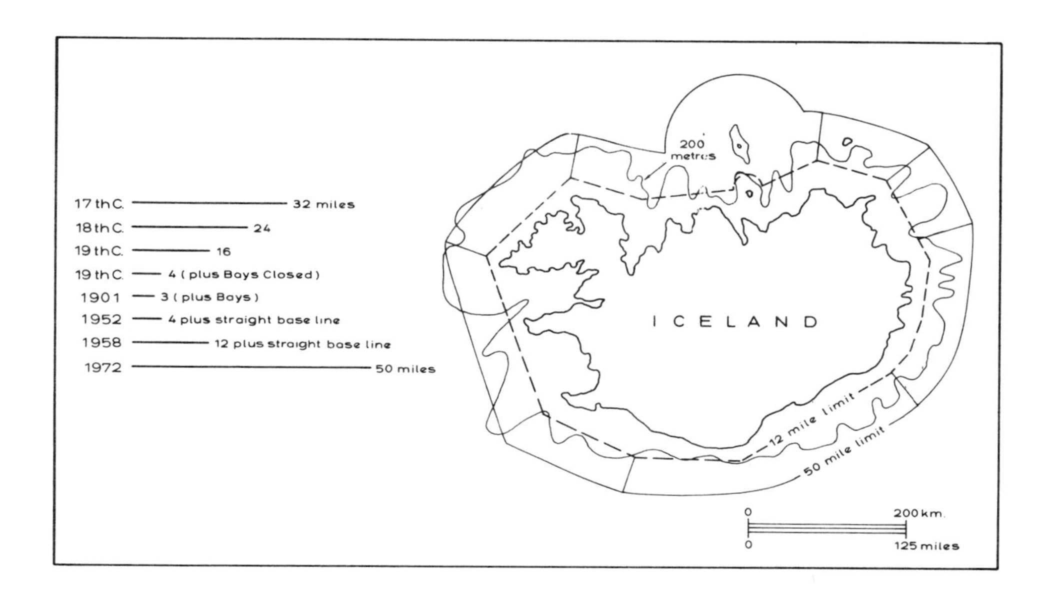

Figure 8–5 Fishing zones around Iceland. The graph shows how Iceland's fishing limits have changed through time. Use of base lines between headlands has extended the fishing limits of many countries. Like others, Iceland now claims control of an 'economic zone' extending to 200 miles.

There has been much disagreement between the nations on the matter of the extent of a coastal state's rights over fishing off its coasts. Until recent times, it was possible to see a sharp contrast between the great powers, which upheld that a 12-mile zone was sufficient to provide shelter for coastal fishing, and the less industrialized and more economically vulnerable countries which have tried to secure the largest possible area for their exclusive economic exploitation. Some of the most bitter disputes exemplify this relationship: between the USA and the Pacific states of South America, between Spain and Morocco; between Great Britain and Iceland.

The problem of fishing limits has been particularly acute in the case of Iceland, where fish and fish products make up more than 75 per cent of all exports. To Icelanders, the case for extending control over all fishing effort to embrace a zone very much larger than the internationally-accepted 12 miles was self-evident: a necessity for the conservation of an already-overfished area on which their economy depended. To British trawlermen, however, it was a claim taken without regard to the rights of those nations which had operated in Icelandic waters throughout the era of the steamship and motorship. Moreover, it would give a nation whose entire population was similar in size to Aberdeen or Hull a monopoly control over approximately one quarter of the demersal fish resources of the Northeast Atlantic. Iceland unilaterally declared a 50-mile fishing limit in 1972 (which ushered in the 'cod war' of ensuing years) and has since followed many of the Latin American and African countries in claiming a protected fishing zone around her coasts of 200 miles (Figure 8–5). Latterly, the more economically powerful nations have done the same, including the USA (1976) and Canada (1977). In Western Europe, the issue is complicated by the varied fishery commitments of the nations. But on external policy, the EEC Commission proposes a 200-mile zone which may exclude the fishing vessels of non-EEC countries like Russia, Poland, East Germany and Finland, yet pave the way for reciprocal fishing concessions with Iceland and Norway.

The whole question of fishing limits has become increasingly merged with the larger issue of the rights of the state over *all* economic resources off its coast, whether in the sea, on the sea-bed, or beneath it. During the last half century, it has been held that coastal states have rights to sea-bed resources on their adjacent continental shelf. We have seen (Figure 8–2) how median lines were drawn up amicably in the North Sea as a prelude to the nation-by-nation development of oil and natural gas. But on the 'open'

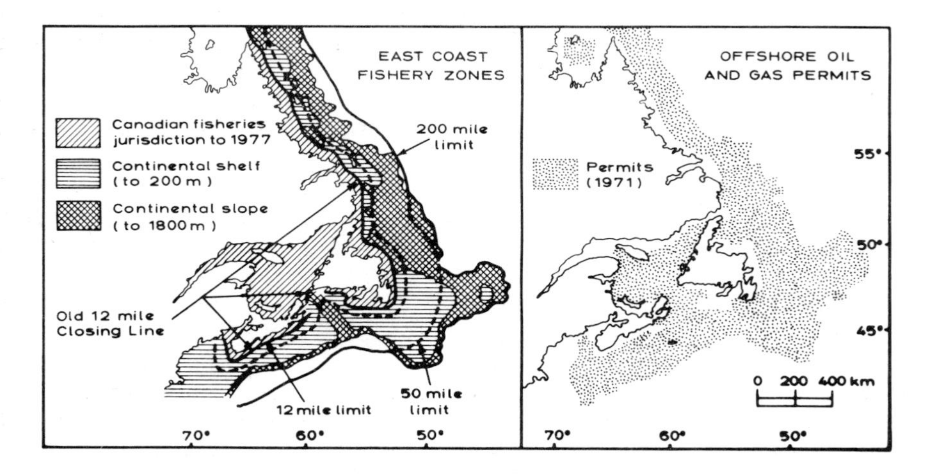

Figure 8–6 Fishing limits and offshore oil and gas permit areas, eastern Canada. The old 12-mile fishing limit 'closed off' the Gulf of St Lawrence. Canada has declared a 200-mile fishing limit and awarded oil and gas exploration permits on the continental shelf and slope to a depth of 1800 metres. Permit zones extend beyond 200 miles to take in major submarine extensions of the Grand Banks.

ocean fringe it has been a difficult matter. While it has been accepted that the coastal state may control the continental shelf to a depth of 200 metres, there is a strong trend among many maritime states in favour of very broad 'economic zones' inside which they can claim all resources for themselves.

While the problems which arise from these claims go beyond considerations of fishing, they are worth noting here because of their very geographical nature. Thus successive international Law of the Sea conferences have been faced with such questions as: Where does the natural prolongation of the land terminate? At what stage does a rock or sandbank become an island? How can we define a strait or a passage? Are functional relationships more significant criteria for measuring rights (and obligations) in marine space than proximity to land and physical criteria? Figure 8–6 illustrates some of the problems in the case of Canada.

FURTHER NOTES ON FORESTRY AS AN ECONOMIC ACTIVITY

Forests cover an area of the earth about equal to that devoted to agriculture. Of the total area under forest, however, only about 30 per cent is considered productive enough to warrant exploitation, and of this percentage two-thirds are dismissed as being inaccessible for use. This raises three questions of paramount geographical importance. (1) What must a forest be like to be classified as productive and hence commercially exploitable? (2) What does the pattern of inaccessible forests look like and how is this pattern derived? (3) How are the potentially exploitable forests utilized?

Before these questions are answered, however, it might be well to note that some of the world's best forests in terms of productivity and accessibility have been destroyed. These former forests account for about 10 per cent of the naturally forested areas of the earth and, as might be expected, were primarily located in what have become highly populated and industrialized areas. Major denuded areas include most of Western Europe, large parts of the US midwestern states such as northern Michigan, southern Indiana and Ohio, and most of China and India. In all areas but India the forests removed were of mixed coniferous and deciduous types. It is debatable whether such forests should have been preserved inasmuch as more profitable uses were made of the land, but more might have been done to maintain forests in some of these areas. In fact, reforestation is undertaken at great time and expense in many parts of Europe and China, and afforestation, the planting of trees on land not formerly covered with forests, is also occurring in many places.

Three major factors determine whether a forest is of sufficient quality for commercial use. The first is the general density of the particular type of tree desired. In far northern and high mountain areas, tree stands are generally thin because of climatic conditions, so that the board metre of lumber per hectare in such areas is considered too low to be harvested. The same factor operates in many tropical forest areas where the tree desired, for example, mahogany or teak, is often widely interspersed among many other tree varieties. Thus a thick forest alone does not necessarily ensure a profitable forest resource. However, the lack of stands in tropical areas is somewhat counterbalanced by the size of trees. On the other hand, some trees are so dense that they will not float and must be hauled overland or in barges to reach ports. Yet large trees tend to compensate somewhat for thinness of stand or generally lower densities. Third, trees that are more scarce, and thus in shorter supply, bring higher prices than do those more widely found. The general size of the forest in terms of area covered is a final factor. A pure stand of good and valuable trees will nonetheless be considered

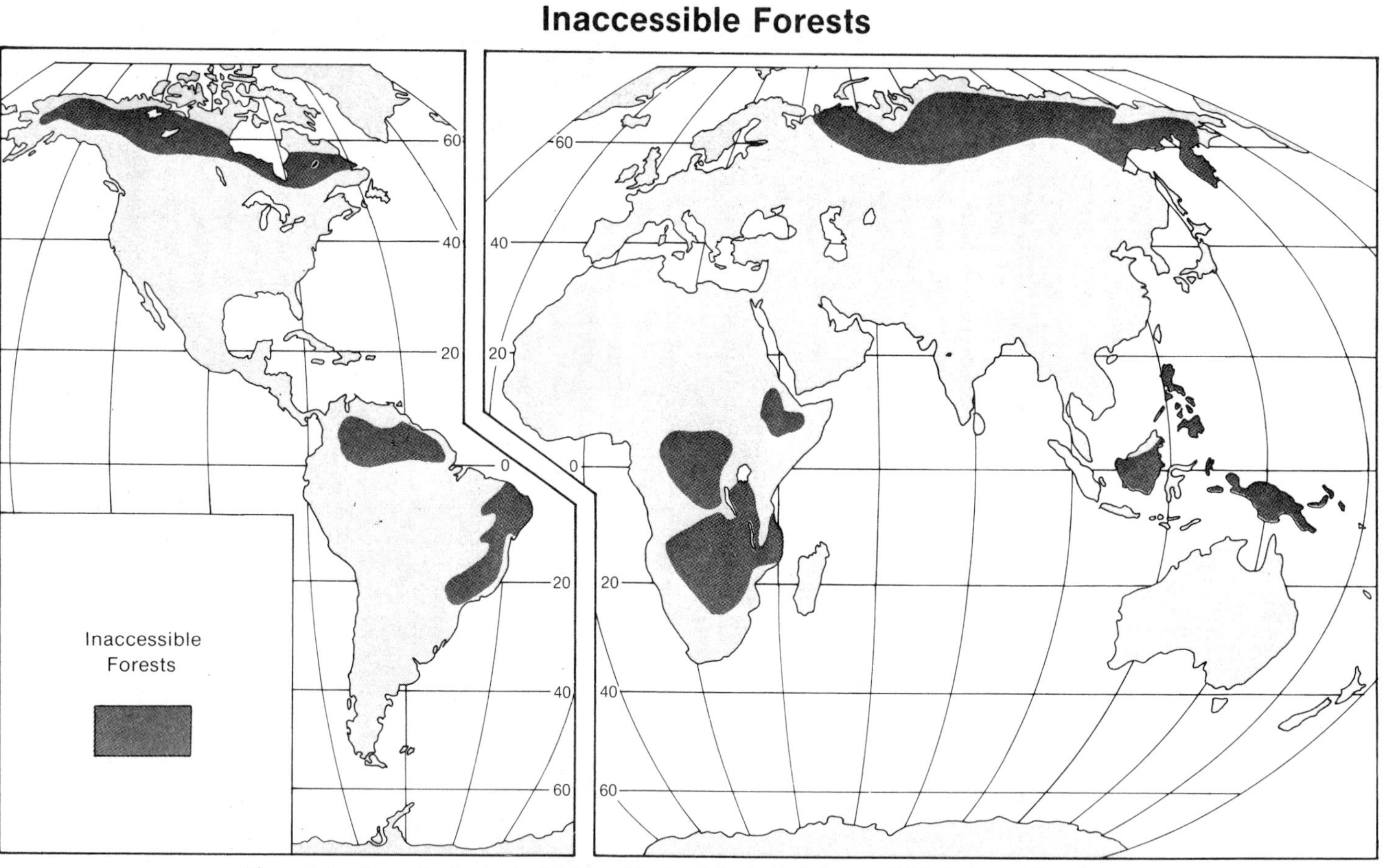

Figure 8–7 Generalized extent of inaccessible forest lands. Such inaccessibility is in part based upon better forest lands closer to the markets.

unprofitable if such a group is not of sufficient extent to justify the equipment and other features necessary for its removal.

The determination of accessibility to the market, or lack of it, is based on the same general features just discussed. Generally, areas are inaccessible if they are distant from major rivers that can be used for transportation and if they are not served by roads and railways. Thus many forests in tropical areas, those in high mountains, those in nondeveloped continental interiors, and those located in very hot or very cold lands present accessibility problems. In this regard, Figure 8–7, the inaccessible forest lands, should be compared with Figure 6–1 (page 73) which shows world transportation regions. Lack of transportation facilities in an underdeveloped area raises costs inasmuch as the timber company must bear the expense of providing them. To define something as inaccessible is thus another way of saying that the area has not been found profitable for other uses and therefore basic transportation facilities have not been provided.

For example, before the transcontinental railways came to Washington and Oregon, in the western United States, and to British Columbia in Canada, forests in those areas away from the coast were considered inaccessible. The costs of building railways for forest purposes alone were clearly too high. But after railways were built on a subsidized basis, great amounts of timber from these areas became profitable to use. A modern example is provided in Brazil, where the trans-Amazon highway is being pushed through the hitherto inaccessible rain forests. Here, valuable trees will be marketable for the first time, although the plan to clear-cut ninety-six kilometres of forest on each side of the highway may also have very damaging effects.

Undoubtedly the reason so much otherwise usable forest land is considered inaccessible is that forests are generally found on land unfavourable for agriculture, industry, and other uses. Thus forests truly support frontier activities. The limitation of inaccessible forests thus provides some measure of the urban-industrial intrusion boundary into underdeveloped lands.

It should be apparent from the foregoing discussion, however, that questions of tree quality and accessibility are interrelated; moreover, a certain degree of substitution exists. This is evident from the fact that so much otherwise marginal forest land in and very close to major markets has been exhausted, while inaccessible but otherwise excellent forests remain unused today.

In terms of future supply of forests, the picture looks fairly good. Certainly, the inaccessible forests will increasingly become more accessible and therefore more usable. Increasing scales of operation, characteristically in the western countries undertaken by very large, tree-owning corporations, are becoming the norm. Many such 'timber empires' benefit through complete management control over all activities, from planting, thinning and felling to transporting, milling and manufacturing timber and paper products. Their responsibilities, however, are considerable. In the leading forest nations, like the USA, Canada, and Sweden, they are guided by legislation framed within a general philosophy which aims to encourage sustained yield and the protection of the environment. Two examples of corporation forestry activity are shown in Figure 8–8.

Classification of trees and forest types are also numerous. Many of these types have very specific and nonsubstitutable uses. For our purposes, however, we might recognize three main groups of trees: (1) coniferous evergreen, (2) deciduous broadleaf, and (3) broadleaf evergreen. The coniferous trees are generally softwoods, and the deciduous and broadleaf evergreen are usually hardwoods. Among the softwoods are such trees as fir, spruce, pine, hemlock, cedar, larch, cypress, and redwood, which are heavily used in the lumber industry, in the pulp and paper industry, and for general fuel use. The hardwoods include a wide variety of trees. In midlatitude areas, oak, maple, poplar, chestnut, beech, birch, elm, ash, hickory, walnut, cherry, and sycamore are used. In tropical areas, the hardwoods that are commercially exploited are mahogany, teak,

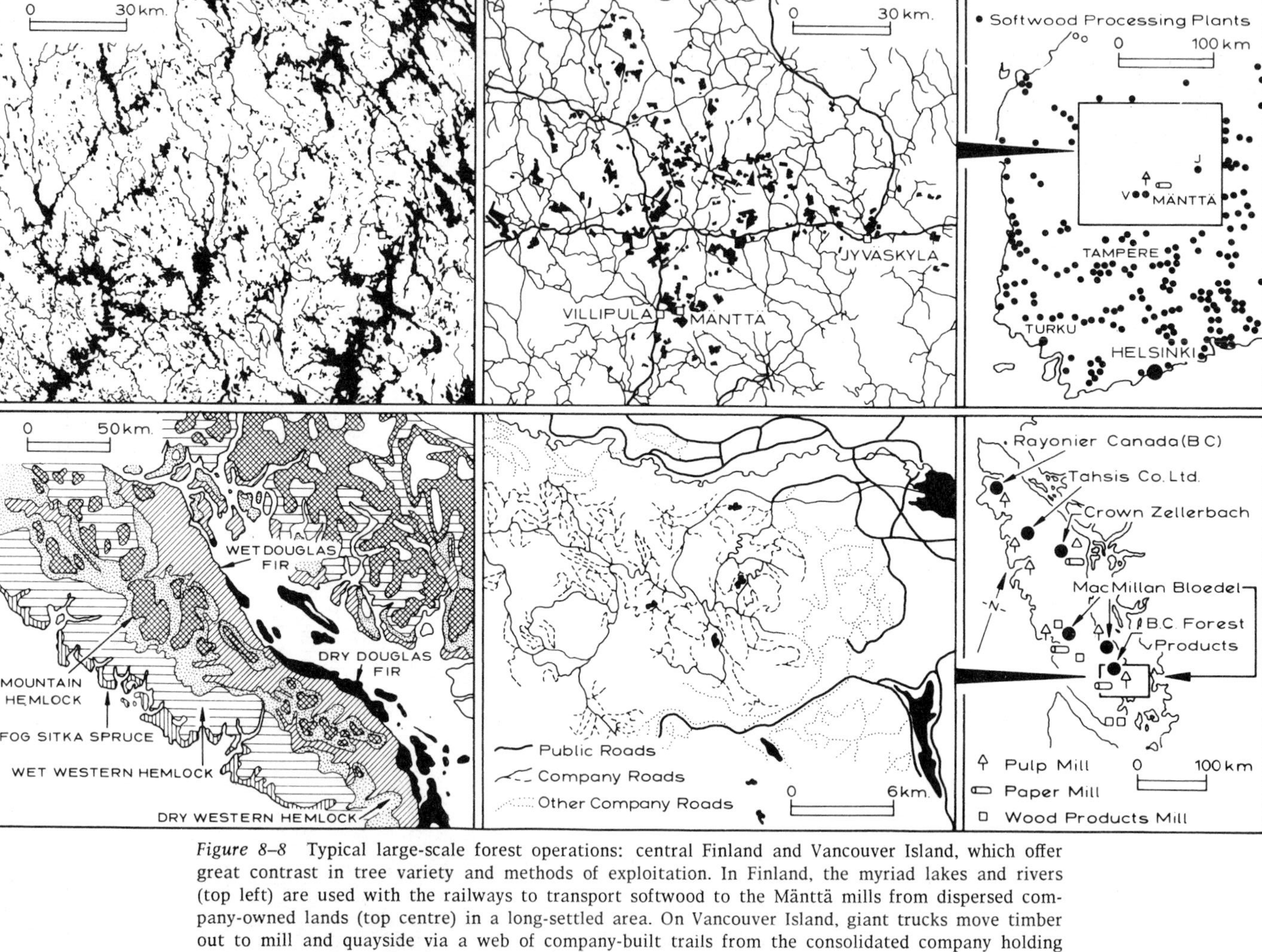

Figure 8–8 Typical large-scale forest operations: central Finland and Vancouver Island, which offer great contrast in tree variety and methods of exploitation. In Finland, the myriad lakes and rivers (top left) are used with the railways to transport softwood to the Mänttä mills from dispersed company-owned lands (top centre) in a long-settled area. On Vancouver Island, giant trucks move timber out to mill and quayside via a web of company-built trails from the consolidated company holding (bottom centre) in the uninhabited interior (after W. R. Mead, *An Economic Geography of the Scandinavian States and Finland*, London: University of London Press, 1964, pp. 245–253, and from information supplied by Macmillan-Bloedel Ltd., Shawnigan Division, 1974).

ebony, and rosewood. With the exception of these, most tropical trees are used more for forest products than for lumber purposes.

Geographically, four types of forest areas can be noted: (1) tropical or broadleaf evergreen, (2) coniferous evergreen forests of far northern areas, (3) midlatitude mixed coniferous and deciduous forests, and (4) montane forests, which are generally coniferous.

These four forest zones are strikingly different in terms of the use to which forests are put. In fact, there are a number of forest uses such as recreation, wildlife reserves, watersheds, and the like, but only the following uses will be discussed here: (1) forest products, (2) lumbering, (3) fuel, (4) pulp and paper, and (5) chemical products.

Forest products are particularly important in tropical forest areas. Here, such products include nuts (copra, oil, palm, ivory, kapok, and coir), leaves (coca), bark (abaca, quinine, and cinnamon), and sap (rubber). In midlatitude areas, forest products include turpentine, tars and pitches; bark (cork) and tanning materials; and such items as molasses.

Both hardwood and softwood trees are used in lumbering. Of these, the softwoods account for about twice the lumber volume of hardwoods. Softwoods are primarily used in the pulp and paper industries like timber or roundwood, in the building of homes, barns, bridges, and so forth. Hardwoods are more expensive and are used for better quality items of a more manufactured nature, such as bowling pins, furniture, pencils and railway wagons. Tropical hardwoods are used for very fine furniture, for carvings, and for highly valuable items.

Fuel use accounts for about half the wood used from all the forests of the world, in terms of volume, although in terms of value, it accounts for only about 10 per cent. Much of the wood used for fuel is inferior or is characteristically used in many underdeveloped areas, where coal, petroleum, and other fuels are absent. In much of Latin America, Japan, and other parts of Asia, charcoal is very inefficiently prepared as a supplement to coal. Great use has also been made of wood for fuel in transportation: along some major tropical area rivers (for steamboats) and beside the tracks of forest-zone railways (for locomotives) in such areas as Eastern Canada and Siberia.

In highly industrialized countries full of 'paperopolises', the demand for paper is increasing very rapidly. Newsprint, magazines, books, and vast new uses for packaging materials have caused the pulp and paper industry to flourish. Such activity is most concentrated in the coniferous softwood areas of the US Pacific Coast, in Canada, and in Northern Europe. Spruce, pine, hemlock, and fir are all heavily involved. Some of the main chemical products derived from wood include acetic acid (perfumes, plastics, rayon, solvents), acetone, (solvents, acetylene, explosives), butadiene (synthetic rubber), charcoal (explosives, fuel, medicines), methyl alcohol (antifreeze, paints, shellacs, varnishes), ethyl alcohol, and various oils, tars, and solvents. Most of the above products are really by-products of wood, used primarily for other purposes, but there is a growing wood chemical industry akin to the tremendous revolution in petro-chemicals.

Thus the 'mining' of the world's forests represents the raw material for many different products. In many cases the site of such raw materials exerts an important locational impact on the next stages of use. Hence an understanding of the major areas of forest commodities provides insights into the location of all related, raw-material oriented activities. Paper-making industries in particular are found fairly close to the forest raw materials. With the use of recycled paper, however, market location may prove more important for the future. Even the tanning industries were at one stage locationally affected by proximity to oak and hemlock trees. Today tannins are made chemically by chromium compounds or from the bark or extracts of certain tropical and subtropical trees.

FURTHER NOTES ON MINING AS AN ECONOMIC ACTIVITY

Although important as a raw material base for industry, forestry is nonetheless small compared with mining. There are well over 200 different minerals in use today, and each has special locations and uses. At least eight categories of mining resources are generally specified: (1) iron ore; (2) fuels such as coal, petroleum, and natural gas; (3) nuclear energy fuels such as uranium, thorium, radium, plutonium, and lithium; (4) construction materials such as clay, granite, sandstone, slate, gypsum, asbestos, and limestone; (5) ferro-alloys, which include manganese, chromium, tungsten, nickel, cobalt, molybdenum, and vanadium; (6) nonferrous metals such as copper, tin, aluminium, magnesium, zinc, and lead; (7) mineral fertilizers such as nitrates, phosphates, and potash; and (8) chemical resources such as salt and sulphur. It can readily be seen that a discussion of all these mining resources would require a book in itself. We will be content to develop some spatial generalizations.

Mineral resources are extremely widespread over the earth. To the casual observer, it might appear that minerals can occur almost anywhere. Nonetheless, mineral deposits are closely associated with particular geological features. For example, mountainous areas such as the Urals, the Appalachians, and various European ranges are outstanding in their abundance of minerals. In the United States, the Appalachians, particularly in West Virginia, are noted for coal. Minnesota and Upper Michigan are important for iron ore, although it is also found at Birmingham, Alabama, and in many parts of the Rocky Mountains. The Gulf states, particularly Louisiana and Texas, are notable for the mining of petroleum, natural gas, and related chemical products. Bauxite is found in the Ozarks. The Rocky Mountains contain a vast assortment of ferrous and nonferrous minerals. By contrast, the Great Plains, the Midwest, Florida, and much of the Pacific Coast show little mining activity. In Europe, few countries are bereft of minerals (Figure 8–9). Most of the coal deposits are situated in the carboniferous strata found close to the surface all around the old Hercynian massives: the Ruhr in Germany, Upper Silesia in Poland, the Donetz Basin in the Ukraine, the Franco-Belgian coalfields area, the Scottish Lowlands, Lancashire, South Wales and the North Midlands in England. Many of these coal deposits have been long-exploited and their mines in decline. But deeper coal in 'hidden' coalfields close by the older mines is being exploited, and wholesale reappraisals of the cost of raising coal are being made in the light of total energy costs. Most of Europe's oil and natural gas is limited to North Sea Europe, but small pockets are found elsewhere, as at Lacq near the Pyrenees. Significant reserves of iron ore are found in Sweden, Spain and France. Other metallic ores are limited in amounts, but widespread. Ores containing manganese, chrome, and tungsten are concentrated mainly in the Mediterranean countries, while the Scandinavian region produces virtually all of the modest amounts of nickel, molybdenum and cobalt that Europe possesses. Nonferrous ores have been mined in small quantities in many European countries, but increasingly, as with other ores, they have been imported from other parts of the world.

The brief lists of mineral resources given above could be considerably lengthened. Indeed, on a world scale, the most striking picture is presented by the great profusion of mining resources in the advanced industrial countries in the Northern Hemisphere. Practically any mining resource one can mention is found somewhere in this area. A possible explanation is that this concentration of resources merely reflects the fact that people have more closely explored nearby areas than those in more remote locations. Such superabundance of mineral resources in the industrialized nations does not appear to fit any neat, logical, physical explanation. If people have primarily found

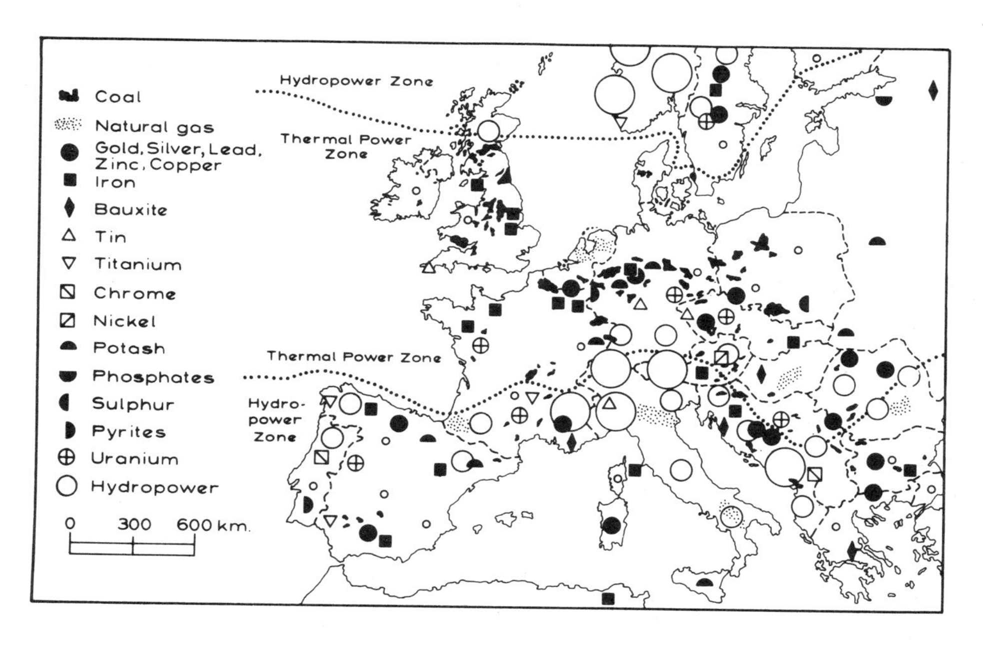

Figure 8–9 Distribution of mineral resources in Europe. Many European nations possess a wide range of useful minerals, but economically workable deposits are today mostly small and scattered. Only in coal, iron ore, potash and salt do the European countries produce more than 10 per cent of world production. Iron ore, lead, zinc, copper, nickel, tin and bauxite are all imported by the industrial nations. Inside the 'thermal power zone', Europe produces 20 per cent of the world's coal and still possesses large reserves (see Figure 8–2 for North Sea oil and natural gas).

minerals where they have looked the most, that is, in areas near them, it would follow that there are considerably more mineral resources on earth than one would imagine simply by looking at present distribution maps.

That this is probably the case is evidenced by periodic discoveries of extremely large and rich mineral resources in other parts of the world. Petroleum exploration has been very successful in many areas outside industrialized countries. In fact, some of the most productive fields in the world are in places like the Middle East, northern Venezuela, and more recently, in southern Nigeria and northern Alaska. Vast mineral resources, including copper, tin, zinc, manganese, cobalt, and others, have been found in the African areas of Katanga, Zambia, and Rhodesia. In short, it appears that to date vast areas of the earth have received little exploratory attention.

Nevertheless, most advanced nations import a number of minerals from other areas. This occurs either because outside resources are richer and more economical to use than are national resources, or because of an outright scarcity of minerals locally. Most imports are the result of the former condition, in which it is more profitable to use foreign mine commodities than perhaps less-accessible or lower-quality resources at home. This is particularly the case where the local resource is in the interior of continents. For example, for steel production along the East Coast of the United States, it is cheaper to bring iron ore from Labrador than to use Minnesota iron ore. On the other hand, Canada finds it more advantageous to import American iron ore, via the Great Lakes, even though the iron ore in eastern Canada is richer. Thus, both Canada and the United States import and export iron ore, a practice illogical to an accountant but sensible to a geographer.

In fact, many advanced countries today could survive on their local mining resources but at a higher cost. Only a very few resources are not available without imports. Nevertheless, various countries have certain critical resources that make local supply almost, if not completely, impossible. Even if the resource is available somewhere within a country, it can still be quite inaccessible and clearly less economical than importing the resource by water. Some classic cases include the relative lack of coal in much of South America and Africa. Although coal is found in a few places there, it is not in a good location relative to iron ore deposits, which by contrast are impressive. Large deposits are accessible to the sea in Venezuela, Peru, and Chile. In fact, considerable iron ore is exported to the East Coast of the United States where massive steel production is occurring. Moreover, Western Europe at present is still highly dependent on the closest major source of petroleum, the Persian Gulf area of Southwest Asia (although petroleum discoveries in the North Sea area (see Figure 8–2) may change this situation). The United States, although having great oil reserves, is a heavy importer of Venezuelan petroleum.

From our review of all three varieties of extractive activity we may come to interim conclusions which once more bring them together. Firstly, we might say that the extractive activities, in the spatial context, are in one sense among the easiest to understand; fishing, forestry, and mining must, by definition, occur where these resources are found on the face of the earth. People select the resources they will use on the basis of proximity of the resource to the market, quality and size of the resource, and price of the resource in the market-place. In another sense, however, that is, in dealing with questions of when and how a resource is to be taken, the extractive activities are made more difficult to understand because of the numerous public-policy questions with which they are plagued. Inasmuch as most of the extractive activities are depletable, especially if abused, questions often occur as to whether or not the resource should be used now or in the future. Moreover, if the resource is not taken carefully, waste often results. Thus questions of how best to extract the resource are highly relevant. Finally, since most extractive activities occupy considerable territory, they are in conflict with nonextractive uses of the resource. Often the overriding concern is what might loosely be called conservation, that is, the wisest and best use of resources.

BIBLIOGRAPHY

Alford, J. J. 'The Chesapeake oyster fishery', *Annals of the Association of American Geographers*, June 1975, pp. 229–239.

Andersen, H. G. 'Iceland extends its fishery limits to 200 miles'. *EFTA Bulletin*, Vol. XVI, Dec. 1975, pp. 45–46.

Bernarde, M. A. *Our Precarious Habitat*. New York: W. W. Norton, 1970.

Coull, J. R. *The Fisheries of Europe*. London: Bell's Advanced Geographies, 1973.

Couper, A. D. Chapter 2 'The economic geography of the sea'. In *Marine Policy and the Coastal Community*, Ed. Johnston, D. M. London: Croom Helm, 1976, pp. 37–63.

Franci, C. T. *An Alternative Arrangement for Marine Fisheries: An Overview*. Washington D.C.: Resources for the Future, 1973.

Fullerton, B., and Williams, A. F. Chapter 18 'Northern Sweden', and Chapter 19 'Finland—an economic introduction'. In *Scandinavia*, 2nd ed. London: Chatto and Windus, 1975, pp. 216–231 and pp. 232–246.

Gillespie, G. J. 'The Atlantic salmon', *Canadian Geographical Journal*, June 1968, pp. 186–199.

Gonzalez, R. J. 'Production depends on economics—not physical existence', *Readings in Economic Geography*. Ed. Roepke, H. G. New York: John Wiley & Sons, 1967, pp. 353–361.

Hare, K. F., and Ritchie, J. C. 'The Boreal bioclimates', *Geographical Review*, July 1972, pp. 333–365.

Helin, R. A. 'Soviet fishing in the Barents Sea and the North Atlantic', *Geographical Review*, July 1964, pp. 386–408.

Hjul, P. 'World fish stocks on a delicate balance', *The Geographical Magazine*, October 1977, pp. 27–40.

John, B. 'Fish for survival in Vestfirdir, Iceland', *The Geographical Magazine*, October 1978, pp. 63–66.

Luard, E. *The Control of the Sea Bed*. London: Heinemann, 1974.

Marts, M. E. 'Conflicts in water use and regional planning implications', *Regional Development and the Wabash Basin*. Ed. Boyce, R. R. Urbana: University of Illinois Press, 1964.

Minghi, J. V. 'The problem of the conservation of salmon with special reference to Bristol Bay, Alaska', *Land Economics*, November 1960, pp. 380–386.

Mitchell, B. 'Politics, fish, and international resource management: the British Icelandic cod war', *Geographical Review*, April 1976, pp. 127–138.

O'Dell, P. 'Europe sits on its own energy', *The Geographical Magazine*, March 1974.

Sheskin, I. M., and Osleeb, J. P. 'Natural gas: a geographic perspective', *Geographical Review*, January 1977, pp. 71–85.

Smith, D. M. *The Practice of Silviculture*. New York: John Wiley & Sons, 1963.

Smith, G.-H., ed. *Conservation of Natural Resources*. New York: John Wiley & Sons, 1958.

Thomas, T. M. 'World energy resources: survey and review', *Geographical Review*, April 1973, pp. 246–258.

Thomas, W. L., ed. *Man's Role in Changing the Face of the Earth*. Chicago: University of Chicago Press, 1956.

9

Bases of Agricultural Location

In this chapter and the next, we focus on agriculture activities, their location and pattern. Such focus hardly needs justification:

> '*All materials for living come directly out of the soil or crust of the earth. The man in a ship at sea or in a steel skyscraper in a modern city gets his sustenance from the soil just as surely as does the farmer who takes potatoes from the furrow*' (J. Russell Smith, *Industrial and Commercial Geography*, New York: Henry Holt & Company, 1925).

The locational understanding of agriculture ranges from the simple to the difficult. In some respects, agriculture can be treated as a simple locational problem much as the extractive activities were in Chapter 8. For example, some crops are highly restricted to particular locales because of such physical restraints as climate. A valid differentiation can be made between agricultural activities that require a tropical environment, such as growing bananas, and those that require a cool climate, such as dairying. The potential areas of cultivation for highly selective crops like cotton, rice, bananas, and coffee can thus be reasonably determined on the basis of physical requirements. In this manner, as with the extractive activities, the locations of the resource areas are given.

Another simplistic approach to agricultural location is to assess crop potential in terms of accessibility to the market. This procedure applies well to regions that have fairly homogeneous physical features, such as similar climate, soils, and topography. In this manner agricultural types can simply be ranked according to the intensity with which they occupy the land: those crops that have the highest profits per hectare, or which for reasons of high cost transportation due to perishability must be close to the market to survive, bid highest and thereby get the land closest to the market. Activities with less rent-paying ability are squeezed outward from the market in a series of zones of decreasing intensity.

As we progress into the discussion, however, these simplifying assumptions will be systematically removed until a fairly realistic approximation of actual patterns of agricultural use is mirrored by the theories. At this latter phase the true difficulty of the agricultural location problem will emerge. For the spatial understanding of agriculture is in reality much more difficult than that of the extractive activities previously examined. Specifically, one more degree of freedom in locational choice is evident in agricultural location than was previously the case. In agriculture, there are two possible questions rather than one. The familiar query is, given the land (resource), what crop, or crops, will best be produced on it? (In the extractive activities, the question is merely whether or not the resource should be used.) A second question is, given a crop (resource), where is the best place to produce that crop? This choice is not available to extractive operations; thus there is considerable interplay in the agricultural location problem among land, crops, and location. Different crops can be and are produced on similar kinds of land. Conversely, similar lands are often used for different crops. By contrast, in the extractive activities, the resource and the place of production are always bound together.

There are three major locational determinants in agriculture: (1) inputs, or procurement costs, (2) outputs, or distribution costs, and (3) on-site production costs. The input costs include items not available on site, such as seed, fertilizer, machinery, and other items required to produce a crop. These inputs involve capital outlay as well as transportation costs necessary to get the 'raw material' items on site. Fortunately, these costs are often of such a nature that they are not decisive in the determination as to whether any given crop will be grown: they do not exert much locational significance within any given region.

MARKET ACCESSIBILITY AND AGRICULTURAL LOCATION

The major locational determinant of the type of commercial agriculture practised in any given area is proximity to market. Unless the farmer can get his produce to market, that area is useless for commercial agricultural purposes. Some parts of the earth that might be physically suitable for a particular crop remain unused simply because of their remoteness from markets. The development of land transportation and the opening up of farm lands have always occurred together. The technical name for such market accessibility needs is distribution costs and will be symbolized by the letter *T*.

The other major cost, and hence locational determinant, with respect to agriculture is the on-site production costs. These relate to the physical quality of the land, to labour costs, and to general operating costs during the production period. Because most of these are variable as a result of differing physical environments, in the following analysis they are labelled on-site costs and are symbolized by the letter *E*.

In order to see how these two costs (*T* and *E*) relate to crop production patterns, the ceteris paribus technique will be used. Initially, all costs associated with production, except *T*, will be held constant. After this effect is presented, the on-site costs (*E*) will be allowed to vary spatially in an attempt to see how site and transportation costs affect the profitability of various lands. Such relaxed assumptions will also set a more realistic framework by which actual agricultural patterns can be nominally assessed.

The analytical procedure discussed below is usually referred to as a Von Thünen's model. It is named after its originator, Johann Heinrich Von Thünen, who in 1826 developed a concentric series of agricultural zones or rings around a central market, each with its crop production pattern determined mainly by differential transportation costs in which distance, weight and perishability of product were the important considerations. His assumptions, which will be used here initially, include first of all a single market (city) on an isolated, homogeneous land surface. Thus all surplus commodities from the surrounding areas must go to this one market. In addition, the tributary area is considered to be homogeneous in terms of the physical environment and in terms of productivity per unit of land for any given crop. Von Thünen's second assumption is an agricultural hinterland traversed only by a single mode of land transportation. All areas within a particular distance zone from the market have equal accessibility to it. In other words, all transportation lines are radially focused on the market town.

Moreover, transportation costs, which are directly proportional to distance, are to be borne directly by the farmers, who must ship all commodities to the city in 'fresh' form. Von Thünen's final assumption is that we are dealing with rational, economic people who desire to maximize their profits and are completely flexible in the type of crop they produce so that they can change production immediately in order to maximize profits.

The above are, of course, assumptions or premises that do not hold in the real world. Nevertheless, the importance of market accessibility is of such great significance for many crops that if one looks for them within the framework of this kind of model, examples of such influence can be seen in the actual crop patterns of the world.

The general result of applying these premises in such a landscape laboratory is that the number of profitable crops that can be grown decreases as distance from the market increases, Thus there is an outer limit beyond which a crop cannot be grown because it would be unprofitable, and there is an inner limit for a crop because choices of more profitable crops exist. In Von Thünen's study the inner zones were devoted to market gardening and fresh milk production. These uses were primarily located here because of perishability considerations (high transportation costs) and the consequent need to get these commodities to the market fresh. Of course, with today's refrigerated transportation, fresh milk can be brought to centres from distances of 500 km or more, and fresh vegetables can be brought in from across continents. The second zone also reflected the technology of the day and was occupied by farmers specializing in producing firewood, then the main type of fuel for heating homes. Its bulky nature made transportation costs high, thus putting heavy emphasis on production in the inner zones. Also, the heavy demand kept the market price fairly high so that timber cropping (silviculture) yielded greater returns to the farmer than did any other crop except fresh vegetables and milk. The third, fourth, and fifth zones were occupied by grain-type crops of decreasing profitability. Zone 6 was used most profitably, according to the premises above and the prices at that time, for livestock production. Von Thünen considered the transportation costs of livestock to be quite low inasmuch as cattle could be driven to market. Another activity in the sixth zone was cheese production, a commodity not highly perishable and one that could easily stand the high transportation costs. Finally, the zones beyond were considered unprofitable for crops and remained unused wilderness areas. It should be apparent from the above discussion that three variables show up as decisive: (1) price or value of the commodity at the market (V); (2) cost of producing the commodity at the farm site (E) (in this case, assumed to vary

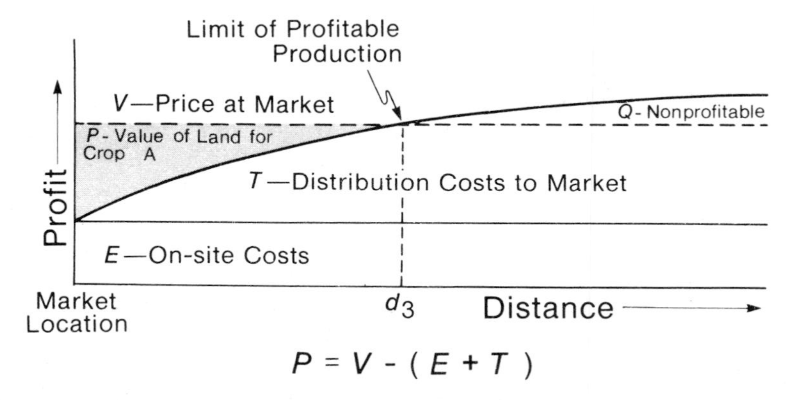

Figure 9–1 Profitability of cropland in relation to distance from market. Note that the value of land for crops (P) decreases with increasing distance from the market because of increasing distribution costs (T) to market. On-site costs (E) are considered equal for all locations.

only among commodities and not among areas); and (3) cost of transporting the commodity to the market in fresh condition (*T*).

The method in which the above variables operate is demonstrated in Figure 9–1. These three variables are considered as follows. The on-site production costs (*E*) are assumed to be equal throughout the market hinterland inasmuch as the physical environment is presumed to be homogeneous. The price at the market (*V*) is given on the basis of going prices. And the distribution or transportation costs (*T*) are assumed to increase for any particular crop in a linear manner with distance from the market. (In Figure 9–1 the distribution costs are shown in an exponential fashion in order to make the example more realistic.) It is evident that the crucial variable in this model is transportation costs, or proximity, to the market. The variation over distance of this variable thereby makes the profitability for producing any one crop (*P*) decrease with increasing distance from the market location. Thus, all crops would be more profitably produced in zone 1, but inasmuch as some crops are more profitable than others, less profitable crops are squeezed out to less advantageous locations.

Note that for the example crop shown, the limit of profitable production is at d_3. Beyond this distance, the production of that crop will lead to a deficit—the farmer will lose money by growing it. The extent of this loss in any given location is shown by space Q.

The critical curve is the line enclosing space *P*. Note that this space shows decreasing profitability as distance from the market increases. Such a curve is called a *rent gradient* and is a spatial representation of the profitability of any given crop for any given area. The rent gradient is generally presented in the manner shown in Figure 9–2.

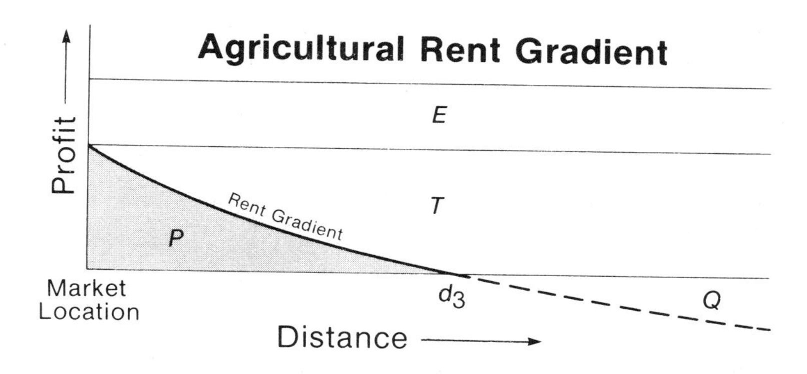

Figure 9–2 Developing an agricultural rent gradient. Based on Figure 9–1.

Note that the rent gradient (*P*) is merely the curve that has been derived from two constant features (*E* and *V*) and receives its distinctive shape from the transportation cost line (*T*). In formula form, therefore, $P = V - (E + T)$, where *E* is held constant.

The application of this in the Von Thünen context is shown in Figure 9–3. This model contains one market and three crops: A, B, and C. Note that crop C is the most profitable crop in the market, but that it also has the steepest rent gradient. Thus its maximum limit of profitable production is at distance Q_c. However, it is not produced outward to that distance inasmuch as crop B is a more profitable crop at distance d_1 before Q_c is reached. Therefore, crop C will be produced to distance d_1, and crop B beyond that point to distance d_2, where crop A becomes the highest potential rent-payer. This ranking of territorial profitabilities among crops leads to a series of concentric zones outward from the market as shown. These concentric zones, it might be noted, are unlikely to be similar in width.

This same procedure can also be applied to multiple markets, as shown in Figure 9–4. Note that the crop with the highest rent-paying ability gets first choice for pro-

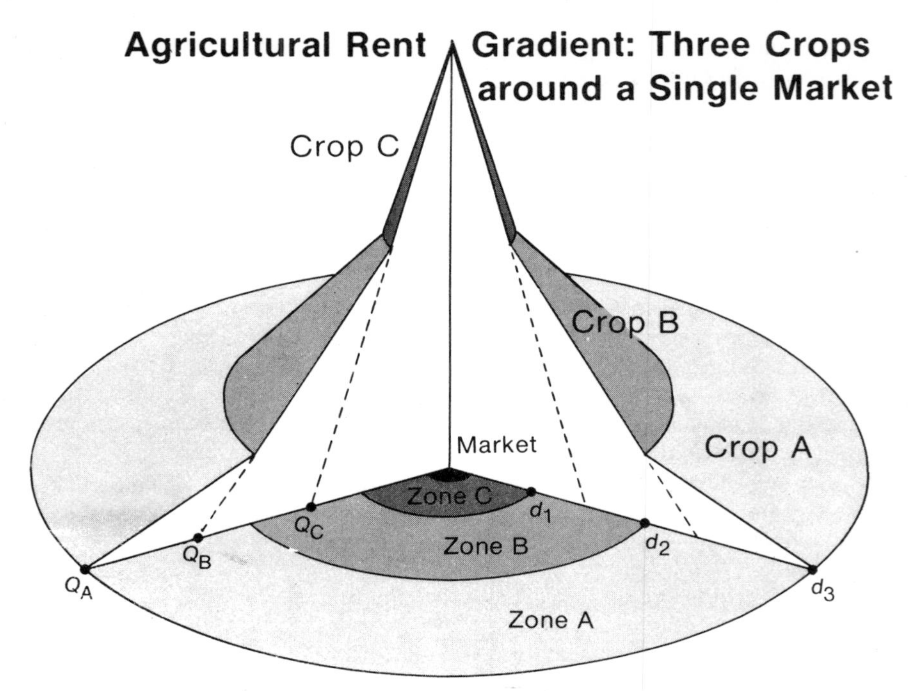

Figure 9–3 Agricultural rent gradients for three crops about a single market place and the territorial result. Note that crop C would be grown outward to distance d_1 to dominate zone C, crop B would be grown between distance d_1 to d_2 and would dominate zone B, and crop A would dominate the area from d_2 to d_3 in zone A. Crop A would not be grown beyond d_3 because it becomes unprofitable beyond that distance.

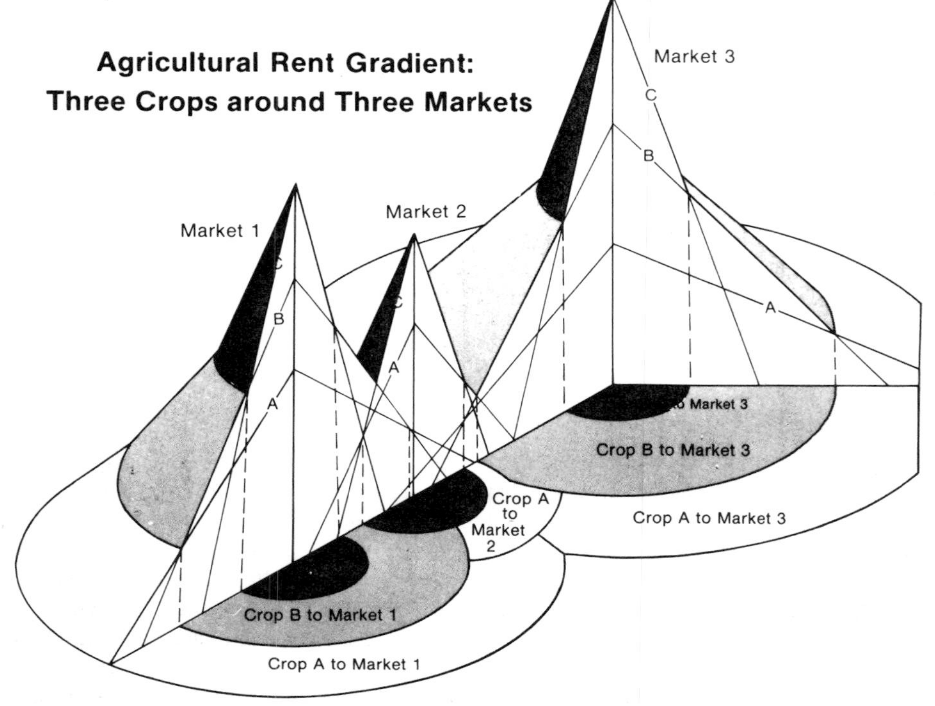

Figure 9–4 Agricultural rent gradients for three crops about three markets. Note that the subterranean lines (dashed) are ignored in the determination of surface crop responses as shown for the areal distribution of crops.

duction in any given area. Thus the crops to be grown along a traverse connecting markets are along the top rent gradient lines. Other possible, but less profitable, crops are shown beneath this line in dashed form. Such rent gradients, in reality, represent those crops next in line for production should the market price decrease for the crop currently most profitable, should transportation costs for it increase, or should the next-order crop increase its rent-paying position. The markets for the most profitable crops are demonstrated by the extension of the rent gradient to its highest point at a particular market place. The profile of crops and markets is shown directly below the rent gradient curves. Note that it is possible that some markets may not be markets for all crops. For example, a town may purchase wheat, but have no stockyard. Thus it is possible that crop C might be shipped to market 3 even though markets 1 and 2 are closer. This principle begins to approach reality in those areas that are fairly physically homogeneous. Thus this model, despite its ceteris paribus assumptions, provides a fairly good first-order approximation of the distribution of many actual crop patterns.

ON-SITE COSTS AND AGRICULTURAL LOCATION

Perhaps the most blatantly questionable premise of the model above is the assumption that we are dealing with a flat and featureless environmental surface. This is a particularly serious variable to hold constant inasmuch as agriculture is so much affected by on-site factors like climate and soil. Therefore, in Figure 9–5 the on-site costs (E) have

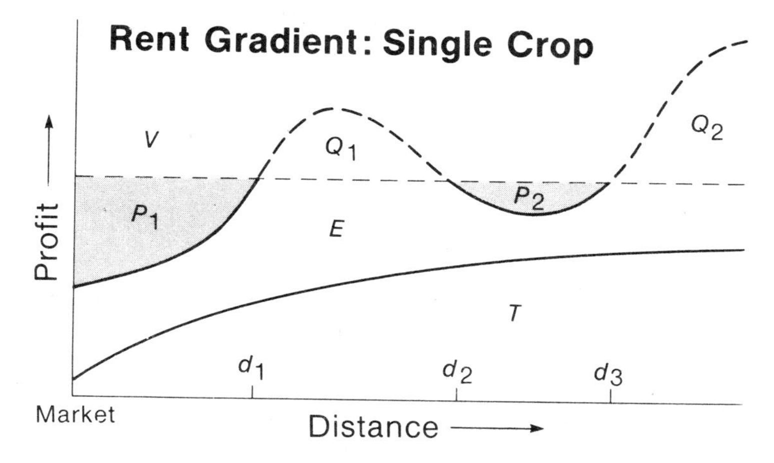

Figure 9–5 Profitabilities of land for a single crop with variability in on-site costs (E). This is but one of many possibilities inasmuch as the on-site costs are based on peculiar topography, soils, farm and field size, and many other variables which are not directly related to distance from market. Compare with Figure 9–1.

been allowed to vary concomitantly with transportation costs. In this case there are two zones of profitability, P_1 and P_2. Thus, as with many actual agricultural patterns, particularly on a local scale, environmental features cause an otherwise smooth rent gradient to become spatially disrupted. An example of how this works is demonstrated by going outward from an imaginary town where the land near the town is relatively flat and fertile, but the land slightly farther out is hilly and rocky. Beyond this zone is a recurrence of the fertile soil like that near the town. Crops might be found in the good soil lands, but perhaps grazing, or some less intensive use, will be found in the hilly and rocky areas. Nonetheless, the general gradient affected by transportation still exerts some effect inasmuch as the profitability further out (say, at P_2) is much less than identical land would be closer to the market. Likewise, the second zone of unprofitability, Q_2, is much less profitable than the nearer zone, Q_1.

The same method of assessing highest and best use of the land can nonetheless be used with this model, as shown in Figure 9–6. Note that crop B is identical to that shown in Figure 9–5. Two other crops with their respective rent gradients are also introduced. The principle of determining which crop is best produced in an area is identical to that used earlier. In the case shown, crop C appears to be least affected by the hypothetical hilly and rocky land and thus is able to use such land at a profit. However, there is some land, perhaps swamp land or very rough land, which is not profitable for any crop. Thus, despite its proximity to the market, this land is left unused for agriculture.

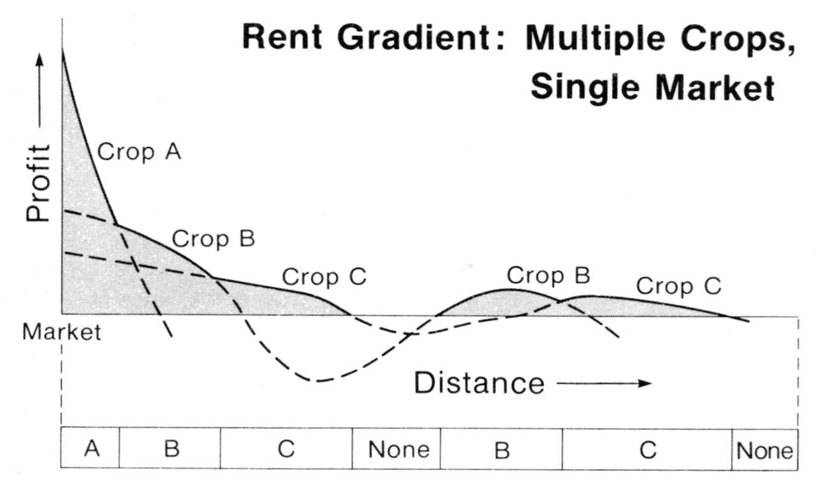

Figure 9–6 Hypothetical rent gradients about a single market with on-site costs considered. Compare with Figure 9–5.

DISTANCE-RELATED PRODUCTION ZONES IN THE MODERN WORLD

So far, we have established a valid form of analysis which identifies some vital aspects of the agricultural location problem. It is salutary to be reminded, however, that in its original intent and proper use, Von Thünen's model depicts an idealized, isolated region of perfect economic competition, land rent maximization and rational behaviour on the part of farmers. These *are* model conditions: what, then, of its application to reality?

Even in 1826, Von Thünen recognized that his assumptions needed to be relaxed to approximate actual conditions. In particular, he noted that in real countries soils vary in quality (to affect fertility), few major towns were without a navigable river or canal (to affect transportation) and capital cities were accompanied by a scatter of other towns (to complicate the market). We have shown, in Figure 9–6, how varied terrain might disrupt the ordered pattern of concentric land-use zones, and it could be shown readily how a navigable river would do likewise. Von Thünen assumed a water transportation rate of only one-tenth of that by land: farmers near this cheaper axis of movement would benefit in such a way as to distend the zonation. As to the existence of smaller towns, they would compete for food supplies, their own small production zones skewed away from the larger markets.

While these relaxed assumptions may bring us a little closer to conditions as they might have been in Tellow, Mecklenburg, at the beginning of the nineteenth century, we are all aware of the enormous changes which have taken place in agricultural and transportation technology, in the growth of world population and in rapid urbanization, all of which may affect the location and intensity of agricultural production in the urban-industrial parts of the world. Some modern geographers have found that Thünen-like relationships were discernible within the agricultural territories which once surrounded some of the great European cities in the pre-industrial era. Many more have observed that the intensity of agricultural activity declines with distance from city, town or village in many present-day remote or economically-backward parts of the world (notably in Africa and India). But for very many reasons, we may not expect to find Thünen's rings neatly disposed about particular cities and towns in the developed countries. Indeed, given only that we have established a valid form of enquiry, it might be more productive to go out into the agricultural countrysides around such cities and towns to find out why we do not find such a zonation. We draw attention below to two of the most important influences. One relates to the rent-bidding power of agriculture close to rapidly-expanding cities; the other to the realities of modern transportation in relation to the market.

In the urban-industrial nations, agriculture is in competition for land with many other uses. In such places, the basic allocating force governing land-use may well be economic or location rent, but the major force influencing such rent may not be simply the transport cost to the market. Space-consuming housing, industries, roads and many other facilities and activities compete for space with agriculture. Almost invariably, such urban-based land-uses command a higher location rent than rural land uses. And rural land which is under 'threat' of being converted to these urban uses also has a higher anticipated value than rural land which is not. Thus, it might be maintained, far from intending to be farmed at the highest intensity, land close to fast-growing suburbs tends to be neglected: the farmer will reduce his inputs of capital and labour if he anticipates the sale of his land at a price per hectare which reflects urban, rather than agricultural, valuations.

Observations such as these led an American geographer (Robert Sinclair, *Von Thünen and Urban Sprawl*, see Bibliography) to suggest that the value of land for agricultural purposes is *lowest* close to an expanding city and *increases* with distance as the 'air of urban encroachment' is dissipated (Figure 9–7). As before, the competitive position of each land-use type is governed by the steepness of rent slope, which in turn depends upon the intensity of agricultural investment. But the extended zones of land use indicate the reversal of Von Thünen's analysis.

A scenario such as this seems possible to sustain on the edge of rapidly developing cities in North America, where new 'subdivisions' engulf the adjacent farmland. Sinclair found backing in the American Midwest where there was much former farmland now vacant within and close to the suburban sprawl (Zone 1 in Figure 9–7). Here, the characteristic agricultural activities which still remained were space-limited, like greenhouses and poultry-keeping. At a further distance (Zone 2), where city-encroachment was considered imminent, the farmer's inputs to agriculture were affected by the anticipation of sale. Much land here was leased temporarily for grazing or recreation. Beyond this (Zone 3) city influence, although weaker, still depressed field cropping and grazing in a number of ways. It was a zone of transitory agriculture outside of which (Zone 4) milk production was oriented to the urban market. Finally, where all local big-city influence had disappeared (Zone 5), the highly intensive and well-integrated crop and livestock associations became dominant—those which we recognize as characteristic of the Midwest and commonly label the 'Central Corn Belt' or the 'Meat Belt'.

Insofar as it incorporates an important feature of our times—the space-hunger of

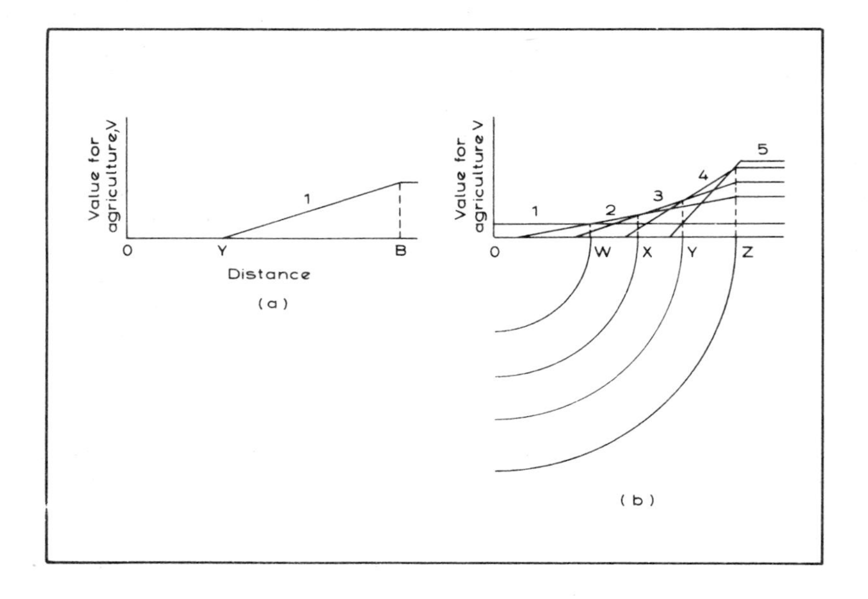

Figure 9–7 Sinclair's reversal of Von Thünen's analysis. According to Sinclair (see Bibliography), the value of land for agricultural purposes is *lower* very close to an expanding urban centre and increases with distance as the likelihood of urban encroachment declines. The relationship for (a) one type of agricultural land use and for (b) five agricultural land use types.

the growing city—the 'inversion' analysis may be considered to advance our understanding of the location of agricultural activities. Like other models, however, it requires further testing to establish its validity in a wider context. In Western Europe, for example, agriculture close to great cities and conurbations may be said to fight a 'rear-guard action' against encroachment by land-uses with a higher bid-rent status. But in many countries this force is not only recognized, but also fairly successfully countered. The establishment of 'Green Belts' around London, the West Midlands and other conurbations and cities in Great Britain is one example; the protection of the agricultural 'Greenheart' of Randstad Holland is another.

If competition with many other land-uses has an important effect on the location and intensity of agriculture in the modern world, so does transportation. It will be appreciated that not only was transportation not so simple a matter in Von Thünen's time and place, but also that in the time elapsed since, the various transportation modes have revolutionized access to market. Indeed, the market for some of the agricultural products of the 'Meat Belt' in the American Midwest may include not only the whole of the USA, but also much of Western Europe as well. Extreme in distance and important in world agricultural trade, the products of Australia and New Zealand find their market on the other side of the world.

These circumstances suggest that in the twentieth century, Von Thünen patterns are most clearly perceived at the 'macro-level', the level of the whole continent in which the great urban-industrial cities and conurbations of northwest Europe and northeast North America are great 'Thünen-Towns'. Indeed, many will go further to suggest that the isolated state has become the world. These last ideas will be expanded as we proceed to examine types of agricultural patterns.

TYPES OF AGRICULTURAL PATTERNS

By now differences will have been noticed with regard to the terms crops and agricultural activities. This occurs largely because agriculture is a highly diverse economic pursuit: in some places, a single crop may dominate the farming system (monoculture), whereas in other places such a crop may be but one of a complex combination of crops. In fact, the question of agricultural classification is surely one of the most perplexing in geography.

Indeed, the first question to be answered is: what is agriculture? Technically, the word 'agriculture'should refer only to so-called cropland pursuits, but in popular usage it has come to refer to any kind of farming activity. In this chapter agriculture will include all commercial farming activities and will encompass *bioculture* (the husbandry of animals), *aquiculture* (fish farming and the farming of such as oyster beds), *silviculture* (timber cropping), and *horticulture* (the raising of flowers, fruits, and vegetables). The primary focus, however, will be on crop and animal farming and various combinations of each.

One system for classifying farming types is according to their general degree of land use intensity (that is, rent-paying ability) as shown in Figure 9–8. This diagram shows the following trend of agricultural activity extremes: (1) areas dominated by a single

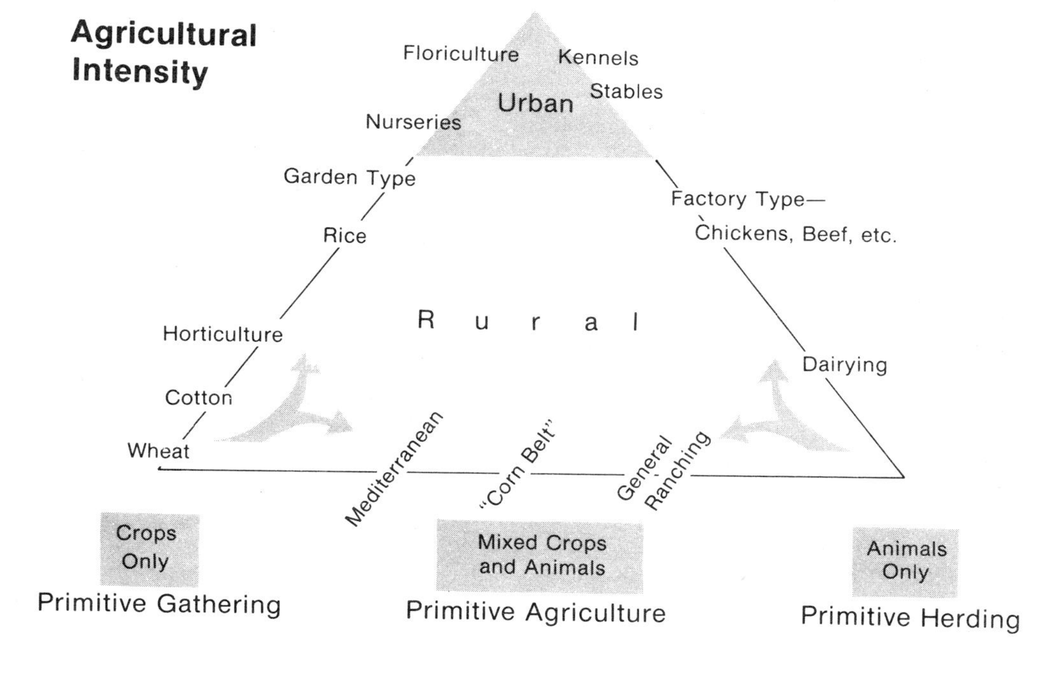

Figure 9–8 Scheme of agricultural intensity rankings according to types of crops and agricultural systems. Note that the most intensive crops are those found closest to cities.

crop such as wheat or cotton, (2) areas dominated by animals such as ranching and dairying, and (3) mixed animal and crop farming. Note that the intensity of land use also generally increases as one moves from the ranching and single crop types to mixed farming types. Intensity also increases, however, as crop- or animal-oriented farming activities become more heavily dependent on proximity to urban markets. Thus garden-type agriculture, primarily vegetables, demands a more intense use of land than does cotton farming; dairying is more intensive than is ranching; and the newly emerging factory-type production of chickens, milk, and beef on very small areas is perhaps the most intensive of all rural activities.

It should be evident also that there is a slippage among land, labour, and capital within the animal-crop dichotomy shown here. Generally, as agricultural types move diagrammatically towards the more intense urban realm, they become more parsimonious users of land. However, the 'crop only' functions appear to change most with regard to labour, and the 'animals only' functions appear to become more capital- than labour-intensive. By contrast, intensity is achieved in the mixed crops and animal combination through an elaborate fusion of these two activities. In the case of agriculture in mediterranean climates, however, such efficiencies are also achieved through irrigation and especially by a system that capitalizes on different crop and animal combinations at different seasons of the year. Thus the mixed systems appear to have been able to achieve a certain amount of efficiency through careful combinations that provide flexibility over space and time.

A general ranking summary of the intensity of land use is shown in Table 9–1. Intensity generally refers to the degree of profit derived per hectare of land. As will be

Table 9–1 Comparative intensity of land use among selected types of agriculture in the United States

Farming type	Degree of land use intensity: High	Medium	Low	Very Low	Estimated average farm size (hectares)
Garden	X				8–60
Mediterranean	X	–			120–240 plus
Mixed	X	–			64–140
Irrigated	X	–			32–400 plus
Rice	–	X			120–240
Cotton		X			120–240
Dairying		X			64–140
Horticulture		X	–		16–60
Wheat			X	–	250–600
Ranching				X	Over 1000

demonstrated in Chapter 10, this does not directly correlate with the extent of total profit made per farm. In fact, some of the less intensive land uses, such as ranching and wheat farming, are among the most profitable on a per farm basis; by contrast, their profit per hectare, and hence their ability to bid for land on a unit area basis, is of a very low intensity. Another measure of intensity, therefore, is reflected in size of farm: the larger the farm, generally the less intensive the land use. However, with the development of large-scale irrigated lands, there are increasing exceptions to this rule.

Of course, Table 9–1 is very general. The exact degree of intensity varies greatly from place to place. The general variation from the average figure (X) is shown by the dash (—). Nevertheless, if physical factors are generally ignored, this seems to be a fairly reliable approximation.

Agricultural Pattern of the United States

The importance of crop intensity in providing spatial understanding for the distribution of crops is demonstrated for the United States in Figure 9–9. Although this map is highly schematized, the general decrease in crop intensity with increasing distance from the major United States 'market' is most evident. Note that with the exception of mediterranean agriculture, all high-intensity crops appear near the major American market. In fact, the rise of southern California as a major supplier of vegetables and citrus products for the main market of the United States is a marvel in large-scale production and marketing efficiencies, and strongly reflects the general revolution in transportation. The medium intensity areas are next farthest removed, except that dairying, an activity dealing with a highly perishable commodity, has in some areas been able to occupy land close to the market. It is also an activity that fares well in the climate to the north, with its shorter growing season, an area generally unfavourable for many other crops. Finally, the low-intensity agricultural pursuits of wheat farming and ranching have been squeezed to outlying areas relative to the market. These very low intensity activities also occupy the land that is physically less desirable for other kinds of agriculture. In the Far West, either because of accessibility problems or environmental difficulties, most of the area in agricultural production is limited to irrigated areas. The primary activities in this most distant zone are forestry and general grazing—even lower-intensity uses than ranching. However, where irrigation is practicable, quite intensive production occurs: these areas are thus able to compete with much closer-in areas.

Thus proximity to markets is clearly a powerful force. But there are several conditions that operate against accessibility providing a full explanation. First, most types of agriculture are affected by physical conditions. Obviously, mediterranean-type agricul-

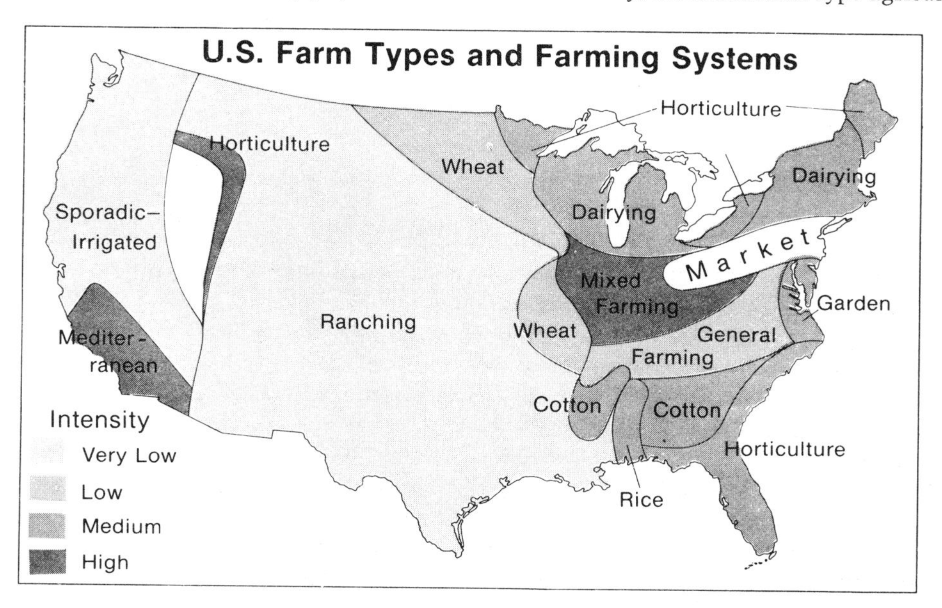

Figure 9–9 Schematic map of various farm types and farming systems within the United States. Compare the proximity of these types and systems to their levels of intensity, shown in Figure 9–8, as they relate to proximity to the major United States market.

ture can occur only in a mediterranean climate. Even if areas with so favourable a physical environment are far removed from the market, they still have potential for fairly intensive production. On the other hand, such areas may be so distant that accessibility problems largely prohibit their intensive use, as is the case with the mediterranean climate belt in central Chile. Second, some areas close to the market are so difficult to use for high-intensity agricultural purposes because of climates slope, or soil conditions, that only very general types of agriculture can be practised. Such is the case in much of the Appalachians and the Ozark plateau.

Finally, a dilemma in interpretation is posed. For example, is wheat farming an extensive use of the land because it is in a poor bidding position for land closer to the market and hence must occupy land farther away, or because it can do better in more difficult lands than most other crops? Certainly, wheat does fairly well in drier areas and is a commodity that is relatively less perishable than most others. Both conditions are involved, with the latter a more likely explanation. This is demonstrated by the fact that the mediterranean agricultural practices in southern California are highly intensive, whereas other nearby areas are used in an extensive way. In fact, we might use a reverse argument: in order to use good lands far removed from the market, they must be used more intensively than otherwise to compete with similar crops grown closer to the market.

Agricultural Pattern of the World

The world pattern of selected agricultural types is shown in Figure 9–10. Note that a pattern similar to that for the United States emerges. Those agricultural types most associated with the urban-industrial heartlands (especially the northeast part of the United States and the northwest part of Europe) are characterized by mixed farming. This type of farming activity is complex and, in fact, is a tightly intertwined system based on a profitable blend of crops and animals. The major food and animal feed crop in the United States heartland is corn, but wheat and other grains are also grown. Lately, soybeans, an industrial and feed crop that enriches the soil, has become widespread in the Midwest of the United States. In Europe sugar beets and potatoes are important crops, but the general crop-animal mix is somewhat similar to that in the United States. Such crops are fed to animals like beef cattle, pigs, and poultry in a manner that provides a year-round income and work pattern for the farmer. In this regard, farmers in midlatitude areas are feeding animals in order to give them some of the same advantages that farmers enjoy in subtropical areas where multiple cropping is possible. Thus animals provide a major source of profit during the off-growing season of the year. It must also be stressed that such mixed crops are also the result of rotation. Of special interest too are the broad bands of dairying activity. This activity, close to the major market source for reasons already mentioned, is primarily associated with advanced industrial and highly urbanized economies.

Dairying is not a homogeneous activity; the commodities produced change relative to distance from the market. Hence areas nearest the market produce primarily fresh milk. The next zones are engaged in the production of butter and cheese, which are commodities less valuable than milk. This arrangement into dairying bands also occurs because such areas are beyond the milkshed zones and because butter and cheese are less perishable than milk. The extreme market orientation of dairying is also shown by the fact that milksheds are found within overnight commuting distance of almost every large urban centre in the world. In fact, less than one-fifth of all the world's milk comes from the dairying regions as marked in Figure 9–10. Dairying is carried on

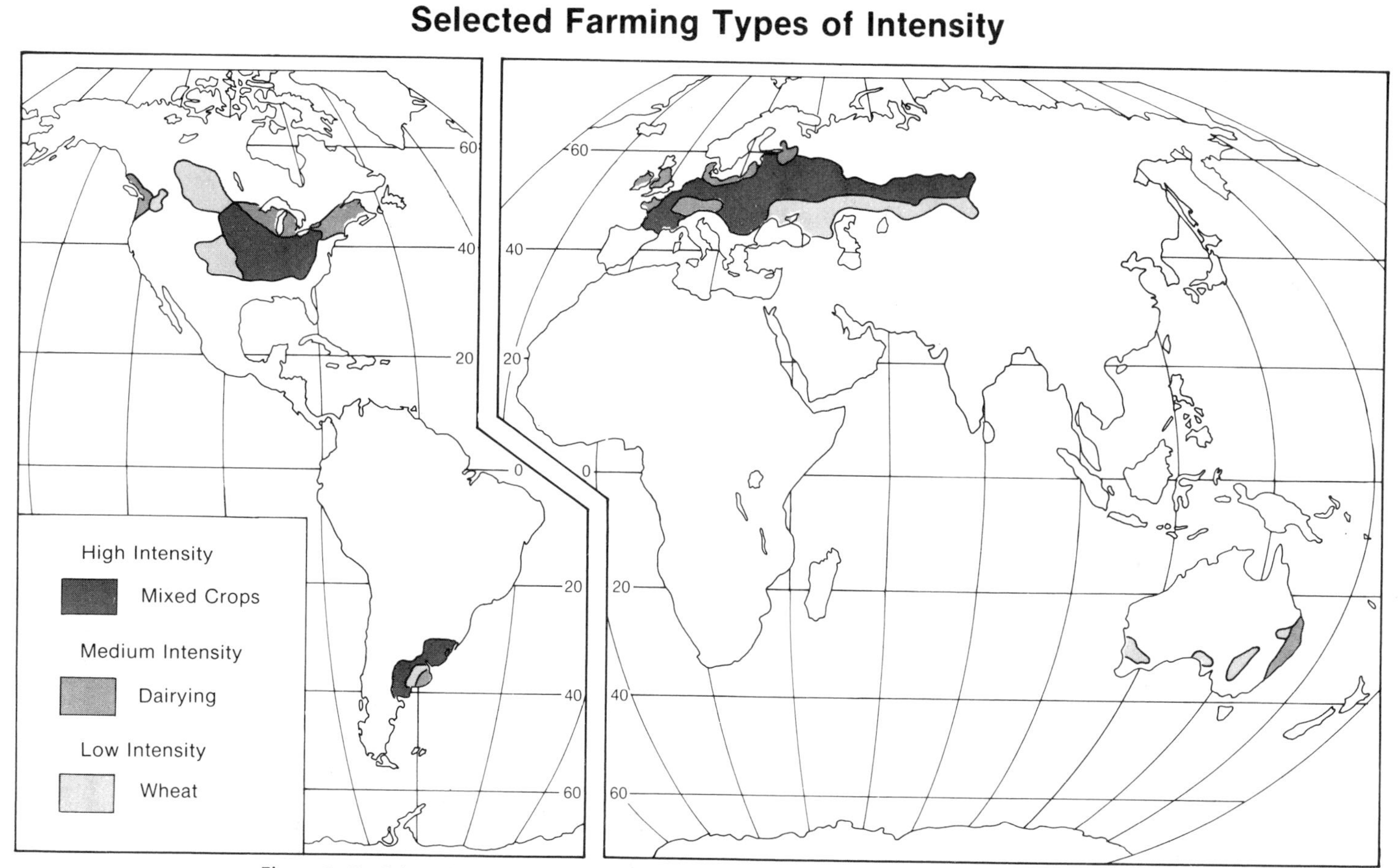

Figure 9–10 Selected commercial farming types by intensity over the world. Compare the level of intensity, shown in Figure 9–8, as these farming types relate to proximity to the European and US markets.

in conjuction with many other farming activities. Perishability keeps dairying very close to even small markets as does the great weight and volume of dairy products. Despite such market proximity, dairy products account for more millions of tonnes than does the output of any other 'crop'. In fact, dairy products amount to almost 400 million tonnes annually as compared with some 250 million tonnes of wheat and about 200 million tonnes of corn. (However, wheat tonne-kilometre volume is greater than that for milk.)

As we move outward from the various world markets, the pattern progresses, as anticipated, into less intensive agricultural pursuits. First, various crops of medium intensity are encountered, followed first by wheat farming and then by ranching, a progression that is particularly orderly in the United States, Argentina, and Australia. Regarding cattle production, it should be noted that nomadic herding and the raising of cattle in conjunction with other farming activities is not specified on the map in Figure 9–10. Thus India, which has over one-sixth of the world's cattle, is not included inasmuch as cattle play a role of little commercial importance there.

Two classes of agricultural activity stand out as peculiar in Figure 9–10. One is the potentially high-intensity mediterranean system, found only rather near the market in Europe. The other is the medium-intensity plantation system, which appears to be located in some of the remote places of the earth. Such spatial peculiarity is obviously the result of highly restrictive physical features. However, mediterranean agriculture located nearest the European and United States markets is by far the more developed, thus perhaps showing somewhat the effect of accessibility.

The physical features of the mediterranean climate are perhaps the best on earth for agriculture. First, the temperature in the mediterranean climate is generally above freezing the year around. This allows it to grow crops that other temperate areas cannot grow. Its full-year growing season also allows it to capitalize on the production of crops calculated to enter temperate markets during the high-price winter period. Finally, through irrigation, the land can be worked in a more intensive fashion than in almost any other area of the world. Consequently, its three major crop types include: (1) those that do well on the winter and spring precipitation, such as grains; (2) those that can withstand the hot summer drought (primarily tree crops, such as oranges, olives, grapes, figs and nuts); and (3) speciality irrigation crops, such as garden vegetables and fruits, which are grown at strategic times of the year to yield the highest market prices. Thus the mediterranean agricultural system represents a monumental commercial symbiosis with nature and is a year-round endeavour. As might be expected, animals are subordinate to crops.

The main problem with agriculture in mediterranean climates appears to be over-success rather than failure. The climate in California is considered so attractive for urban settlement and recreation that considerable invasion of agricultural land has occurred. The same trend, though on a smaller scale, is observable in Mediterranean France. The urban intrusion, however, has also had its sparking effects on extremely intensive gardening.

The other physically contaminated system is plantation agriculture. It should be noted that this system is limited to tropical areas where Western people, primarily European, have attempted to augment tropical commodities through a commercial farming system. Most plantations are thus islands of Western intrusion into an underdeveloped world. All commodities are heavily dependent on native labour, based at first on slavery-type conditions in many areas, and all commodities are prepared for export to developed countries. Typical commodities include rubber, coffee, tea, bananas, sugar, pineapples, and cacao. Initially, such plantations were made possible by the development of the steamship, railways, and other improvements in transportation which, when used in conjunction with cheap water transportation, opened many tropical areas to commercial agriculture.

Plantations are characterized by large-scale production under a centralized management corporation. One of the largest is the United Fruit Company, whose operations are so extensive in Central America that countries in this area are often derisively referred to as 'banana republics'. Some plantations are of such an extensive corporate nature that they are almost completely vertically integrated. These integrated plantation-based companies, therefore, control not only the production of the commodity but its transportation, manufacture, and distribution to commercial outlets.

Despite their large-scale nature, plantations occupy only small portions of the tropics and their location is highly generalized on the map. They are thus mere pinpoints of commercial agriculture in a vast sea of subsistence economies and largely unoccupied territory. In fact, small-scale native farming is on the increase in tropical areas. Farmers there are preparing many plantation-associated commodities for export. Of particular note are cotton from Uganda, cacao beans from Ghana, rubber from Malaya, and bananas from Ecuador. With the sovereignty of many countries in Africa and Asia, the future role of plantations under external control is much in doubt.

CONFRONTING REALITY: AN INDIVIDUAL CROP APPROACH

The true difficulty, and perhaps delight, of the locational problem in agriculture is manifest when one examines the distribution of a particular crop. An individual crop is usually not limited to a single area or farming system but found in a wide variety of agricultural settings. Thus explanation for its distribution is challenging indeed and provides an exploration into the intricate workings of the farm operation itself—intricacies to be discussed in Chapter 10.

The examination of a particular crop reveals many other locational factors crucial to full distributional understanding. These include: (1) seasonality, (2) changing market demand, (3) problems of disease and crop pestilence, (4) particular varieties, (5) possibilities for mechanization and for achieving economies of scale, and (6) many other variables like the spread of such a crop over the landscape and the residual patterns left from former times of profitability. Principles like this are perhaps most clearly and easily demonstrated with regard to one of the most ordinary of crops, the potato. Over the world some 250 million tonnes of potatoes are produced annually. This tonnage compares evenly with that of rice, and is superior to wheat and corn tonnages. In terms of area, it occupies some 24 million hectares as compared with some 200 million hectares in wheat. It is a high-calorie crop and one that provides the staple diet for many Europeans and Americans. An examination of the changing pattern of the common, or Irish potato in the United States presents a fascinating example of the way fore-mentioned principles operate, and it provides as well an inkling into the real nature of the agricultural location problem.

The spatial dynamics of potato production and acreage patterns in the United States from 1900 to 1964 are shown in Figure 9–11. Note that potatoes were heavily market oriented in 1900 but have since shifted to outlying areas. In 1900 there was almost a perfect correlation between state population and the relative amounts of potato acreage and production in each state. By 1964 some of the least populous states accounted for sizable productions—for example, Idaho, Maine, Washington, and North Dakota.

Also evident is the tremendous decline in area despite the fact that production has remained almost constant: fewer hectares are doing more of the producing through an increase in yields per hectare. Average yield has increased from 7.6 tonnes per hectare in 1900 to well over 25 tonnes per hectare today. However, inasmuch as population has grown tremendously, annual per capita consumption has dropped from over

Figure 9–11 Changing Irish potato acreage and production patterns, 1900–1970. Note the general proximity of potato production and acreages to the United States market in 1900 and the great lack of proximity to market in recent years.

90 kg in 1900 to slightly over 50 kg today. The rapidly rising potato-chip, processed-potato, and frozen-potato industries are largely responsible for recent gains in per capita consumption, but fresh weight consumption is still dropping and is now about 35 kg annually per person.

Three major factors are behind the shift in production patterns away from the market. First, transportation has improved to the point that this highly perishable commodity can be shipped to distant markets. Second, vast irrigation areas have been developed in the western part of the United States and have been put into intensive potato production. The effect of irrigation on California and Washington production is clearly evident in Figure 9–12. The potato yield of irrigated lands averages more than

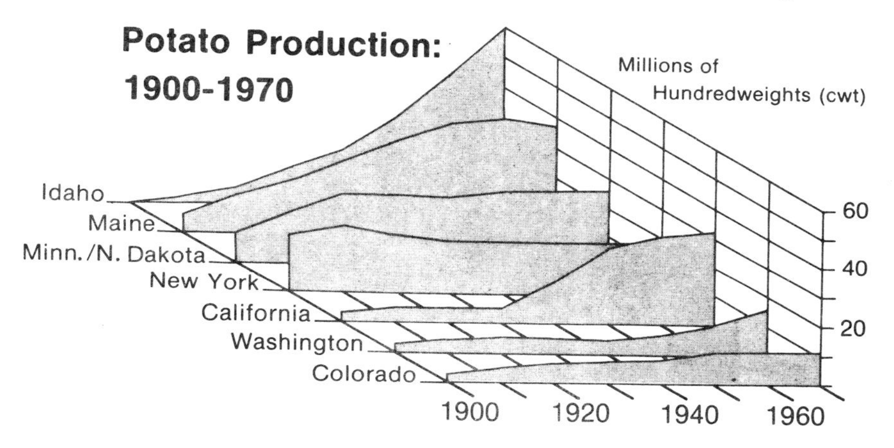

Figure 9–12 Potato production by selected states, 1900–1970. Note the tremendous long-term rise of Idaho potato production throughout this century.

twice that of most nonirrigated land. Finally, the newer areas are the most mechanized and generally the most efficient because of large-scale production. Many of the older areas have been unable to mechanize (smaller fields, poorer soils) and have consequently gone out of production. Potatoes are a heavy-labour crop, especially without mechanized harvesting, so that the scarcity of this kind of labour has been a critical factor in many areas. In the realm of increased production, Idaho has been the agricultural wonder. As shown in Figure 9–12, Idaho was an extremely small producer of potatoes in 1900 but has increased its production steadily until it is now the major producer in the United States.

Concentration trends are, however, offset by seasonality (Figure 9–13). Particularly significant is the late spring and early summer production response from areas in the southern parts of the United States, namely, southern California. This response, in fact, amounts to some 15 per cent of total production. These areas capitalize on their climatic advantages in a manner calculated to enter 'early' potatoes in the market at a time of highest prices. Finally, the pattern is made possible by the development of particular varieties. Some are good, all-round potatoes, such as those produced in northern areas (for example, the russett of Idaho and the katahdin of Maine). Others are early potatoes, developed to mature in a short growing season (for example, the white rose of California and the sebago of Florida). Still other potatoes, such as the kennebec, are grown primarily for the potato chip industry.

This chapter has explored some of the major factors that affect the location of agriculture on a large scale. It has attempted to specify those bases that are most vital to the interpretation of broad agricultural activity patterns. We have progressed from a fairly simple and abstract approach to an exposure of the major complexities in the geography of agriculture.

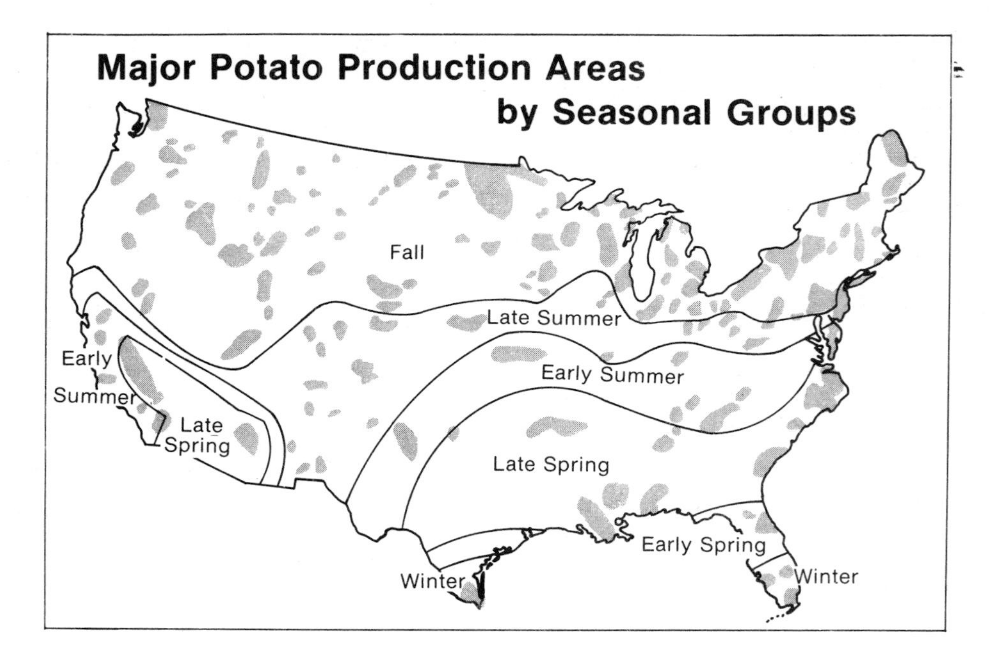

Figure 9–13 Major potato-producing areas by seasonal groups. Note the widespread nature of potato production: every state has some commercial production. Note also the seasonal effects on production.

To appreciate these bases fully, it is necessary to spend considerable effort applying them to new areas and to new patterns. We have attempted only to provide frameworks for analysis, not much of the analysis itself. We can experiment with the models in various ways and pursue locational facts about farm activities with the goal of interpreting them within a theoretical context. Finally, the goal is not so much to understand the present distribution of things but to anticipate and make inferences about probable future patterns, given new developments in transportation and farming technology as well as changes in demands among commodities. This predictive quality is clearly one of the hallmarks not only of a specialist in geography but of any educated citizen.

However, the specific nature of those variables that affect the profitability of an agricultural undertaking at the on-site or farm level must be attended to. The discussion in this chapter has been on a fairly general and rather high level. To appreciate the agricultural location problem fully, it is necessary to get right down to earth at the farm-scale level.

BIBLIOGRAPHY

Baker, O. E. 'Agricultural regions of North America', *Economic Geography*, April 1931, pp. 109–153.

Blaikie, P.M. 'Spatial organization of agriculture in some North Indian villages', Part I, *Transactions of the Institute of British Geographers*, Vol. 52 1971, pp. 1–40; Part II, *Transactions of the Institute of British Geographers*, Vol. 53 1971, pp. 15–30.

Boyce, R. R., ed. *Regional Development and the Wabash Basin*, pp. 22–39. Urbana, Ill.: University of Illinois Press, 1964.

Chisholm, M. *Rural Settlement and Land Use: An Essay in Location*. London: Hutchinson, 1972.

Dunn, E. S., Jr. 'The equilibrium of land-use patterns in agriculture'. *Spatial Economic Theory*. Ed. Dean, R. D., Leahy, W. H., and McKee, D. New York: Free Press, 1970, pp. 233–250.

Durand, L., Jr. 'The major milksheds of the northeastern quarter of the United States', *Economic Geography*, January 1964, pp. 9–33.

Found, W. C. *A Theoretical Approach to Rural Land Use Patterns*. London: Arnold, 1971.

Garrison, W., and Marble, D. F. 'Spatial structure of agricultural activities', *Annals of the Association of American Geographers*, June 1957, pp. 137–144.

Gregor, H. F. 'The changing plantation', *Annals of the Association of American Geographers*, June 1965, pp. 221–238.

Griffin, E. 'Testing the Von Thünen Theory in Uruguay', *Geographical Review*, October 1973, pp. 500–516.

Grigg, D. 'The agricultural regions of the world: review and reflections', *Economic Geography*, April 1969, pp. 91–132.

Hall, P., ed. *Von Thünen's Isolated State*, Oxford: Pergamon Press, 1966.

Harvey, D. W. 'Theoretical concepts and the analysis of agricultural land-use patterns in geography', *Annals of the Association of Geographers*, June 1967, pp. 361–374.

Hoover, E. M. *The Location of Economic Activity*. New York: McGraw-Hill, 1948.

Horvath, R. J. 'Von Thünen's isolated state and the area around Addis Ababa, Ethiopia', *Annals of the Association of American Geographers*, June 1969, pp. 308–323.

Jonaasson, O. 'The Agricultural regions of Europe', *Economic Geography* I, 1925, pp. 277–314.

Johnson, H. B. 'A note on Thünen's circles', *Annals of the Association of American Geographers*, June 1962, pp. 213–220.

Leaman, J. H., and Conkling, E. C. 'Transport change and agriculture specialization', *Annals of the Association of American Geographers*, September 1975, pp. 425–432.

Morgan, W. B. 'The doctrine of the rings', *Geography*, November 1973, pp. 301–312.

Muller, P. O. 'Trend surfaces of American agricultural patterns: a macro-Thünian analysis', *Economic Geography*, 49, 1978, pp. 228–242.

Peet, J. R. 'The spatial expansion of commercial agriculture in the nineteenth century: a Von Thünen interpretation', *Economic Geography*, October 1969, pp. 283–301.

Rutherford, J. 'Agricultural geography as a discipline', *Jerusalem Studies in Geography*, 1970, pp. 37–105.

Sinclair, R. 'Von Thünen and urban sprawl', *Annals of the Association of Geographers*, 1967, pp. 72–87.

Symons, L. *Agricultural Geography*. London: Bell, 1967.

Ursula, E. 'The Von Thünen principle and agriculture location in colonial Mexico', *Journal of Historical Geography*, April 1977, pp. 123–134.

Whittlesey, D. 'Major agricultural regions of the earth', *Annals of the Association of American Geographers*, December 1936, pp. 199–240.

10

Farming Patterns

In the preceding chapter the general effects of on-site production costs and transportation costs to markets were related to agricultural patterns. It was found that proximity to market, when interpreted in light of differential site productivity among crops, provided a fair approximation of general location patterns. Nonetheless, on-site production cost variables were taken as given within any climatic or physical environment. Moreover, the procurement costs, or costs of getting needed material to the farm site, were largely ignored.

These costs will be examined in some detail. For example, why should production costs, and hence intensity of land use, be higher for potatoes than for wheat? Under what conditions do procurement costs affect relative profitability of the same crop grown in two locations? These questions will be answered from the vantage point of five factors: (1) factors that affect ownership and occupancy patterns, (2) determinants of farm site, (3) determinants of field size, (4) rotation characteristics, and (5) possibilities for achieving farm efficiencies through farm co-operatives and vertical integration.

These features are vital to the full understanding of how on-site production costs vary among crops and among agricultural regions. Knowledge of these factors provides the basis for determining the intensity of uses among crops and farming systems; this treatment thus provides the foundation material necessary for more than superficial use of the models in Chapter 9. Many site costs are primarily spatially based, for example, size of farm, shape and size of fields, and rotation patterns. However, we will not attempt to understand fully all the various on-site production costs but only those derived from geographical factors. The other costs will still be, for the most part, accepted as given.

This approach raises a long-standing debate about how much geographers should know about the subject they study, whether it be agriculture, manufacturing, or cities. What should they take as given and what should they try to explain? Of course, they should know something about those factors that affect spatial distributions, but how much? Should they know what causes the factors that cause the patterns to exist? Undoubtedly, their geographical understanding should include some knowledge of phenomena one step removed from the item on which they are focusing. Beyond this, however, things become more uncertain: for full understanding, all things may have some relevance.

The foregoing is perhaps best put in methodological terms. In statistics there are often one dependent and many independent variables. The dependent variable represents the phenomenon one wishes to understand. The independent variables include those factors that correlate with the dependent variable in a meaningful manner and thus provide understanding of it. This procedure is shown in Figure 10–1 with regard to an understanding of agricultural location. Note that our approach through Chapter 8 has been to take the independent variables as given. However, the independent variables can also be studied as dependent variables in another context, so that explanations for their characteristics might also be pursued.

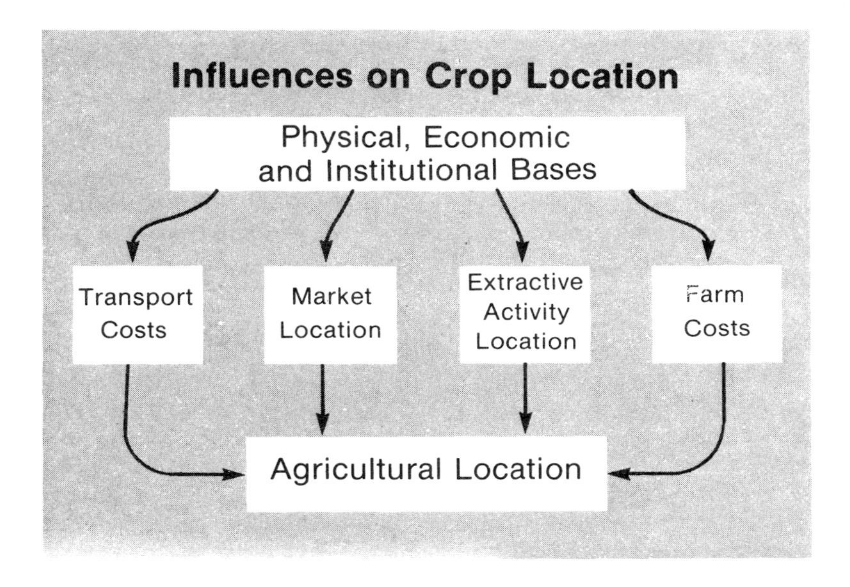

Figure 10–1 Influences on agricultural production. In Chapter 9 crops were explained in terms of location and distance of farm from market, on-site costs, and other variables. In this chapter farm costs are of paramount interest and other variables are held constant.

More formally, the order of removal with reference to agricultural location can be shown as follows (see Figure 10–1). Assume C to be the dependent variable, the spatial pattern of agriculture. This pattern would then be a function of procurement and on-site production costs (E) as well as distribution costs (T), both of which would be labelled the independent variables. But T and E can also be individually treated as dependent variables, just as transportation cost variables were treated in Chapter 6 and as E is treated in this chapter. In this chapter it is assumed that procurement and on-site production costs (E) are dependent variables and that such things as farm and field size, rotation systems, and vertical integration are the independent variables. Moreover, we are attempting to explain variation in such things as farm and field size. Now working forward sequentially, much more can confidently be said about the relationships discussed in Chapter 9.

DETERMINANTS OF FARM AND FIELD SIZE

Few variables affect farm production costs more than do the general patterns of farm ownership and occupancy inasmuch as they correlate with farm and field size. Most cultures have developed standard procedures for settling the land for agricultural purposes. Figure 10–2 shows a variety of land-allocating systems, drawn from different places and different times. Today, we can suggest that four basic strategies are evident. First, there are the large-scale, government-owned patterns evidenced in Communist countries. Second, there are patterns characterized by large-scale estatelike private holdings in which only a few persons own most of the land. Third, there is the strategy of allowing the maximum individual ownership of land through a freehold system. Finally, there is the tenant system, a compromise that is becoming more and more prominent.

In all but the state-owned patterns, agricultural settlement forms have generally resulted in a single farm being composed of noncontiguous units. One such fragmented field system was practised in much of medieval Europe, where the farmers settled on fiefdoms. The basic pattern there was the three-field rotation system, whereby one field was allowed to lie fallow every third year. According to the prevailing philosophy, each farmer should have a plot in each of the three major fields. This caused no great difficulties in medieval times but has resulted in a diffusion of plots for each farm—a highly inefficient system when the plots are of similar physical characteristics.

Another widely practised land-allocation system was the long-lot farm system practised by the French along many rivers in North America, particularly in Quebec. It was rationally thought that each farm should have access to the river, the only reasonable form of transportation at the time, and that each farm should comprise various types of land, including river bottom land, upland, and others. To implement this thinking, strips of land perpendicular to the river were plotted. This system, however, prevented any one farmer from being able to achieve the scale necessary to specialize in any one crop. In fact, today these long-lot farms have been split into several parcels, subdivided along roads that have since paralleled the river at various distances. This has further reduced farm size. Because of the initial ownership pattern, it is difficult for one farmer to acquire just one physical type of land (say, bottomland) in any contiguous pattern. Consequently, most large farms here are composed of widely spaced fields.

A strategy similar to that of the French was characteristic of initial Mormon settlements in Utah during the 1850s and 1860s. Here too it was deemed most appropriate for each farmer to have some of each kind of land. Inasmuch as diverse physical land types were situated in various places around the farm village settlement, each farmer ended up with disjointed fields. This mattered little in pioneer times, when the village was centrally located to most of the different fields and each field was of an economic size. Today, however, any one field is of such small consequence that much cost is involved in moving from field to field. Because of the initial ownership pattern, it has been almost impossible for any one farmer to assemble a contiguous farm unit. Many of these farming areas have practically disappeared today as a result of such inefficiencies. With a new farm pattern, many of these former farming areas might once again become profitable.

Seemingly the most sensible basis of settling an agricultural area was the system of homesteading used in the United States. Here, the land parcels were highly compact and of sufficient size for an efficient operation at the time. This compactness was primarily the result of the rectangular grid land-survey system, in which the land was divided into square-mile sections that could be combined or subdivided into half sections and quarter sections. The size of the homestead was modified in various environments so that it amounted to a full section in many areas, but only a quarter section (160 acres; 64 hectares) in many areas the government thought irrigable. Nonetheless, the farm was of more than sufficient size initially, and had the potential of being a viable farm pattern arrangement.

Subsequently, two conditions arose that have caused difficulties even in these areas. First, as many farms have become more specialized in crop types, and as mechaniza-

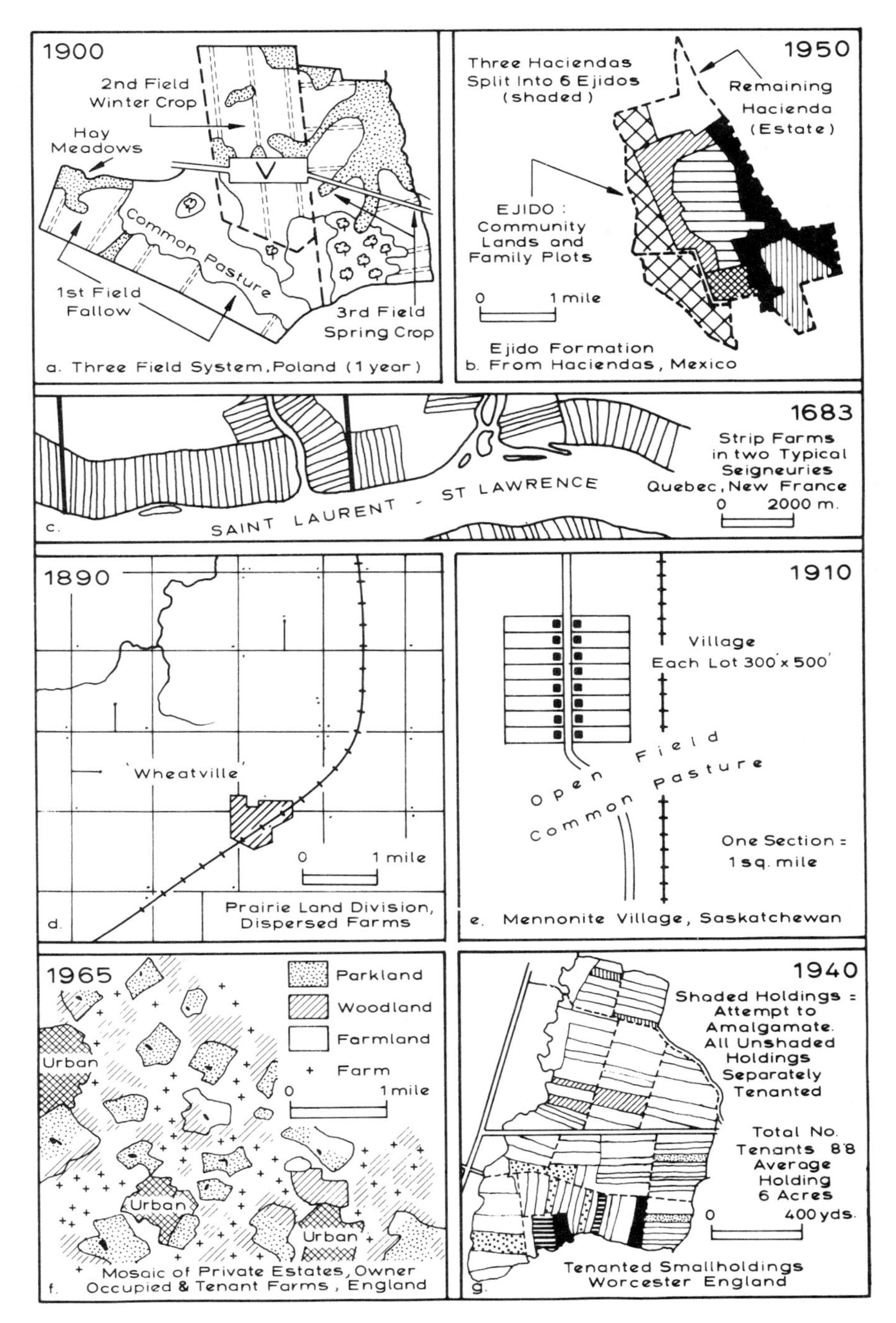

Figure 10–2 Land allocation systems: (a) medieval survival in Poland, with traditional commonages and fragmented arable (one farmer's strips shown by dotted lines); (b) feudal hacienda (latifundia) in Mexico, split into ejidos, whose communal lands are legally granted by government (tillable lands in family parcels 100–300 h. each); (c) linear pattern of rented holdings under feudal landlords (seigneurs) in Quebec, river-front homes giving access and security; (d) Canadian prairie sections with dispersed farms and isolated homesteads; (e) adaptation of (d) by Mennonites with community spirit and co-operative tendencies. Saskatchewan; (f) typical mosaic of private estates, owner-occupied and tenant farms, central England; (g) rentable smallholdings achieved by breaking up large farms, Bretforton, Worcestershire, England.

tion has allowed greater acreages to be farmed by a single farmer, many of the original homestead plots have proven too small. This might have been overcome by the purchase of nearby farms if the scourge of subdividing homesteads among the offspring of the pioneer could have been circumvented.

Size of Farm and Field

At this point, a brief look at the laws governing land inheritance would be helpful. In parts of Europe the law of primogeniture, under which a man's entire landholding passes to his eldest son, still applies. Designed to keep holdings intact, this law has kept farms a viable size and allowed for considerable response to changing times. In the United States, however, Thomas Jefferson sponsored a bill making illegal the rule of primogeniture and entail (limiting the passing of property to a specified line of heirs). Jefferson was primarily concerned about the large slaveholdings in Virginia, which he wished to break up so that the land might be distributed to a greater number of owners. His bill was thus called a 'land reform' bill. Another reason for repealing the rule of primogeniture was that only men who owned land (freeholders) had the right to vote. Thus Jefferson acted to broaden the suffrage.

The long-term result of this law in the United States is the 'share and share alike' principle whereby all offspring of the landowner hold equal land-inheritance rights. Over time, homesteads and many other farms of viable size were cut up among the children. This worked well initially, inasmuch as the homestead plots were perhaps too large. Then inefficiency in farm size became prevalent: unable to assemble more land, farmers sold or rented their plots and left the farm for the city. A major rural migration occurred, but only after much hardship and many futile attempts to survive on the part of would-be farmers. Many farmers found their income to be marginal. In this position many survive today at a level not much above the poverty level. Many traditional family farms, the so-called backbone of America, are of this type.

The general cycle of average farm size in the United States is summarized in Figure 10–3. Note the four main stages of farm size: (1) the initial 160-acre homestead, (2) the fragmentation of the farm because of inheritance, (3) the consolidation of fields by the fortunate few in their long climb toward a viable farm unit, and (4) the current period, in which additional farms have been assembled.

A not-dissimilar sequence of events as described above has been experienced in many agricultural areas in Western Europe, albeit at a slightly different time and on a

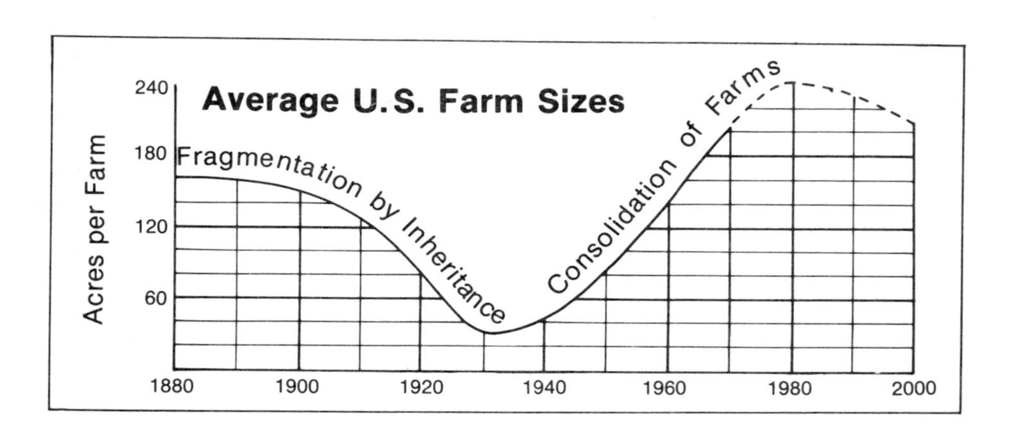

Figure 10–3 The result of fragmentation by inheritance and reconsolidation of fields by the more successful inheritors. The data, based on Idaho, reflects typical farm size change in the US.

different farm size scale. The most notable case is France. Here, no less than 51 per cent of farm holdings (over 950 000 in all) were under 10 hectares in size in the early 1960s. Many of these small farms were badly fragmented, could not be mechanized (or, in the south, could not be successfully irrigated; see later in this chapter) and stuck to traditional low-yielding crops and livestock.

In the case of the United States, the 'shake-out' of inefficient farmers and the subsequent trend to larger, consolidated holdings (Figure 10–3) is mainly achieved by free market forces. In France, however, while the open agricultural land market has operated in the same way, the movement towards consolidation and enlargement of farms has been slow and needed to be bolstered by official policies. Thus rationalization programmes begun in the early 1960s involved both retiring farmers with state pensions and government intervention in the agricultural land market. Within ten years, ageing farmers left a total area of over six million hectares, or 20 per cent of the national agricultural area. Their fields and farms have mostly been absorbed by existing farmers anxious to enlarge their holdings. But government-backed 'Companies for Land Rearrangement and Rural Establishment' have also been able to purchase land that comes on the market to create new farms. The whole process, usually known as 'remembrement' still continues.

The plague of farm fragmentation, however, is not peculiar to the United States and Western Europe. In fact, in most areas occupied by Catholics, Buddhists, Muslims, and Hindus primogeniture is rarely practised. Consequently, in some parts of Asia a two-hectare holding may be subdivided into some 20 different strips. Undoubtedly, the emergence of smaller and smaller farms (and the farm fragmentation resulting from the attempt to assemble larger holdings) is one of the major problems in agricultural production today.

Paradoxically, problems of inefficiencies can also result when large landholdings are somehow kept intact. The *latifundia*, or estates, in much of southern Europe and Latin America well demonstrate the difficulties caused by large landholdings. Here many estates have been piece-rented to farmers. But in many parts of the world, large landowners are in about the same moral position as typical large-city slumlords; they milk the land as much as possible and give very little in return. Moreover, tenant farmers are in a very precarious position if they can rent only on a year-to-year basis.

Even so, where tenants' rights are carefully spelled out, the system can be a fairly beneficial one for the tenant farmer. Tenancy is particularly high in England, perhaps because tenants' rights are so well protected and so fairly established. But tenancy is not just the result of scandalously extensive lands controlled by an absentee landholding class. In the United States large holdings have resulted in vast regions worked by sharecroppers, as in the cotton areas of the Mississippi Valley and eastern Texas, and tobacco areas along the east coast Piedmont. Tenant farming is also particularly prevalent in the Midwest. In fact, about one-third of the farmland in the United States is tenant-operated. Here some of the richest and most profitable land in the world has gone into tenant status. The reasons for this appear to be primarily social in nature. Farmland holds great status among urban midwesterners, perhaps because of their rural upbringing. Many invest in farms which are immediately leased to tenants who provide the landowner with 'outside' income.

The most successful farms appear to be those which have reached a size large enough to permit economics of scale. Such farms have been assembled in many areas of the United States, particularly by family corporations. Some of the largest assemblies in the world include ranches in Texas and, of course, the banana and rubber plantations owned by large corporations in Latin America and South East Asia. But other very profitable farms have been developed in most kinds of crops, such as cotton, wheat, and potatoes. In the latter case an extensive, irrigated farm might be only a thousand hectares in size, but would be massive in comparison with the very small farms nearby.

Other large-scale farms occur in Communist countries and in those places that have nationalized much farm land. The *sovkhozes* (collective farms) in the Soviet Union often amount to over 20 000 hectares each, and several are well over 40 000 hectares in extent. Nevertheless, perhaps for political and social reasons, many of these farms have not been as productive as might be expected. Other variations on the communal tenure patterns include the *ejido* of Mexico, the *kibbutz* of Israel, and the *kung-she* (collective farm) of China. Ironically, many of these countries initiated collective farms as a reaction against the ills of large estates and latifundia. Other collectives, of course, are attempts to consolidate multitudes of small landholdings. An attempt to show how these and earlier-mentioned land holding systems are related is made in Figure 10–4.

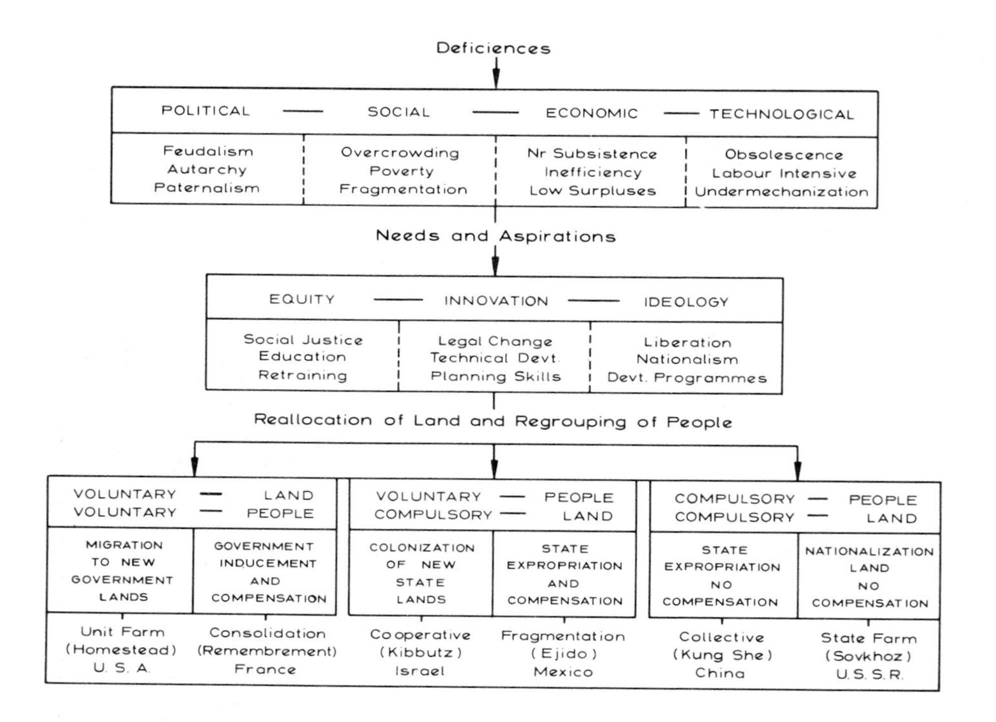

Figure 10–4 Land holding systems and their reform: characteristic deficiencies ascribed to older land holding systems; reform needs and aspirations; reallocation of land and regrouping of people, with examples.

Determinants of Field Size and Shape

The determinants of field size are somewhat different from the determinants of farm size. One of the primary determinants of field size is natural or physical features such as small streams and variations in soil and topography. Such physical features cause irregularities in shape but do not reflect an ideal pattern. Nonetheless, small and irregular fields are often put to less intensive uses than those in nearby fields.

Other, more functionally related factors that determine field size and shape are the nature of the agricultural system and the degree and type of mechanization. The nature of the farming system is strongly correlated with field size. For example, the medieval system was based on cumbersome ploughs pulled by oxen—ploughs that were difficult to turn at corners. Consequently, in the typical three-field system each man's property was in long, narrow fields. Each field was about one furlong (220 yards; 201 metres) long. Thus in this system the length of the field, not its width, was the critical factor.

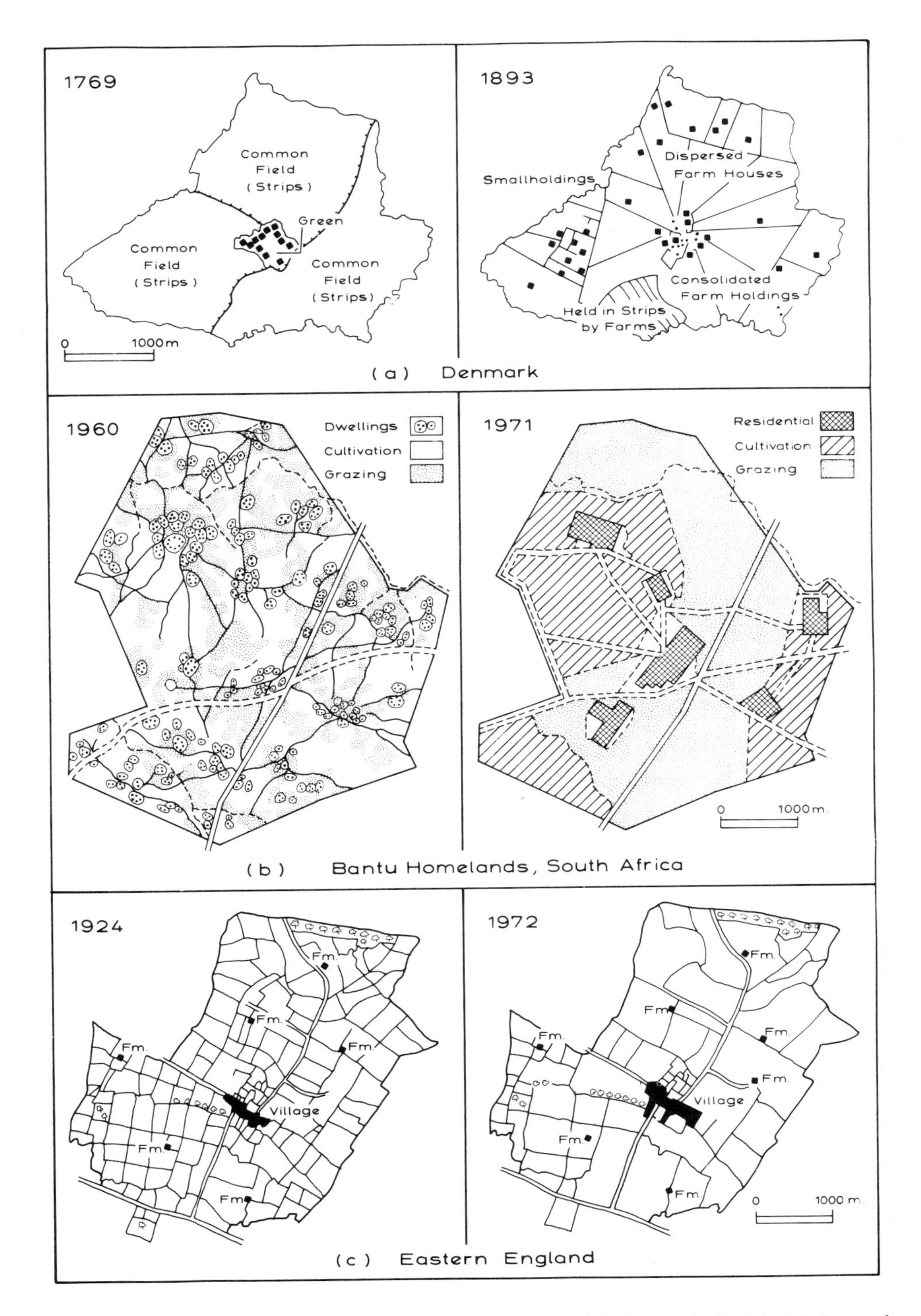

Figure 10–5 Farm and field rationalization. Three examples of 'before and after' in (a) Denmark: the medieval parish and its common fields (in strips) consolidated into owner-occupied farms and small holdings; (b) Bantu Homelands, South Africa: tribal area with scattered cultivation and communal grazings consolidated by tribal consent; (c) England: field enlargement in a Huntingdonshire parish (achieved mainly by hedgerow removal).

Field plots were rectangular, with a length-to-width ratio of about 15 to 1. Freedom from feudal bonds demanded change, especially if necessity or opportunity dictated the specialization of agricultural activities. Thus in Denmark the long, narrow strips within the communally farmed, three-field village gave way to fenced fields and paddocks consolidated about the house and barns of the isolated family farm, geared to a dairy-livestock regime (Figure 10–5). By contrast, modern grain farming, in such areas as Eastern England and Northern France, requires large machinery, which is best used in a circular fashion. Inasmuch as circles leave unoccupied land between fields, the next best thing, the square, is commonly found. Here the ploughing and harvesting is done by going about the field in a series of smaller and smaller 'circles'.

The major problem in the use of machinery, however, is fields that are too small. Such machinery is often difficult to turn around and requires much space to be operated efficiently. Consequently, many farms with small fields have been placed at a cost disadvantage, inasmuch as hand labour or less efficient machinery must be substituted for the more mechanized operations. This is a major reason for the widespread enlargement of fields (and consequent loss of hedgerows) in many parts of Southern England and especially in East Anglia (Figure 10–5).

On surface irrigation farms, the size of field is particularly critical. If the fields are too small, potential cropland will be taken up with numerous irrigation ditches. If the fields are too long, the soil near the irrigation sources will be soggy before the water reaches the far end of the field. Another problem, largely the result of increasing field size, is the deterioration of the soil through the buildup of alkalinity. Vast areas in the American West have been made largely unproductive for this reason alone, a good example being the basin of the Colorado near the US–Mexico border.

CROP ROTATION: ASSET OR DEFICIT?

Rotation is another characteristic of farming that affects the cost of production considerably. Basically, it is practised in order to build up nutrients in the soil that have been lost through cropping. Inasmuch as many cash crops are particularly hard on the soil, some way must be found to restore lost elements. The traditional restoration procedure is to plant crops that replace removed materials. Today, however, these lost elements can be resupplied by the use of fertilizers. It might thus seem that rotation is an archaic system and no longer need be practised.

Certainly it is an old system. As previously indicated, it was practised in the medieval three-field system where one-third of the land was left fallow each year. A rotation system was postulated by Von Thünen in his third, fourth, and fifth rings outward from the market (see Chapter 9). With each outward ring the intensity of use decreased, as did the number of years between fallowings. Thus in the third zone the land was made fallow only every six years, whereas in the fifth zone the land was fallowed every three years. Today, leaving the land fallow as part of a rotation system is practised primarily among less intensive crops like wheat as it is grown in North America or the USSR in areas of low rainfall. Here, land is left fallow every other year, or sometimes every third year, not necessarily to restore nutrients lost in grain growing (actually grain is a grass-type crop not too hard on the soil), but to clean-cultivate in order to eliminate weeds and to restore moisture content.

The more common practice is to maintain a rotation scheme that provides for some crop on all fields each year. Thus crops like corn, tobacco, or potatoes, which remove vital nitrogen from the soil, are followed by crops that tend to restore nitrogen, for example, various hay crops (alfalfa, clover, timothy), peas, or beans. For many crops, nutrients and trace elements like iron, manganese, boron, copper, zinc, iodine, and cobalt, which occur in very minute but critical amounts, can be fed directly back into the soil through fertilization.

Nevertheless, there ar two major reasons for continuing the common rotation practised with many crops, namely (1) the disease dangers inherent in the growing of the same crop year after year, and (2) the dangers of extensive erosion. With continued cultivation, many disease organisms build up in the soil. In potatoes, for example, there may be serious blight, root and tuber problems like wireworm and nematodes, or various potato skin diseases such as warts. Even with a careful rotation certain crops have had to be eliminated in the United States because of disease organisms. Examples include the Colorado potato beetle in northern Colorado and the boll weevil in much of the South.

Crops that are open-tilled and grown on sloping land must also be rotated in order to prevent extensive erosion. Cotton, tobacco, potatoes, and corn are examples. In such cases, other crops must be grown periodically in order to hold the soil in place. In some places like the Great Plains in the United States and the steppe areas in the Soviet Union, vast areas have been set aside as 'shelter breaks' in order to keep the wind from blowing topsoil during parts of the year. In other places, strip farming and contour farming are practised. All rotation schemes, though necessary, add up to a considerable cost factor or an overall lowering of productive intensity. First, rotation requirements prevent ultimate specialization in any one crop. Other crops must also be grown, separate machinery must be purchased for these (often less profitable) crops, and farmers may be prevented from taking maximum advantage of high market prices.

In some cases the rotation patterns are not sequential over a decade or less but take place over a generation or more. The slash-and-burn type of agriculture practised by primitive agriculturists in the tropics extends over such a long period that formerly cleared lands are once again completely overgrown with trees and vegetation. Although this is not rotation in the usual sense, it does represent a sequential use of the land.

An example of the more common type of rotation pattern is that found in southeastern Idaho. This flood-irrigated cropland has passed through various six- or seven-year cycles with changing crop emphasis since it was first settled in the late nineteenth century. In each case the cash crop has been hard on the soil, so rotation is an absolute necessity. Moreover, the rotation pattern has already passed through about four main stages. Generally, such crop patterns have progressed from fairly extensive to a highly intensive production during the period studied.

In the pioneer period most of the land was grazing land or pasture. Cultivated cropland was used for experimental crops such as peas, sugar beets, corn, and potatoes. These cash crops were heavily supplemented by nitrogen-building legumes in such a way that a typical rotation pattern on cultivated land included a cash crop about once every four years.

A number of rotational patterns were practised during the remaining stages. In order of intensity, these included (a) sugar beets, potatoes, feed grains, alfalfa, alfalfa, alfalfa, and (b) wheat, alfalfa, alfalfa, peas, potatoes, potatoes. Another common pattern was wheat, alfalfa, alfalfa, potatoes, wheat, and potatoes. In the last stage, potatoes had become *the* crop and the pattern was highly intensified, as illustrated by the following rotation scheme: wheat, potatoes, potatoes, wheat, potatoes, potatoes, wheat, alfalfa, alfalfa. Note that all but the last rotation pattern covered a six-year cycle.

The most prominent sequence is from a grain crop like wheat to a hay crop like alfalfa. This occurs because alfalfa takes one full year to reach maturity. Consequently, it is usually seeded along with a grain crop. Thus a harvest of grain is received during the first summer. Also, the 8- to 10-inch-high alfalfa after grain harvest is excellent for pasture purposes, The grain stubble is cut high so that the alfalfa is left on the land. In fact, alfalfa is not so extensive a crop as it might first appear. It does triple duty. First, it returns needed nitrogen to the soil during growth and then, as green manure, it is ploughed under to prepare the soil for a cultivated crop. Equally important, inasmuch as the hay is fed to animals, alfalfa is indirectly valuable as barnyard manure. Finally, alfalfa is not just a one-crop harvest during the summer—two and even three hay crops are common.

It can therefore be seen that rotation is not just a simple matter of keeping the soil at constant fertility. It is a fairly elaborate system that is subject to change and it is, moreover, greatly affected by the market prices for various crops. In places where the land is extensively used, rotation appears to be more leisurely. In places and times that present great opportunities for profit in cash crops, the rotation pattern, indeed the whole system of farming itself, can be drastically changed.

FARMING CO-OPERATIVES AND PRODUCTION COSTS

The possibilities for achieving farm efficiencies through farm co-operatives and vertical integration are considerable. These features will be exemplified below in case studies which also involve other aspects of farm efficiency already noted. First, however, we draw attention to farming co-operatives and production costs as a general consideration.

One way in which small farmers affected by high production costs can greatly increase efficiency is by forming co-operatives. Consequently, co-operatives are widespread in those agricultural areas that have achieved distinction in a particular commodity. Co-operatives are found in dairy, livestock, tobacco, wine, wool, and poultry areas, to name a few. The system operates through joint ownership of critical off-farm facilities by co-operating farmers. Such farmers may co-operate in the marketing of crops, the consumption of goods and services, or in making available farm loans through credit unions. Marketing co-ops, in which the farmers attempt to set grading standards for their crops and to process and sell produce, are the most widespread. Some marketing co-ops are heavily involved in packing, storing, trucking, advertising, and even in testing activities with regard to the crop in question.

Figure 10–6a illustrates the structural relationships of the range of these co-operative activities in the case of Denmark, where such co-operatives were pioneered in relation to dairy products. Co-operatives have also been long-established in other European countries, an outstanding example being the village wine co-operatives of southern France. There are, too, notable examples in Britain and the old Commonwealth countries and in the western part of the United States.

In all these cases, the producers act in concert to lower their costs and to enable their products to penetrate greater and more distant markets. In extending the demand for their crops far beyond what otherwise would be the case, co-operatives have enabled small farmers to receive prices higher than those likely to prevail in a free market economy. They have also achieved, through their communal entry into the processing and marketing channels, a reduction in the costs of delivering their produce to market. Moreover, they are able to procure their tools of production (fuel, feed, seed, machinery, and so forth) at less cost than otherwise. They are also able to get loans at a lower interest rate than that obtainable through normal financial channels. Finally, as shareholders in the co-operative, small farmers may even make some money on their investments. These benefits all have importance in reducing procurement, production, and distribution costs for farmers so that they are able to achieve higher profits under this system than they would by working independently.

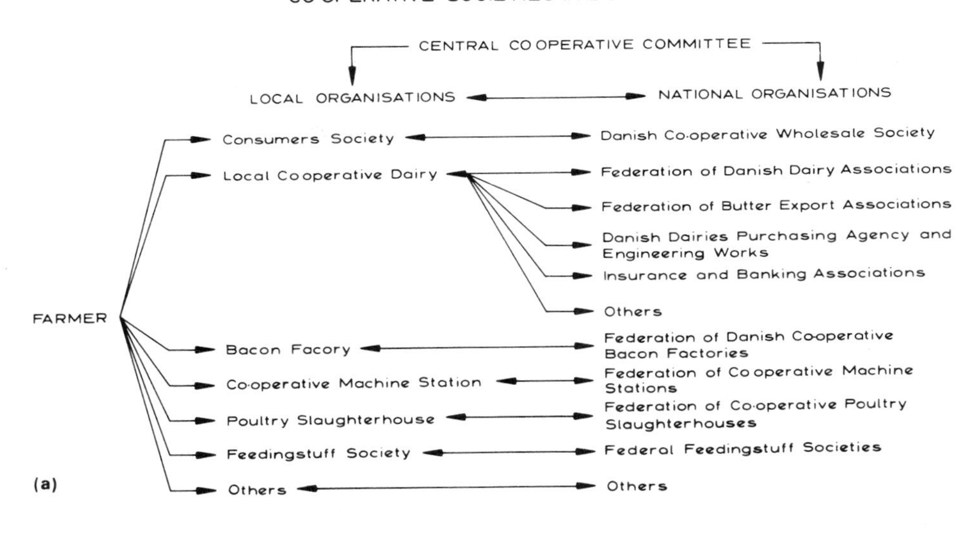

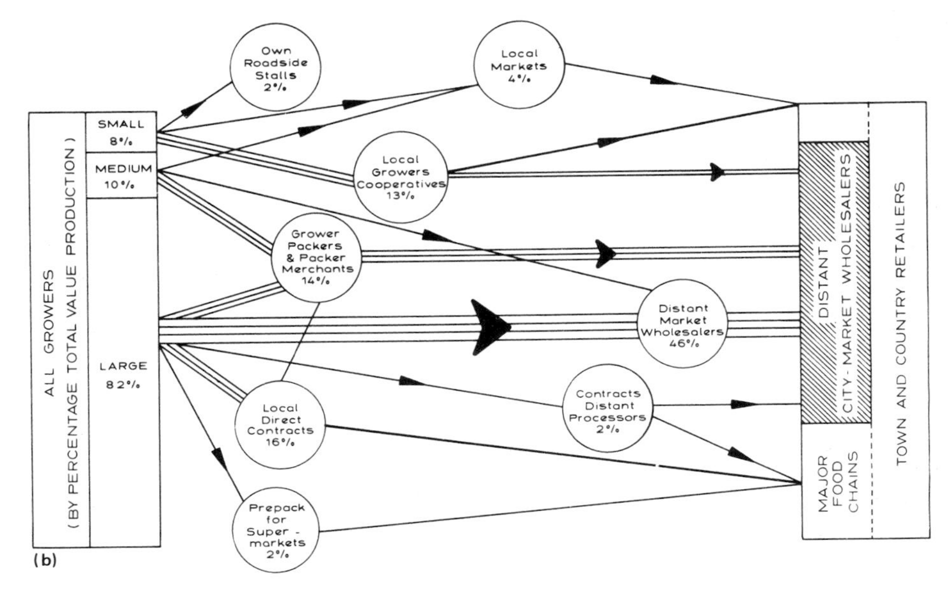

Figure 10–6 Co-operation and production costs. (a) the farmer's relationships with Co-operative agencies in Denmark and (b) systems of marketing in the Vale of Evesham, England for horticultural producers. Note that despite the alternatives available to the English growers, most produce still passes through the hands of distant market wholesalers before reaching consumers (see also Figure 10–7).

The Impact of Co-operation in an Area of Old-Established Agricultural Smallholdings

Many of the general features of co-operation might be illustrated by reference to one or other of the crop-growing or livestock-producing regions in the British Isles. Here, we choose a relatively small-scale example of such enterprise in the Vale of Evesham Worcestershire, England. This is a part of the Midlands particularly well-endowed for a varied agriculture on mixed farms of moderate size (60–120 hectares)—but it is especially well known nationally for its horticultural produce from hundreds of smallholdings whose average size is under 20 hectares.

Lying between the limestone Cotswold Hills and the flood plain confluence of the Avon and Severn Rivers, the area is one of gently rolling terrain. The flatter lands in the Vale consist of either well-drained river terraces with a friable, loamy soil, or a much less tractable clay. The earliest 'Garden-croppers' were probably the worker-monks of the great Abbey of Evesham, who undoubtedly recognized the value of the terrace soils. By the eighteenth century, the old feudal systems dominated by Church and State had vanished, the broad open fields had been enclosed and the farms were held by an assortment of 'landed gentry', owner-farmers, and tenants. The chief advantages of the area at this time, apart from its position in the middle of England, were its lower rainfall and longer sunshine hours than most of the surrounding districts. To these circumstances we can add the skills which, with spade and hoe, year by year enabled a rising surplus to be marketed beyond the local towns.

In the early nineteenth century, the produce of the Vale of Evesham travelled regularly by cart, canal and river to markets in a 40–100 kilometre range: to Birmingham, Gloucester and Bristol. With the coming of the railways, the farmers took their produce by cart to one of a dozen or more local sidings where their small bundles, sacks and crates were bulked in wagon loads to reach a wider market in the Midlands and the South. The growth of the industrial cities increased demand, with more land converted from mixed farming to orchards, bush fruits and vegetables. Potatoes, cabbage, lettuce, asparagus, peas, beans, rhubarb, blackcurrants, strawberries, apples and plums (the local Pershore variety) were all produced in greater quantities. The success of very labour-intensive production, and the services it generated, brought about both immigration and a natural increase in the local population. In many parishes, three-quarters of the active work-force was engaged in agricultural labouring. Very few so employed could hope to rent or own their own farms, and even basic needs like housing were in short supply.

These unrelieved pressures were recognized after World War I by Parliamentary Acts which enabled the County Councils to purchase large farms and break them up into 30 acre/12 hectare rented smallholdings. A main aim of the new legislation was to create the first rungs of a 'farming ladder'. Through the years, most success attended those 'County Council Smallholders' who could either rent or purchase more land to create viable mixed farms, or else intensify production on their original grant by turning to horticulture. Neither course could be achieved easily. For most new growers, their main advantage was that they were located among neighbours who shared an established horticultural tradition. But the established freeholders occupied the lighter soils, and many of the new smallholdings were located on the clay soils, the worst of which were heavy and cold. Moreover, the new homes were not on the new smallholdings but built in rural 'terraces' which might be some distance away.

As they faced these problems, many would-be growers failed to prosper. They left the land, to be replaced by others with more enterprise, skill and dedication. But even this was not enough where the small scale of production could only be achieved with high-priced supplies and much time and money needed to be spent in getting small quantities to market. In most situations, small quantities meant low prices from the middlemen in the business, and roadside stalls (still a feature of the Vale) could only partly relieve the problem. The answer for the majority was to join those independent growers who had pooled their resources (by shares) to establish co-operatives, the first of which had appeared in the area by 1908. The aim, of course, was to achieve some of the economies of scale of the larger producers (Figure 10–6b). By the 1960s, two major 'growers co-operatives' had over 1400 members. These purchase the growers' supplies in bulk (sold at a discount to members), collect produce from the growers, grade, pack and load on the premises, sell by telephone and telex, and despatch by road to distant markets (Figure 10–7).

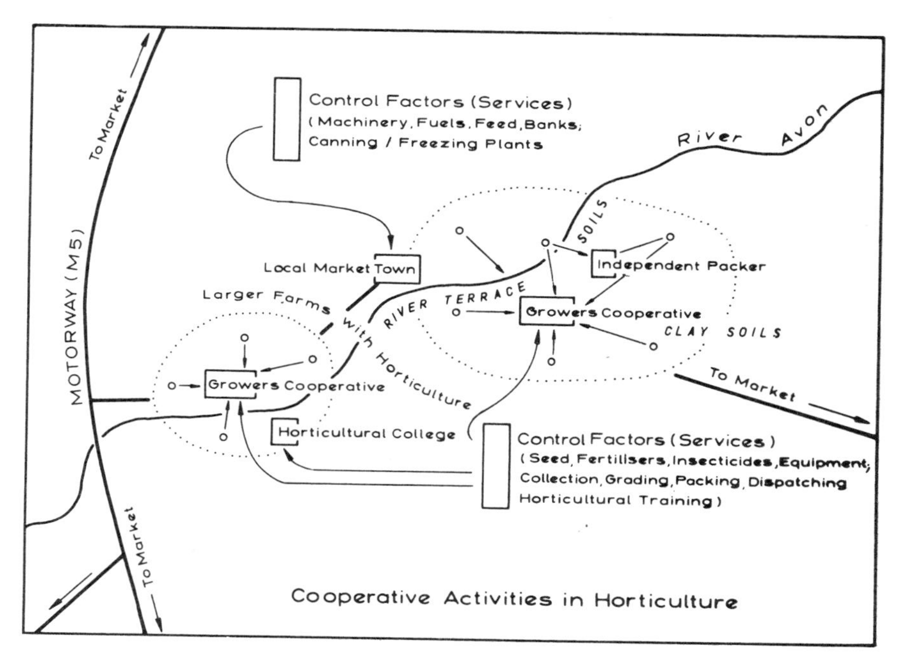

Figure 10–7 An example of co-operative activities in horticulture. Within the ringed zones, most smallholders belong to the leading co-operative agencies owned by growers. Larger farms market their crops independently.

In this example, the co-operative system ensures that the small grower benefits from the best prices prevailing, since the co-operative is in daily contact with wholesale-retail fruit and vegetable markets in places as far apart as Cardiff, London, Nottingham and Newcastle-upon-Tyne. The annual turnover of the largest Evesham co-operative in the 1970s is in excess of £2 million. It is undoubtedly true that many—if not most—of the smallest growers would be in poor economic circumstances without the co-operative. But it is also important to keep their co-operative opportunism in proper perspective. Growers in this area keep abreast of competition in many other ways: by new methods of 'forcing' early crops under polythene or glass, by 'phasing' their indoor and outdoor crops in a sequence of continuous operations that spread labour inputs, and in a host of other ways.

Finally, we may make some more generalizations about co-operatives. They do have an important role in many regions—they are noted again below in the USA, and in the Midi, France. Yet for both farm goods and horticultural produce, there are usually alternative methods of buying and selling. There is no compulsion to trade through the co-operative, which relies only on the loyalty of its members for its existence. A strength of many co-operatives may be their ability to arrange forward contracts with members who then supply a given crop of agreed quality by a given date and at a guaranteed price. Such arrangements are characteristic of produce which can be canned, bottled or frozen. But co-operatives need to be very well supported to maintain operations of this kind; they are likely to be outbid by specialist organizations (preserve manufacturers, frozen pea companies and the like) which can pass on some of the benefits of their economies of scale to the farmer-producer.

FARMING EFFICIENCIES AND VERTICAL INTEGRATION

In the example just employed, co-operation appeared to be a necessity for the survival of small producers. However, large-scale farms in the Vale of Evesham, as elsewhere, may produce horticultural and field crops in such quantity that all activities in the purchasing-growing-selling system are internally integrated in order to optimize their benefits of scale.

Economies of scale are achieved in two ways. First, as discussed in Chapter 4, large producers can use their machinery more effectively and can generally save in ways not open to small producers. Second, large-scale farmers often get discounts on the basis of volume for most of their procurement inputs.

The economies achieved through vertical integration are formidable indeed for small farmers to compete with. Such integration is possibly the most effective way in which large-scale operators truly make their operations pay. Basically, the aim is to control both backward and forward costs associated with the farm enterprise. The backward costs are those associated with inputs to the farm activities, like machinery, fuel, seed and fertilizer. The forward costs are those related both to the production and the marketing of crops and livestock. The scale of on-farm activities may mean that by-products, wasted on the small farm, can provide additional on-farm income by their use in subsidiary enterprises; and the off-farm control of storage, warehousing and processing may bring considerable further benefits.

Examples of economies of scale and vertical integration of this kind are present in British and European agriculture. Notable examples in the United Kingdom are in central and eastern England where pig farmers and poultry specialists control the processing, cold storage, transfer and sales processes to command a market which is as large as the EEC. On a smaller scale, but still notable, farmers in eastern Scotland may control their own outlets for livestock production by owning butcher's shops or dairy rounds in nearby urban areas. Here, however, we demonstrate the full benefits of large-scale production and vertical integration by reference to potato production on the largest farms in Idaho in the USA.

The Impact of Large-Scale Production in an Area of Traditionally Irrigated Potato Farms

We have previously noted the long-term rise of Idaho potato production throughout the twentieth century (Figure 9–12) and considered many of the reasons which have brought this state to pre-eminence in potato farming in the USA. We can now look more closely at the contrasting potato farm activity in Idaho near the Snake River, which illustrates the competitive challenge of new large-scale farming and the benefits of vertical integration.

When they were first settled by pioneers of the Mormon sect in the late nineteenth century, the farms of the area were each of one quarter-section—160 acres (64 hectares).

The farms varied somewhat in quality because of physical features and proximity to market centres, but most held more land than could actually be used for crops for some time. It seems to have mattered little that the soil of the area was found to be a heavy, calcareous loam, often rocky, and with a tendency to be cloddy. More important, perhaps, was that it could be irrigated. Canals were laboriously dug from the tributaries of the Snake, with headgates built to control the gravity flow of water into the new fields.

The farming operation generally prospered during the first thirty years as more and more land was brought under cultivation and as experimentation in various crops occurred in order to find the best crop combinations. The potato was found to do well, but sugar beets, wheat, corn, hay and all manner of vegetables and feed grains were attempted. Unfortunately, in conformance with general custom, the farms were subdivided between second generation sons and daughters in equal parcels, reducing their viability.

Six- or seven-year rotations, too, were necessary in order to keep the land in good heart, with a leading cash crop marketable every four years. While potatoes became dominant, sugar beets and peas were also cash crops of importance. Sugar beets could be sold to a local sugar factory, and peas despatched to local canneries. Peas had additional advantages: they were both good for the soil and, after harvest, provided excellent forage for pigs. Both beets and peas, however, went into decline with the arrival of high-cost harvesting machinery. Earlier, the farmers had employed the free labour of their entire families for the back-breaking tasks of lifting and picking; now they could not afford machinery which, anyway, was designed to be cost-effective on much larger farms.

Fortunately, for most farmers, the market for the potato expanded with the rising population of the United States and its demand for an increasing range of potato-based products. But by the end of the 1960s, the main prizes in potato production in the Snake River area were won by newer and larger farms established alongside the old. Indeed, a veritable revolution was brought about by the transfer of production to the West of the Snake, where dryfarm wheat land suddenly became potentially irrigable. This was the result of enough capital being available in the area for well-drilling and pumps so that deep-well sprinkler irrigation could be attempted. The farmers in the valley had long known about the potential but had not possessed sufficient risk capital to undertake the experiment. By the early 1960s several of the larger and more prosperous potato farmers met the challenge. Some of the first of these farmer-entrepreneurs faced disaster, but soon an entirely new scale of farm and field size was in the making—a scale which not only would affect the rotation pattern but also would modify the entire future of the older farms. In a few years, the large-scale sprinkler-irrigated farm was a reality and a major threat to traditional, smaller-scale, surface-irrigated farms.

The new lands converted from dryfarm to 'wetfarm' had several obvious advantages. First, inasmuch as such wheat dryfarms were of extensive size, having more than several thousand acres each, the land was ready-made for large farm and field use. Second, with the development of sprinkler irrigation, which relies on pipes rather than ditches to distribute the water, levelness of the fields made little difference. Hence no extensive field preparation was involved in converting the land to irrigated use. Third, the soils in much of the dryfarm and desert areas were a lighter, more calcareous, and very fine sandy and silty loam. The Bannock silt loam had excellent drainage characteristics and was perfectly suited for mechanical harvesters. Moreover, these lands were tremendously fertile and, for the first few years at least, had higher yields than did the traditional potato farms nearby. Finally, during the first year or two the entire farm could be put into potato production: if the land had been entirely in grain use, it was ready for potatoes. Thus the new farmers had tremendous opportunities for making a killing the first few years, and several of them made over $1 million annually.

We can now proceec to compare the older potato farm with the new on a financial basis. Reliable figures show that the traditionally irrigated farm still has the advantage on a cost-per-acre basis. These costs are shown in Table 10–1. Average procurement and production costs per acre amount to only $149 for traditional, surface-irrigated farms versus some $174 for the new large-scale sprinkler-irrigated farms (in 1970 dollars). The clue to the competitive advantage of the sprinkler-irrigated farms, however, is seen in their level of production, that is, the amount of potatoes produced per acre.

Table 10–1 Comparative costs: surface-irrigated versus sprinkler-irrigated potato farms in southeastern Idaho

Type of cost	Traditional, surface-irrigated farm (per acre)	New, sprinkler-irrigated farm (per acre)
Procurement costs		
Seed	$44	$44
Water	2	20*
Fertilizer	10	16
Machinery	8	20
Pipes and pumps	None	10
Subtotal	64	110
		*(includes power)
Production costs		
Seed-cutting labour	4	4
Planting and soil preparation	6	4
Irrigation labour	3	3
Cultivation	3	3
Harvesting labour	50	10
Ditch maintenance	1	None
Digging	9	35
Taxes	3	1
Subtotal	85	64
Total	149	174

The profit to be made on a traditionally irrigated farm with 160 acres is only about one-seventh of that on a sprinkler-irrigated farm. The latter is about 1000 acres in size and has some 300 acres in potatoes (during the first several years all of the 1000 acres are in potatoes). By contrast, the traditionally irrigated farm can support only about 40 acres in potatoes in any one year because of its rotation pattern and because of heavy labour requirements at harvest time. If it is assumed that the average yield is 300 hundredweight per acre in each case and the field run price is $1.80 per hundredweight, then the traditionally irrigated farm will clear $1.30 a hundredweight, or $15 640 in all, whereas the sprinkler-irrigated farm, although making only $1.10 per hundredweight, will clear some $82 500. Thus the sheer profitability of the land on a per acre basis is not always a reliable measure as to the ability of various crops to pay rent or to prosper.

However, the real advantage of the sprinkler-irrigated potato farm over the small-scale surface-irrigated potato farm lies in the farmer's access to economies of scale in production and to possibilities for vertical integration. Neither of these were figured in the previous calculations. How large farms can achieve economies of scale has already been explained; also, it will be appreciated that small farms can fight back somewhat through their co-operatives. The economies achieved through vertical integration however, are formidable indeed to small farmers: it is through such integration that large-scale operators truly make their operations pay.

In the Snake River area, the large-scale operator controls both forward and backward costs associated with his potato-producing operation (Figure 10–8). He controls backward items by sharing heavily in the profits of machinery, pipe, fuel, and seed companies, even of financial institutions. He controls forward items by having his own storage cellars and warehouses for shipment to markets, but he also attains considerable benefit by making use of those potatoes for which the small farmer receives little or no payment. These include knotty, too small, and otherwise damaged tubers, which are sent to a processing plant for dehydration and shipment to extensive markets throughout the United States. The wastes from the processing plant are then largely used as feed for beef animals kept in large feedlots. The large-scale raising of beef cattle in this manner affects even Midwest beef growers in a competitive way. Finally, as if this were not enough, the beef operation is sufficient to support a separate meat-packing plant.

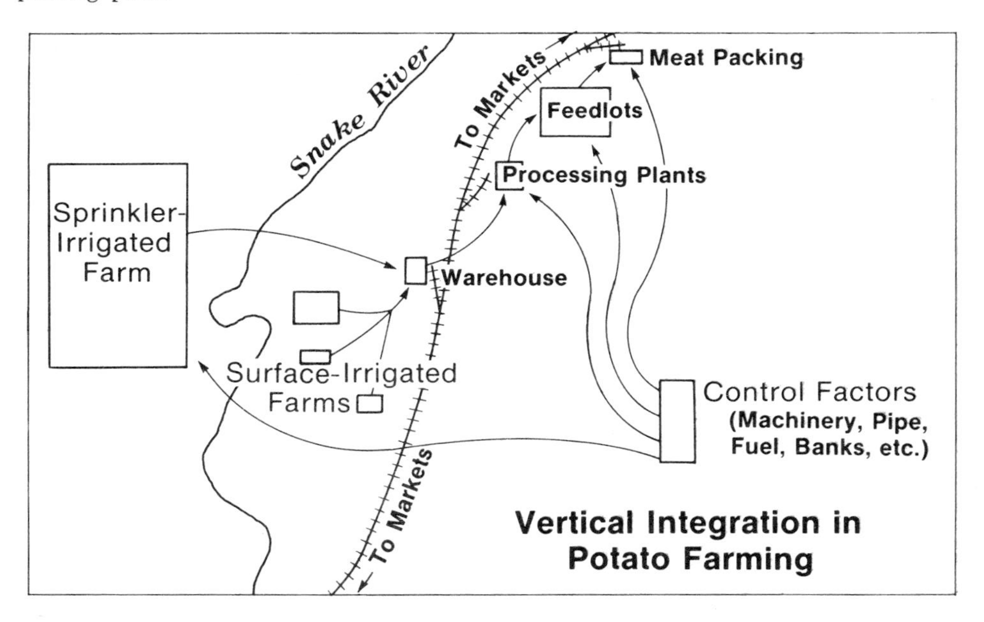

Figure 10–8 An example of vertical integration in potato farming. Note the small size and fragmented nature of the older, surface irrigated farms on the East side of the Snake River in comparison to the sprinkler-irrigated farms on the West side.

With such a system, it is a mystery that the traditionally irrigated farms have survived as long as they have. But complete vertical-integration systems are just beginning to gain full steam in Idaho, and perhaps it will not be long before the older system fades into oblivion. At any rate, the demand for small farms will surely decrease and a new method of agriculture will have to be developed. Given this latest challenge, coupled with the imminent need for sharing the farm among the offspring, the future looks bleak indeed. Thus one of the more interesting revolutions in land tenure and pattern will probably occur in this area soon. Indeed such changes will occur in many areas throughout the United States as large-scale and integrated farming becomes more and more entrenched.

CONFRONTING REALITY: FARMING PATTERNS WITHIN A REGIONAL-NATIONAL FRAMEWORK

We may now attempt to enhance our understanding of farming patterns by stepping up to multi-crop situations within the regional-national framework. The southernmost part of France—the Midi—provides an excellent example for this purpose, for it enables us to bring together many aspects we have so far considered fractionally. Although it is not possible to explain all influences which bring about the farming patterns we observe—some features must be taken as 'given'—a main aim will be to identify the major controls of agriculture development. In the example, crop production is very strongly geared to meet the demands of a market which is national and international in scope and highly competitive. In view of the importance of agriculture in the economy of France, we shall also discern why it is that agricultural developments have not been left to the farmer alone, even if his role is crucial to the success of such developments. In confronting farm production and marketing in this way, the 'other things being equal' approach is largely suspended, but we must still remain mindful of the principles gained from that approach.

THE OLD AGRICULTURAL PATTERNS IN THE MIDI

Within the Midi, we focus on that part which lies on either side of the valley of the lower Rhône, in Provence and Languedoc. Close to the Mediterranean, this is a region with a Mediterranean climate. The detailed character of this climate must remain one of the 'givens', but it is vital to emphasize the extreme seasonality of rainfall, long sunshine hours and the effect of its predominant valley wind, the *Mistral*. This last will be noted later. The most dramatic of these characteristics, which has exercised the strongest control over the agriculture of the area, is the extreme seasonality of rainfall: almost all of it (averaging over 600–700 millimetres per annum) falls between October and March, with the summer period from April to September almost always dry, sunny (over 2500 hours of sunshine per annum) and hot (daytime July 32°C). The principal growing season being bereft of rain, the Old World Mediterranean triad of wheat, wine and olive reigned supreme for centuries wherever cultivation was possible. In other places, the land was useless: its upland forests were steadily denuded by fire, felling and grazing; its limestone hills were bare or covered with garrigue (thorn scrub); and its valleys were either too dry or too wet by season. Moreover, all three of its traditional crops could be grown as well, if not better, elsewhere: wheat in Northern France, olives in Italy and Spain, and grapes in many other parts of France and the Mediterranean.

Indeed, the best quality wines in France have always come from select areas, few of which are in the 'deep south' of France. Yet it was the vine, with its searching taproots and demand for ripening sunshine, which flourished best in the alluvial sands and gravels of the basins, valleys and plateaux of the region, so much so that most of the cultivated area submitted to monoculture of the vine. The arrival of the railway aided viticulture considerably in the mid-nineteenth century, but the whole area suffered grievously from its one-crop dependence as the result of the phylloxera epidemic of the 1880s. The few farmers to escape this disaster were those who, because they could rely on irrigation water, had been able to diversify. Until modern times, the irrigated areas were quite local, the supply being limited to narrow strips of land near to the major rivers. Notable in this regard was the lowest section of the valley of the Durance, especially close to the confluence of that river with the Rhône. Here, with great difficulty, the medieval engineers had irrigated the fields higher than the river downstream but lower than the point of intake, the canal water being kept moving by gravity. The fields gained not only water, but also its suspended load of silt which built up the surface and helped to maintain its fertility. Producing early ripening fruits and vegetables (*primeurs*), this area benefited greatly with the coming of the railway which enabled the produce to be sent to northern markets well ahead of other agricultural regions in France.

Other peasant farmers who might wish to follow this example could only struggle with inadequate or unreliable groundwater supplies. But more than water was necessary for change. In both Provence and Languedoc, farm efficiency—whatever was produced—was considerably hampered by the preservation of an antiquated system of possession and tenure. In valley and plain, the characteristic rural settlement was the nucleated or compact village. Its form and pattern still reflected the demands of the old agrarian system requiring communal effort and co-operation between the village-based peasants tending pieces of land scattered about the fields of the *commune*. Beyond these *parcelles*, large and isolated farm units (*mas*) survived through the centuries or were created in the nineteenth century.

All focused on viticulture, and the actual vinification was carried out at the *cave co-operative* (winery). These undoubtedly helped the peasant wine-grower to survive, but the production of inferior wines often exceeded demand. Poverty and depopulation was rife, yet the Midi could offer few alternatives to its monoculture: there was no coal, few minerals, and few industries or entrepreneurial skills. As the commercial-industrial economy of Northern France developed with these very assets, the backwardness of this most southerly part of France became most marked. Beyond the influence of the major towns, this was a situation which persisted until well after World War II. By the middle 1950s, with millions of hectolitres of wine unsold, the French Government was forced to buy the surplus and convert it to industrial alcohol (which

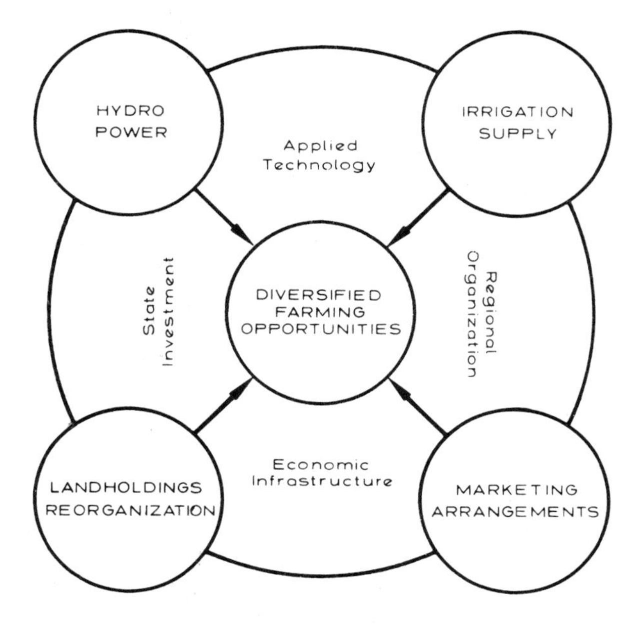

Figure 10–9 Organizational inputs (capital, energy, water, land reform, and marketing arrangements) leading to opportunities for the diversification of agriculture in the Midi, France.

could be made more cheaply from the northern French farmer's sugar beet). There came the realization, therefore, that up to one-third of the two million hectares of vines in France should be turned over to other crops, and that the Midi should be in the forefront of this change.

Basically, the region was deficient in (i) developed sources of energy, (ii) facilities for the widespread distribution of water for irrigation, (iii) a modern system of landholding, and (iv) capital, organization and infrastructure. These deficiencies are now being overcome to permit a diversification of agricultural activity in marked contrast to the old (Figure 10–9). The key to development was the control of water, so it is to the nature of this control that we turn first.

CONTROL OF WATER AND OTHER FACTORS IN AGRICULTURAL DEVELOPMENT

The major rivers of the area both flow into the Mediterranean part of France from distant highlands where precipitation is abundant. Flowing from Lake Geneva, the Rhône becomes the largest river in France as it picks up many tributaries. The discharge which reaches the Midi has a mean flow of 1700 cumecs (cubic metres per second), but the maximum flow can be six times that of the mean. Such variability of flow is related to seasonal or local conditions of flood or drought within the wide catchment. In spite of this problem, over the centuries the Rhône was improved for navigation and to ensure water supplies and local irrigation. The Durance was quite another matter because of its wildly erratic flow and steep gradient. Fed by Alpine snowmelt, in its middle course its maximum discharge could be 4800 cumecs (a broad, raging torrent) while its late summer minimum could be as low as 27 cumecs (a string of pools in braided channels with barely perceptible flow). With none of the Durance problems of pebble and silt, its major tributary, the Verdon, has clear waters derived within a wild and unproductive limestone hill country.

Since World War II French engineers have almost completely transformed these natural conditions on all three rivers by extensive works which are multi-purpose in design and amount almost to complete control (Figure 10–10). The Rhône has been invested with barrages, raising the water for all-year navigation, and the difficult river sections have been by-passed with ship canals. At these places, the fall has been utilized to produce hydroelectricity. The water has also been used to supply a reorganized irrigation canal system and to supply sprinkler irrigation, using electric pumps, on the higher terraces where only dry farming could be practised before. In the Durance Valley, hydroelectric stations, storage basins and their associated canals have enabled the reorganization and redistribution of water for agriculture, again over a wide area. Even on the valley floor, where the old 'Scourge of Provence' once waxed and waned uncontrollably, new irrigated farms and fields have been created in the flood-plain.

West of the Grand Rhône in the delta below Arles, the main problem is excess water. Here, it was necessary to drain the 'wetlands', but the new control of the Rhône enables rice fields to be flooded as necessary and later dried out. In strong contrast, overlooking the lagoon-filled Camargue to the west of the Rhône delta, is the Costière, a dry, pebbly plateau which requires irrigation for agriculture to be possible. Here, water deficiencies have been made up by the supply of water from the Rhône. Using the hydropower generated by the river upstream, water is pumped up to the plateau, which then irrigates a system of new canals which reach westwards to the southern parts of Languedoc.

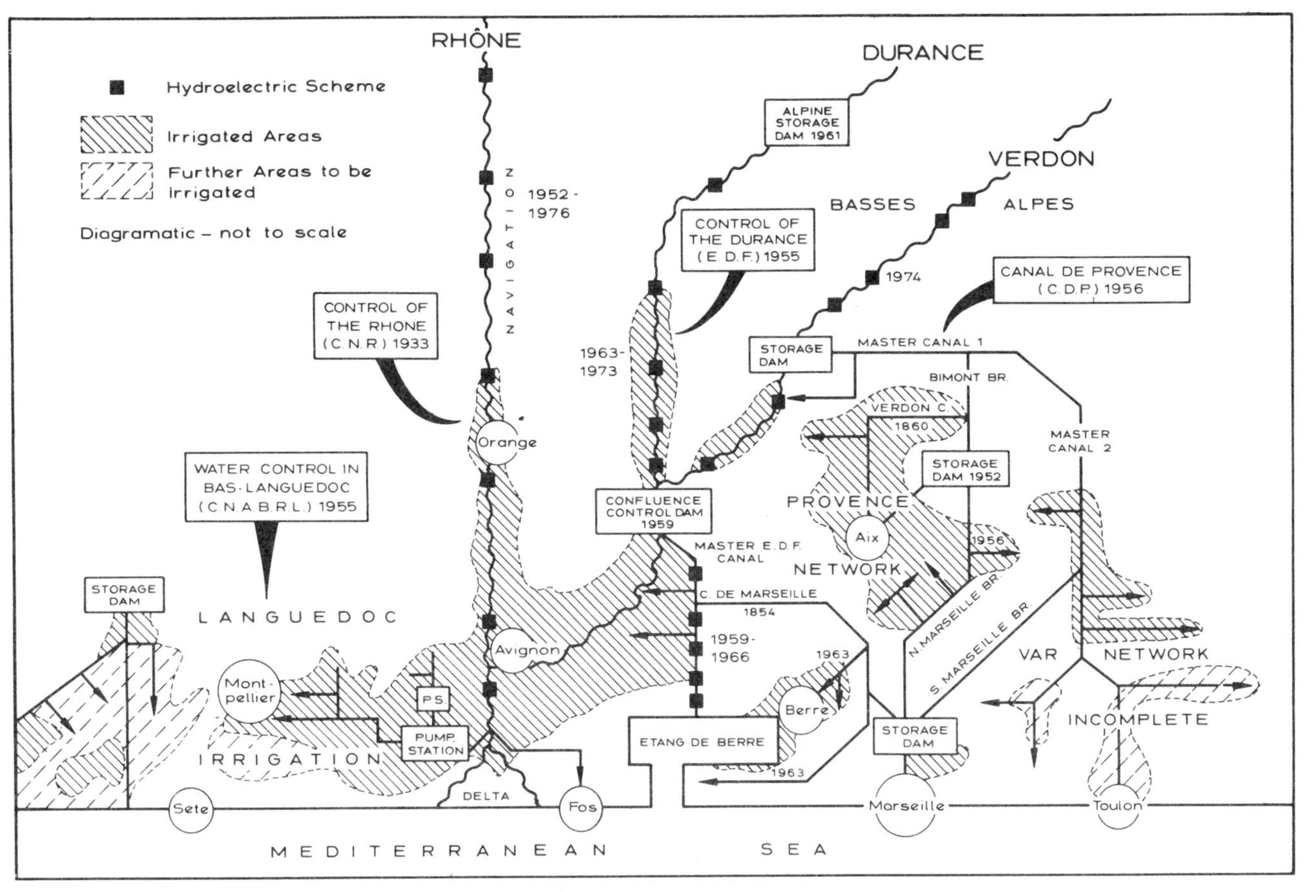

Figure 10–10 New canal systems and irrigated agricultural zones in Provence and Languedoc areas of the Midi, France. Effective control of water has been an essential prerequisite for agricultural development within an area of summer drought. To show the master distribution networks clearly, the diagram is not to scale.

ORGANIZATIONS CONTROLLING WATER AND ITS DISTRIBUTION

These works, as dates in Figure 10–10 indicate, have taken over twenty years to complete and each input had to be carefully phased. Major elements were funded through the Central Government and actually undertaken by Mixed Economy Companies, with their own directorates representing the state, local water and electricity users, local communities, and local and national commercial and industrial agencies. In the case of the River Durance, overall control is exercised by a nationalized industry, Electricité de France. EDF not only holds back a year's supply of water in store in the uppermost reaches, but also guarantees flows to the local canal systems still controlled by local communities. In the case of the Rhône, the Compagnie Nationale du Rhône (CNR) was committed to the improvement of navigation, to produce and sell electricity to EDF, and to provide irrigation water on demand for the lands adjacent to the river and its major canals. Another agency, the Societé du Canal de Provence, was brought into existence (1956) to organize the distribution of the new major source of water made possible by control of the Verdon. This agency has constructed a new network of interfluve-crossing canals which serve domestic and industrial users in the Mar seille Metropolitan Region and, on the way, provide irrigation water for the adjacent farms and fields. Finally, the supply of water to the Costière plateau and to similar areas to the Southeast is under the control of a regional planning agency, the *Compagnie Nationale d'Amenagement de la Region du Bas-Rhône et du Languedoc* (CNABRL), which has wide initiating powers in relation to land reclamation and can offer advice on agricultural methods and soil conditions. One of its most important responsibilities has been in the area of landholding. Many small, fragmented and uneconomic smallholdings have been consolidated (*remembrement*) and many others re-grouped into larger units. Old, large estates also have been split up (*démembrement*) and new farm roads, drainage works and farm houses built.

ON-FARM PRODUCTION IN THE MIDI: INPUTS AND INTERCROPPING

As a result of the arrival of power supplies, irrigation water, and many back-up services of the kind just described, many parts of Provence and Languedoc have been provided with the basic infrastructure for the revitalization of its agriculture. The area irrigated has quadrupled, and the opportunity for increased income from irrigated crops is therefore much more widely available. In areas where irrigation is not possible (or has yet to be provided: the Languedoc and Var distribution networks are incomplete), monoculture of the vine prevails, but in other areas the vine now takes second place to a full range of tree and bush fruits, including cherries, apricots, pears, apples and table grapes, field vegetables, and all manner of market garden produce, like salad greens, beans and onions.

Land on thousands of tiny smallholdings and small farms is worked almost continuously. Vegetables, ground fruits and fruit trees are often combined on the same farm, Europe is subordinated to the system of intercropping. Three or even four successive crops are grown each year from the same field. Thus the maintenance of high yield and quality demands prodigious amounts of chemical fertilizers, as well as the application of herbicides and fungicides. Irrigation water is provided every two to four days in the summer: row flooding, sprinkler and 'drip-feed' irrigation methods are all used as appropriate to a given crop. Glass cloches, polythene tunnels and straw matting are cleverly employed to bring on the primeurs. Small tractors and rototillers are the principal mechanical aids, but most tasks, like sowing, transplanting, weeding, lifting and sorting must be achieved by manual labour. The use of labour is well spread, but peaks at the busiest harvest period from mid-April to June. First to be harvested are carrots, peas and asparagus, followed by strawberries, tomatoes, early potatoes, squash and lemons. The fruit picking is spread from May to October, while cauliflower, cabbage, artichokes, beans and onions mature later in the year.

On the farms, therefore, there is much capital, time and effort invested in order to win a better farm income than might be achieved from the same ground if it were devoted to wheat or vines. But further to these things there is the need to protect against wind and frost, especially if primeurs and fruit are to be the basis of the farm's prosperity. In many of the irrigated areas, there is need for shelter fences, hedgerows or tree-lines against the chief danger to growing crops—the *Mistral*. This is a strong relentless wind which funnels down the Rhône Valley from the north. Normally, it is caused by a continental pressure gradient when a depression develops over the Mediterranean as an anticyclone advances from the West over central France. Its main effect in unprotected orchards and fields is to cause leaves and stems to break on tender plants and flower and fruit to be bruised or to fall prematurely. It is also claimed that constant high evapotranspiration in unsheltered fields decreases yields by slowing down photosynthesis (see Daniel W. Gade, 'Windbreaks in the lower Rhône valley', *The Geographical Review*, April 1978, pp. 127–44, for an admirable survey of this problem). A more obvious impact of the *Mistral* is that it can bring dangerously low temperatures to irrigated fields or the flat valley floors. The last killing frosts may occur towards the end of March, a critical time of the year for the grower of primeurs.

MARKET OPPORTUNITIES: BASIS FOR CHOICE

Finally, in our case study, we must attend to the marketing needs and arrangements. The new range of crops has to be sold in the most effective way. Broadly, this is dictated by the size of holding, its location, the type of crops produced and the time of year.

Perhaps of most importance is the size of holding, for this affects the range of crops and their scale of production. Although a great majority of the farms are small (5–10 hectares) by comparison with the size of the average farm in England (40 hectares), we may still appreciate the importance of local differences in scale. The medium size holdings produce the widest range of crops, both because they have space to do so and because they wish to spread their risks as insurance against failure. With adequate production at any one time to make market journeys worthwhile, such producers have a fairly wide range of selling opportunity (see the base of Figure 10–11 and follow the arrows).The small scale producers try to do the same, but with small quantities to sell at any given time they may rely on local retail sales, local co-operatives or forwarding agents. Both the latter will bulk up the produce of the small producer to dispatch it further down the chain of marketing. Lastly, the large growers can boldly attempt to reap economies of scale by specializing in the quantity production of a small range of crops. This may enable them to use more direct selling methods.

Thus in a region where once the wineries of the *mas* and the *caves co-operatives* were

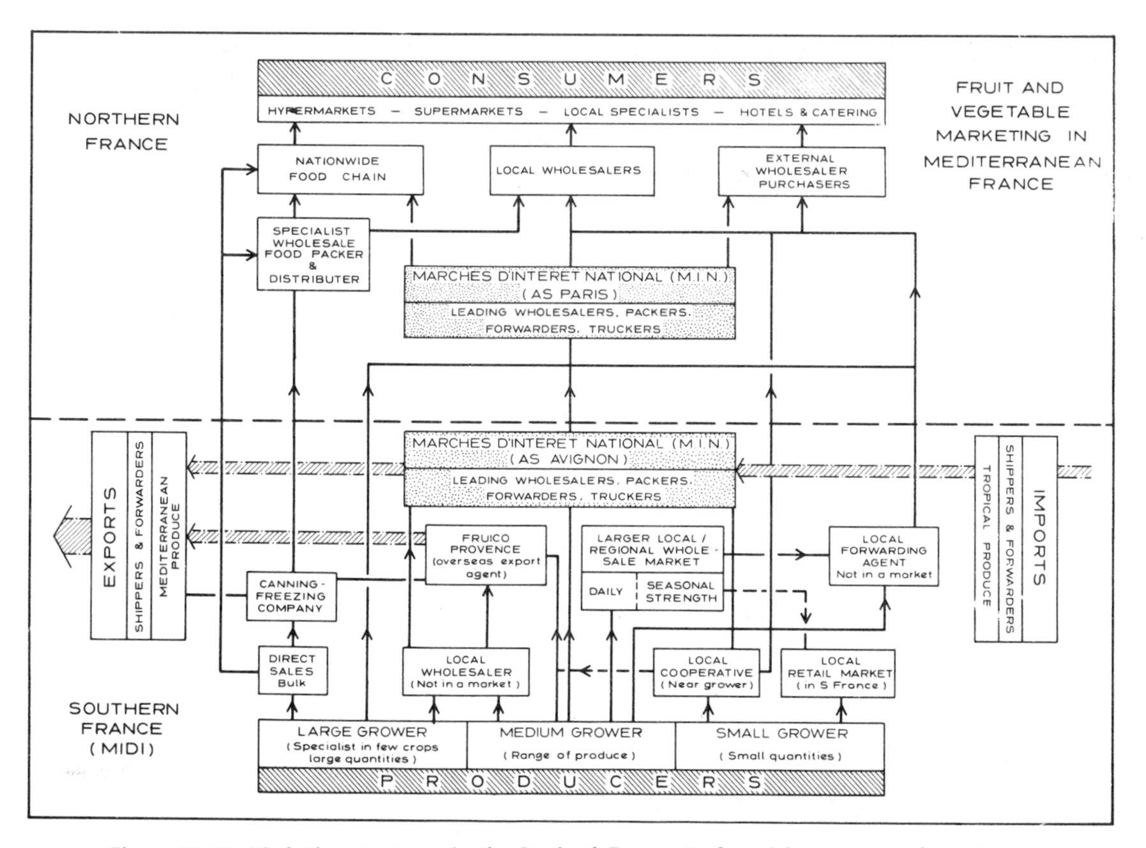

Figure 10–11 Marketing structures in the South of France. At first sight a system of great complexity, this flow diagram logically pursues the most likely movements from farm producers to market consumers. The key to understanding is the scale of production on the farm and the opportunities that this presents. Which producer-group has the widest range of marketing choice?

enough, there is now a complex web of producer-consumer markets of widely varying spheres of influence, together with alternatives which by-pass both market and middleman. The major quantities of farm produce are channelled through one or other of five great markets of national status (MIN), such as those at Avignon and Nîmes. Each contains stream-lined auction systems and dispatching facilities, plus up-to-the-minute price information from all parts of France. Specialist growers may, however, be tempted by good prices in other, regionally-important markets which themselves specialize in, say, tomatoes, cherries or table grapes when in season. Again, local wholesalers may be competitive with local markets, especially if they have direct connection with distant purchasers. As was seen with the case of the larger potato farmers in Idaho, the largest producers may hold the edge in this regard through forward linkage.

Finally, it is also the case that the largest producers are often seen to be the pioneers with new crops and new farming methods. At the forefront, it is they who also bear the greatest risks. Their desire to specialize is related to the demands of a widening EEC market, but their story is not all one of success. Soft fruit specialists are especially vulnerable to frost, hail or wind. And at the other extreme they may face the spectre of glut with a highly perishable commodity. In fact, in recent time there has been overproduction of apples, peaches and melons in the Midi. And competition in the European market may become much stronger as other Mediterranean countries like Spain and Greece become members of the European Economic Community. Other producers, like Cyprus and Israel, are also successfully penetrating the same European market.

Perhaps the greatest advantage of the Midi in regard to such competitors is its position in the Rhône Valley, which gives direct access by high-speed road and rail to the great population clusters in northern and western Europe. Its greatest challenge appears to be in the way its comprehensive range of farm production will affect older established fruit and vegetable producing areas even nearer to the northern urban-industrial market. Indeed, in terms of Von Thünen relationships discussed in Chapter 9, it is not difficult to conclude that this particular outlying zone of the European farming economy, formerly characterized by an extensive and low-yielding farm product, is fast becoming a part of the innermost zone characterized by intensive production and —potentially—high rewards.

We have now arrived at the conclusion of our first major attempt at a comprehensive understanding of one activity. It is perhaps unnecessary at this point to repeat that the agricultural location problem is indeed multifaceted, interconnected, and difficult. Moreover, it is an activity that is spatially highly changeable (albeit not nearly so changeable as the distribution of activities within cities) so that an emphasis here on present-day facts and statistics would be of little use to the student. What *is* important is acquiring the framework for spatial understanding, inasmuch as that framework can be applied, with prudence of course, to any distribution at any time. More important, such a framework should aid the economic geography student in the interpretation of new patterns as they rapidly emerge on the rural landscape.

BIBLIOGRAPHY

Bowler, I. R. 'The common agricultural policy and the space economy of agriculture in the EEC'. In *Economy and Society in the EEC*, Ed. Lee, R. and Ogden, P, E. Westmead: Saxon House, 1976.

Cartwright, D. G. 'Institutions on the frontier: French–Canadian settlement in eastern Ontario in the nineteenth century', *The Canadian Geographer*, Spring 1977, pp. 1–21.

Chappell, J. E., Jr. 'Passing the Colorado salt (Colorado River delta irrigation problems)', *The Geographical Magazine*, July 1974, pp. 569–574.

Clout, H. D. *European Agriculture Since 1945*. London: Macmillan, 1971.

Curry, E. A. 'Land tenures, enclosures and field patterns in Co. Derry in the eighteenth and nineteenth centuries', *Irish Geography*, Vol. 9, 1976, pp. 52–62.
Duckham, A. N., and Masefield, G. B. *Farming Systems of the World*. London: Chatto & Windus, 1970.
Dumont, R. *Types of Rural Economy: Studies in World Agriculture*. London: O.U.P., 1957.
Edwards, A., and Rogers, A., eds. *Agricultural Resources: An Introduction to the Farming Industry of the United Kingdom*. London: Faber & Faber, 1974.
Gade, D. W. 'Windbreaks in the lower Rhône valley', *The Geographical Review*, April 1978, pp. 127–144.
Girtin, T. 'The fertile corner of the world which is the market garden of Europe', *The Geographical Magazine*, 1966, pp. 542–553.
Graves, N. J. 'Une Californie Française: the Languedoc and lower Rhône irrigation project', *Geography*, Vol. 50, 1965, pp. 71–73.
Gregor, H. F. 'Farm structure in regional comparison: California and New Jersey', *Economic Geography*, July 1969, pp. 209–225.
Gregor, H. F. *Geography of Agriculture: Themes in Research*. Englewood Cliffs, N.J.: Prentice-Hall, 1970.
Gregor, H. F. 'The large industrialized American crop farm', *Geographical Review*, April 1970, pp. 151–175.
Grigg, D. 'The agricultural regions of the world: review and reflection', *Economic Geography*, 1969, pp. 117–120.
Hart, J. F. 'Land rotation in Appalachia', *Geographical Review*, April 1977, 148–166.
Highsmith, R. M., Jr. 'Irrigated lands of the world', *Geographical Review*, July 1965, pp. 382–389.
King, R. *Land Reform: The Italian Example*. London: Butterworth, 1973.
Lambert, A. M. 'Farm consolidation in western Europe', *Geography*, Vol. 48, 1963, pp. 31–39.
Lewthwaite, G. R. 'Wisconsin and the Waikato: a comparison of dairy farming in the United States and New Zealand,' *Annals of the Association of American Geographers*, March 1964, pp. 59–87.
Lewthwaite, G. R. 'Wisconsin cheese and farm type: a locational hypothesis', *Economic Geography*, April 1964, pp. 95–112.
Morgan, W. B., and Munton, R. J. C. *Agricultural Geography* (The Field of Geography Series). London: Methuen, 1971.
Stevenson, I. 'Sour grapes from rich harvests', *The Geographical Magazine*, 1976, pp. 262–265.
Tarrant, J. R. *Agricultural Geography*. Newton Abbot: David & Charles, 1974.
Thompson, I. B. *The Lower Rhône and Marseille*. London: O.U.P., 1975.
Warriner, D. *Land Reform in Principle and Practice*. London: O.U.P., 1969.

11

Bases of Industrial Location

Here and in the following chapter, we review the basic elements entering into industrial location decisions, particularly those which relate to manufacturing industries. As with the preceding chapters on the location of agricultural activities, we build from the stance of the components incorporated into a basic, or classical industrial location theory. This theory, we shall find, attempts to describe the circumstances determining successful locations under market place conditions in an idealized world. Later, we attend to more complex circumstances and real world situations.

The economic geographer asks many questions of the 'what, where, when, why?' kind about industrial location. Among them are: Why is the single factory or plant (the individual producing unit) located where it is? What influences bear upon the corporate firm's decision to locate its plant (or many plants) where it does? How do various physical, economic and technical influences affect industrial location decisions? The on-going viability of location as time passes is as important a matter for examination as is the study of location choice de novo. In our search for understanding, we shall find that the study of industrial location is not an exact science and that the locations of most plants cannot be explained in terms of maximum profit or minimum cost. Indeed, all manner of preferences have to be taken into account when considering the reasons for the choice of particular locations. Like farmers and horticulturalists, manufacturers and businessmen are frequently asked why they are located where they are to produce what they sell. The answers most frequently given are intensely personal to themselves, seeming to deny the search for generalizations. Yet all must operate within competitive market conditions and it is these above all else which impose a rationality or regularity to industrial location patterns.

SOME FUNDAMENTAL CONSIDERATIONS

We can begin our analysis by suggesting that the three basic components in industrial location are analogous to those contained in our discussion of agricultural location: (1) procurement costs of raw materials (generally, transportation costs required to assemble the raw materials on site); (2) on-site or processing costs; and (3) distribution costs, that is, costs of transporting the finished product to market. However, unlike agriculture, all three costs may be independently dominating or all may be intermixed to bring about a particular locational response. Especially different is the fact that for a number of industrial firms, procurement costs are of extreme importance. It will be recalled that in the location of agriculture, these costs (for example, seed, machinery, and fertilizer) are of small importance.

Thus the location of manufacturing is a much more complex subject and is more difficult to understand than is the location of agriculture. This is also due to the myriad of locational possibilities and the host of variables that may play a significant role in manufacturing location. Any one of the three types of costs enumerated above may dominate the locational decision. Thus for many types of manufacturing, an additional locational choice occurs, affording a degree of choice not available to activities previously examined. This additional variable is one of the primary ingredients that tends to make locational understanding of manufacturing more complex than that of the primary activities.

There is also much more flexibility and substitutability in the major factors of production (land, labour, and capital) than there is in factors of the other activities thus far studied. For example, labour and capital are highly interchangeable in that machinery can be substituted for labour, and vice versa. Thus a factory situated in a low-cost and relatively unskilled labour area may opt for using much labour and little machinery in the production process. The degree to which such substitution possibilities can operate, however, as with agriculture, is highly dependent upon the type of manufacturing. In comparison with agriculture, however, most manufacturing is highly machinery-oriented. However, furniture and apparel manufacturing are to a certain extent atypical in that they use heavy labour inputs.

Land is likewise a rather flexible item. Certainly, land costs vary greatly; centrally situated metropolitan sites in the United Kingdom often cost more than £100 000 per hectare, while suburban sites are generally under £20 000 per hectare and rural sites still less. This, of course, is important in differentiating among industrial types in terms of their rent-paying needs and abilities. However, it is land costs in the broad, regional context that most interest us here. Manufacturing plants are not prohibited in extremely hot or cold areas. Operation costs in such environments are higher. Thus processing costs might be increased because of the need for air conditioning, insulation, or heating. Mostly, however, other factors are more dominating so that these costs are often not critical.

Manufacturing is more complicated than agriculture for other reasons as well. First, possibilities and practices of integration are more widespread and more extensive. Many industries are involved in controlling, through *vertical integration*, activities that are linked in both a forward and backward direction to their production processes. Unlike agriculture, *horizontal integration* is also possible in manufacturing. Here an industry, through amalgamation or mergers, can control that sector of production in which it specializes. Thus it has much more influence on the competitive framework than does agriculture. No matter what is done in the South of France, say, farmers there cannot control the price of fruit and vegetables in the EEC market: too many other areas are also involved and would independently move in to take advantage of any strategies southern French farmers might adopt. If it were determined that production should be curbed in the Midi in an attempt to raise prices, other areas would increase output and reap most of the price benefits. But in industry, oligopoly (control of the market by only a few major producing firms) is common in many activities so that one firm can have considerable possibilities for manipulation and be far more able to control the destiny of its own undertaking. In fact, the number of firms involved, that is, the extent of competition, has very clear locational implications.

This is evidenced by the fact that most manufactured products, unlike agricultural commodities, are known to the consumer by brand name. Thus a given firm can encourage consumption of its product primarily through advertising, service, and general reputation of quality. This likewise provides greater possibilities for dominating markets than are open to other activities.

Manufacturing is also distinctive in that it is heavily affected by agglomerative economies. This is one of the major reasons for the clustering of manufacturing firms in an area. Agglomerative economies result from three general features: (1) the fact that the manufacturing process of many products occurs in an extensively linked system wherein one firm's finished product becomes another's raw material; (2) the development of plants for the manufacture of by-products from waste materials; and (3) the complementary advantages manufacturing firms achieve by being close to one another, for example, drawing upon a large common pool of skilled labour. Use of the various business services that result from the concentration of plants, and particularly from the development of large cities, is another agglomerative advantage. This latter feature is particularly instrumental in causing those manufacturing firms that otherwise have the choice to locate in or very near large cities.

The effect of a highly linked system, in which one firm supplies another with raw materials, is regional concentrations of such feeder-related industries. Industrial linkage is, indeed, a most important consideration and as such is given special treatment in Chapter 12. For introductory purposes, however, we may note here two famous examples in the automobile manufacturing industry. One is the system which developed in the United States in Michigan and upper Ohio. Here, various firms manufacture parts for the automobile assembly plants in Detroit and the surrounding area. The comparable example in the United Kingdom is the West Midlands, where the assembly plants in Birmingham are fed by hundreds of different companies each involved in producing items for the manufacture of a single automobile (see Chapter 12). Other well-known feeder manufacturing systems include the petrochemical concentrations in and around Houston (sometimes called the 'spaghetti bowl' because of the elaborate interconnections among firms) and, in Europe, in the neighbourhood of such oil-importing centres as Rotterdam and Fos, Marseille. Many other industrial types reveal a strong tendency to linkage, including the textile industries in Lancashire and the apparel industries of several 'world' cities like New York (Manhattan), London and Paris. Many agglomerations offer the situation where a major firm(s) may be feeding 'parasitic' industries that utilize by-products of the parent industry. In many such cases, one firm may achieve tremendous economies through vertical integration. In most instances, these by-products must be processed fairly close to the main plant so that the locational pull is extremely direct and strong.

The combination of all these possible conditions clearly makes manufacturing a more complicated locational problem than are the other activities studied so far. Nevertheless, there are wide differences in the complexity of the problem among industries. Within the secondary sector, the location of some plants is rather simply dictated by the necessity of being adjacent to raw materials; other plants must locate at the market. The most complex problems, however, are presented by those diversely located activities which may be found in a variety of places for reasons not evident by the dominance of any single factor such as procurement, distribution, or on-site processing costs.

PROCUREMENT COSTS AS A LOCATION FACTOR

The major factors in industrial location are shown diagrammatically in Figure 11–1. Any one of these components may be a determining factor. The procurement costs primarily represent the costs of transporting the raw material resources to the factory. For those industries that use a single raw material (for example, the mineral and petroleum processing industries) maximum adjacency to the raw-material resource results when a firm finds these costs to be most critical. However, most manufacturing involves multiple raw materials from a number of different sources. Each raw material has its own weight, volume, value, and transportation cost characteristics. Thus, even if the best location for such a firm is a place where its raw-material costs will be minimized, the determination of that place is no simple matter. Consequently, a number of rather elaborate techniques for minimizing transportation input costs, to be discussed later, have been developed.

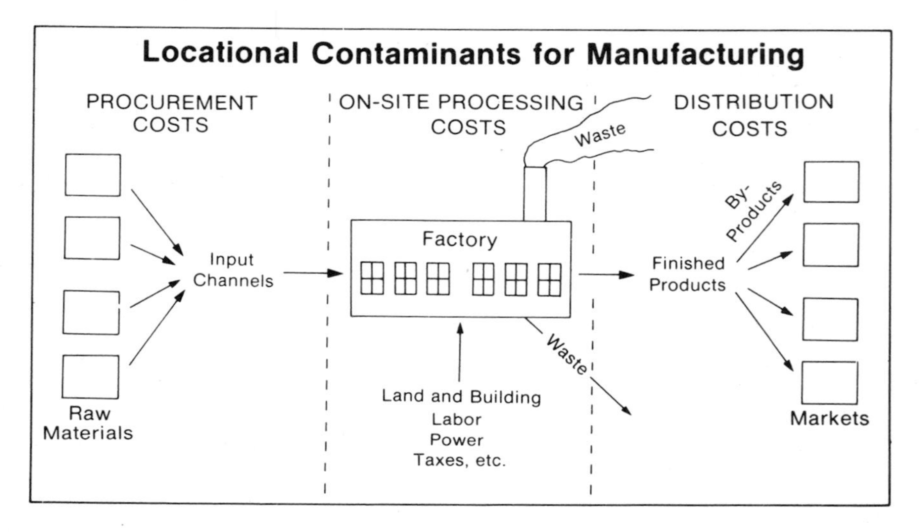

Figure 11–1 Locational contaminants for manufacturing. Note that each of the costs may exert a decisive locational pull on the placement of a factory.

This becomes an especially knotty problem inasmuch as there are usually possibilities for substitution among raw materials. Thus wood can be replaced by plastic or metal in the manufacture of an item. Instead of purely high-grade coal being used for coking (in the manufacture of steel), lower grades can, by *beneficiation*, an enrichment process, be mixed with it in a way that lowers total raw-material transfer costs. Instead of iron ore being used exclusively, scrap can be inserted. And gas can be used as a fuel in place of coke in some parts of the operation. The same kinds of possibilities are apparent in most manufacturing so that, unless the raw-material sources and costs are given, even a determination of the location that minimizes procurement costs is fraught with uncertainties.

Processing costs on the site include not only labour, land, building rent, and machinery costs, but also taxes, power, water, and costs associated with waste disposal, all often critical. Any one of these might be a determining item in the locational decision. Labour is clearly one of the major costs in production, and for many labour-intensive industries, this may be the overriding factor. Small differences in wage rates among cities and regions may make large differences in terms of production costs. Within the United States, wage rates have traditionally been lower in the South than in the North, thus encouraging certain firms to relocate. The textile industry's migration from New England to the Piedmont area in North Carolina is a classic case of how wage rate differences have affected locational patterns. Of course, most industry is limited in its ability to take advantage of such wage rate differences among nations, but some international movement does occur. Clearly, wage rate differences between the United Kingdom and Japan give Japan a major advantage in the manufacture of many items. This model assumes that wage rates and labour are immobile, but in reality this is hardly the case. European workers are free to move among Common Market countries for employment purposes, a circumstance which, for example, has enabled many thousands of Italians to find better paid work in West German manufacturing concerns.

Cheap labour in newly-emerging industrial countries in the Far East, notably Taiwan and South Korea, has seriously affected the ability of manufacturers in the United Kingdom to compete in the mass production of such things as cutlery and footwear. Even so, lower wage rates alone do not tell the full story; productivity of the workers for wages paid is a better measure. Skilled workers are usually more productive and consequently worth higher pay than unskilled workers. More and more are manufacturing plants requiring greater training and skill on the part of the workers. Thus not just any labour will do—the labour must be capable of doing the job. If the labour is unskilled, the wage rates will be lower, of course, but the factory might have to emphasize labour rather than mechanization. In much of the industrialized world the general trend has been for more mechanization to occur concomitantly with increasing skills and wage levels of the workers. Extremely high wage rates thus force industry to substitute machinery for labour.

For other industries, power might be the big cost in the production process. In this regard, industries like aluminium seek out those locations that have the cheapest power rates. In the United States, the earliest aluminium industry used thermal electricity at Pittsburgh, but the development of cheaper hydroelectricity in such locations as Upper New York State (Niagara), Arkansas (Tennessee Valley Authority schemes) and in the Pacific Northwest (Columbia Valley power project) pulled the smelting activity away from market locations. Similarly, capacity in Europe was developed at hydroelectric sites in Norway, Scotland, France and Switzerland. The aluminium industry is so sensitive to power costs that, as cheap hydroelectric sites have become more difficult to find, its newest smelters now also take advantage of coal, natural gas and nuclear energy where these are competitive. In many cases this has brought the processing of aluminium pig back to locations very close to its markets in the manufacturing belts of the USA, the UK and Northwest Europe.

In still other industries the need to dispose of waste products may dictate locational requirements. For example, on a small scale, slaughter yards have been relegated to outlying areas of metropolises. Other obnoxious industries include the pulp and paper industry, which pollutes both air and water, and the iron and steel industry. In many areas such industry is severely discouraged because of these effects. Thus a general principle might be that the more obnoxious an industry, the more it might be prevented from moving into certain urban areas, which would otherwise be optimum locations.

Finally, a number of industries, which otherwise have considerable locational choice, have been lured to areas offering lower taxes. The classic North American case is the heavy industry discrepancy between New York and New Jersey. New York has much higher industrial taxes than its neighbour, with the result that industry abounds on the side of New Jersey near New York but is scarce by comparison in New York proper. Many small communities in the American south and elsewhere have also offered tax advantages for new industries in order to bring in employment. Some firms are even given free land and are offered leases on ready-built structures.

There are many comparable situations in the United Kingdom. Throughout the land, metropolitan authorities, Chambers of Commerce, New Town Development Corporations and the like all compete with one another to attract industry into their communities. But Central Government also encourages relocation specifically in the so-called 'Development Areas' of the North and West. Development Area incentives encouraging the transfer of the firm or the development of new branches, may include help towards the cost of new machinery and employment subsidies enabling it to expand its labour force. To note such involvement reminds us that in many countries the entrepreneur and his profit motive is not alone in locational decision making. In Britain, many students of the problem find it necessary to distinguish three kinds of firms: the private capitalist, the corporate capitalist, and the state. The state may be very instrumental in location decisions and, quite frequently, it may insist that social goals be considered alongside those of profit and growth.

With or without Government encouragement, another on-site 'cost' or asset that has strongly affected industrial location has been the movement of certain kinds of plants to places offering the best overall community amenities. In this case, there may be a marked difference between experience in North America and Europe. In the United States, manufacturing activities whose products can stand high costs of transportation and require a highly skilled high-priced and fairly discriminating labour force are apt to locate in places where such a labour force would like to live. Of particular notice has been the movement of electronics and space industries to places with desirable climates (California and Florida, for example) and to cultural centres, especially those with prestigious universities. Thus a new trend is being set for this footloose type of industry that has a highly mobile labour force and a fair latitude of choice among production areas. In Europe as a whole it may be said that the manufacturer and his labour are apt to be less mobile. Language and cultural obstacles have restricted widespread movements of highly-skilled labour between countries. While it is true that the sunny Southern European countries of Portugal, Spain and Greece have achieved a fast growth rate in industrial production in recent times, they exert only a modest pull on the manufacturers located in the great industrial agglomerations of northern Europe. Within countries, however, there is a noticeable dispersion of the Head Offices of large business corporations and some Government agencies or departments to areas where the built environment and natural surroundings are attractive, living costs are lower than in the great metropolises, and—perhaps most important of all—office rents and rates are lower and communications with the rest of the country are good.

DISTRIBUTION COSTS AS A LOCATION FACTOR

Another major transportation cost component is the distribution of finished products to market. As transportation rates are usually more per tonne-km for finished products than for raw materials, these costs can be considerable. In fact, if there is no reduction in weight between the raw-material inputs and the finished-product outputs, other things being equal, the plant will invariably be best located at the market.

Market location is also critical where perishability is involved. Examples of highly perishable products include baked goods, confectionery, and newspapers. This factor thus dictates that the manufacturing outlet be close in time, if not in space, to the consumption outlets. On the other hand, as is the case with many agricultural commodities, sometimes the raw-material inputs are more perishable than the finished product. In this case, the plant will be raw-material-oriented in order to preserve the longevity of the product through freezing, packing, dehydrating, canning, or the like.

These relationships are shown in Figure 11–2. Note that inasmuch as raw-material transportation costs (PC) are lower than finished product costs (DC), the lowest aggregate transportation costs (TC) will be at the market location. (Costs are less than four at market versus more than five at raw material location.) However, in places where a break-of-bulk operation is necessary in the shipment process, an intermediate location between raw-material source and market may occur in order to save the costs of such an operation. Inasmuch as raw-material transportation costs have been decreasing

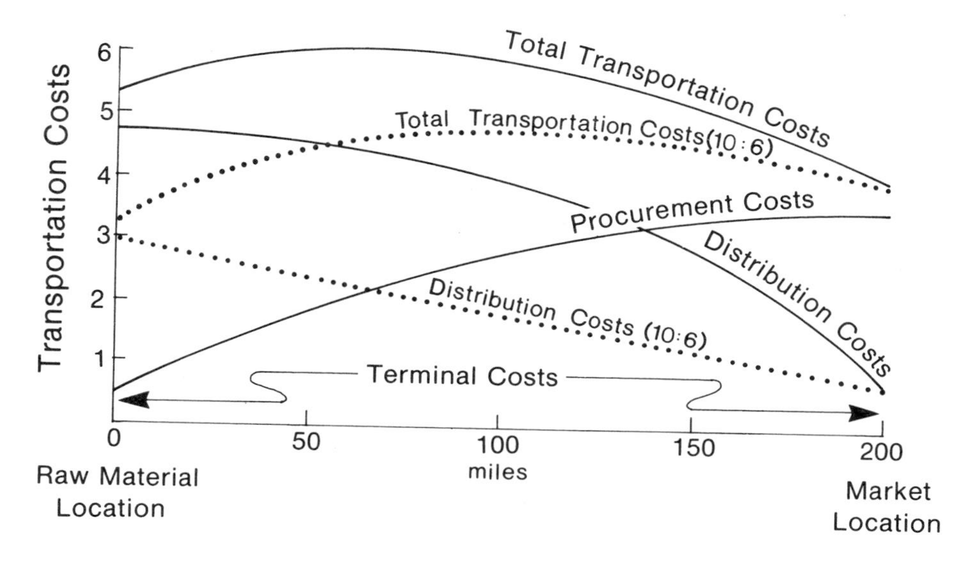

Figure 11–2 Procurement costs versus distribution costs to markets as industrial locational determinants. Other things being equal, the manufacturing plant will locate where total transportation costs are the least. The dashed lines show a changed weight-loss (10:6) and its effect on the best plant location.

faster than finished-product costs over the past several decades, this also makes a market location more advantageous for many activities. It is particularly critical where procurement and distribution costs are not too different between the raw-material and market locations.

But most manufacturing processes generally result in a reduction in bulk between the raw-material and finished-product stage. This can be calculated as a weight-loss ratio. Products that lose more than 30 per cent of volume between input and output stage might be labelled as having high weight-loss ratios, in which case the location of the plant, other things being equal, will be at the source of raw materials. The general formula for computation is to divide the finished-product volume by raw-material volume and then subtract the quotient from 100 per cent. Thus, if it takes 10 units of raw materials to produce 6 units of finished product (10: 6), the weight-loss ratio would be 40 ($100-6/10 = 40$).

Now, note that with a raw-material to finished-product ratio of 10: 6 (dashed lines), the total transportation costs (TC) are lowest at the raw-material location. The total costs remain about the same at the market source but are considerably lowered at the raw-material source. Thus the effect of weight reduction is to favour a raw-material location. Note also that the highest cost point has shifted closer to the market. Nonetheless, the least profitable point in this example, that is, the place of highest total transportation costs, is still intermediate between raw material and markets.

TERRITORIAL IMPLICATIONS

It should be evident from the foregoing discussion that a change in any one of the costs of production may affect the location best suited to a given activity. Such changes might include the costs of transportation, a change in the mix of raw materials, or an increase or decrease in output. These, in turn, very much affect the territory from which a manufacturing plant might receive its raw materials and to which it might send its finished products.

The ability of any firm to compete for the marketing of its products in a territorial context is based on its general costs of production as these are accumulated at the place of consumption. As an example, let us assume the existence of three different plants, each producing the same product and competing in the same region. Let us also assume that, for reasons associated with the necessity of being near the source of raw materials, three plant locations are feasible. But each of the three plants has different processing costs because of different on-site costs such as labour, power, and the like. Because of these varying on-site production costs, they also have differing abilities to bid for customers over a given territory. As a result, the plants will be of different sizes and hence will achieve different scale economies. In turn, this will result in second-round effects whereby the more advantageously situated firm will obtain an even larger market.

This principle of territorial control and territorial progression is illustrated in Figure 11–3. First, assume that the three firms are raw-material-oriented and fixed on location,

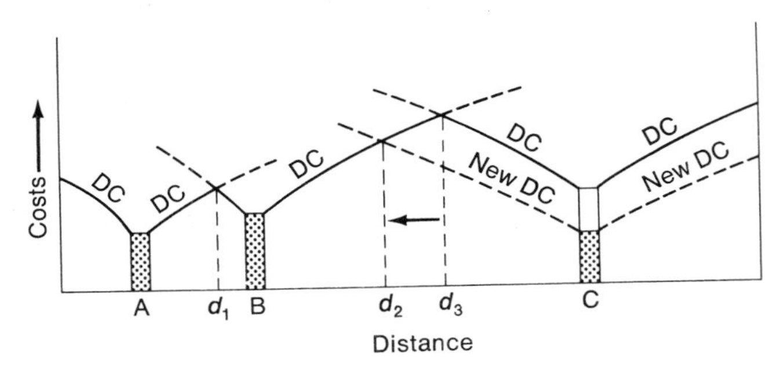

Figure 11–3 Effect of distribution costs (DC) from manufacturing plants, A, B, and C on territorial dominance. The vertical bars at each plant represent procurement costs (PC) and on-site costs. Should on-site production costs be lowered as shown at plant C, the new distribution cost curves (DC) would allow plant C to shift its zone of territorial dominance toward plant B from distance d_3 to distance d_2.

but that they have varying on-site production costs. Plant A actually has the lowest on-site costs, but the closeness of plant B prevents it from dominating much territory. Its limit of price advantage is at distance d_1. Plant C actually has the highest on-site costs, but because of its isolation from the other competitors, it dominates a much wider territory (from d_3 to the right side of the figure). This means that plant C will have a greater volume of output and hence can achieve greater economies of scale (see Chapter 4) than can either plant A or B. This advantage will enable plant C to reduce its on-site production costs considerably so that its costs are now lower than those of its nearest competitor, plant B. The spatial result is that plant C now has a price advantage in the area from d_2 to d_3—territory in which plant B formerly had the advantage.

However, it must be cautioned that because of the brand name features of most manufacturing, these plants are not homogeneous in terms of the consumer market. Some customers may prefer to buy from a particular company even though another company actually has a price advantage in the area. Thus even in those cities where a major plant of a large manufacturer is located, the local population will buy many products from other companies. Price is not a definitive measure of territorial control with regard to manufacturing. Nevertheless, it does indicate which firm has the advantage and which should, with equal competitive spirit and service, provide the product for the majority of customers in an area.

MINIMIZATION OF PRODUCTION COSTS

As might be anticipated, there are a number of partial solutions to the manufacturing location problem. Most such 'solutions', however, are based on keeping several components fixed and given. Basically, the problem can be dissected in three different ways: (1) finding the point of minimum transportation costs for the procurement of raw materials; (2) finding the minimum transportation costs for distribution of the product to markets; and (3) finding that location at which on-site production costs are minimal. In the overall sense, the best location for a given plant is the one that minimizes all these factors.

However, it is soon apparent that such costs can be minimized by merely producing nothing. Therefore, when we talk about minimizing production costs in terms of a locational optimum, we really mean minimizing such costs, given certain levels of output. Perhaps an even better way to state the problem is to find that location that will maximize profits.

When the problem is put in this latter framework, all kinds of factors emerge that are not covered by the theories discussed above. Such variables include notions about plant size, growth potential, long- versus short-run profits, competitive strategies and pricing systems, all of which will, in turn, increase output and perhaps allow greater scale economies. These factors are complex indeed and will be covered mostly by examples in forthcoming chapters.

A SIMPLIFIED, TRANSPORTATION-BASED LOCATION MODEL

The nature of the transportation aspect of the manufacturing location problem is perhaps most easily seen by examining a hypothetical situation involving one market and two raw materials. (The case involving one market and one raw material was just covered.) From this slightly more realistic vantage point, several major axioms can be developed.

Figure 11–4 shows such a situation. This is modelled after the famous locational triangle solution of Alfred Weber, a German economist who in 1909 formulated locational principles for industry along the lines developed by Von Thünen for agriculture at a much earlier period. (Alfred Weber, *Theory of the Location of Industries*, trans. Carl J. Friedrich, Chicago: University of Chicago Press, 1929.) Like Von Thünen, he had to single out certain major elements for attention, and eliminate complications by making simplifying assumptions. He postulated a uniform country, a good or goods of uniform quality, and known positions of sources of raw materials and points of consumption.

Labour was geographically fixed but freely available and with wages predetermined; and transportation costs were considered to be a function of weight and distance. Like Von Thünen, Weber acknowledged that his simplifying assumptions and the analysis which stemmed from them would not explain the industrial locations of the real world. His basic aim was to examine the effect of transportation upon location in a model setting, and in so doing to clarify underlying issues.

In the location triangle (Figure 11–4), given are two raw materials, RM1 and RM2, and one market. In this example, all three are fixed points 200 miles apart. Quite simply, the question is: where is the best place to locate an industrial plant, given these fixed points and the various distance-cost relationships?

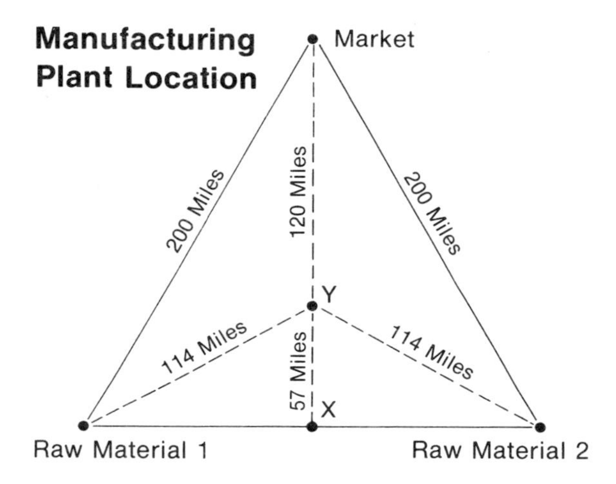

Figure 11–4 A locational triangle for manufacturing plant location given one market and two raw-material sources. The minimum cost solution depends not only on mileages involved but upon weight-loss.

First, if the raw materials are pure—that is, if there is no weight loss incurred between raw materials and finished product—the plant would be best located at the market at a 'transport cost' of 400 unit miles. However, if the weight-loss ratio is 0.50 —that is, if two units of raw materials are required to make one unit of finished product—then the location that will use fewer distance units can be found. The point that minimizes raw material transfer costs is located halfway between the two raw materials at distance X. If the plant is situated here, then the raw material distance units would amount to only 200 distance units (for example miles) and the finished product units would amount to some 177 distance units, or some 377 distance units in all. However, some quick students of trigonometry might have discovered that location Y is the best place inasmuch as it is the centre point between all places. Some calculation shows that total distance units of only 348 (114 distance units for each raw material and 120 distance units for the finished product to market) are needed to serve this location.

As another variation, let us assume that different raw-material inputs are required for production. Let us further assume that the weight-loss ratio remains the same (0.50), but that it takes one and one-half units of RM1 and only one half-unit of RM2 to produce one unit of finished product. In this case, it is seen that the best place to locate the plant would not be at M or X or Y, but at the source of RM1. Here, the total distance units would be only 300. Thus where raw-material mix is involved, the best location is strongly pulled toward the most dominant raw-material source.

This example could be greatly expanded, but it should now be evident that when we say 'raw-material-oriented', we are describing a multifaceted problem. In fact, solution of actual problems of this type require high speed computers. Weber set up what he called a 'locational figure' (the Varignon frame) to demonstrate where an industry would be located under some conditions (see Bibliography).

A MORE REALISTIC MODEL

While we remain still inside Weber's model world, we can apply the same logic to an even more realistic situation (Figure 11–5). In this case we shall keep the schema of two raw-material locations and one market but introduce variable transportation costs among raw materials and finished products, different costs between water and land,

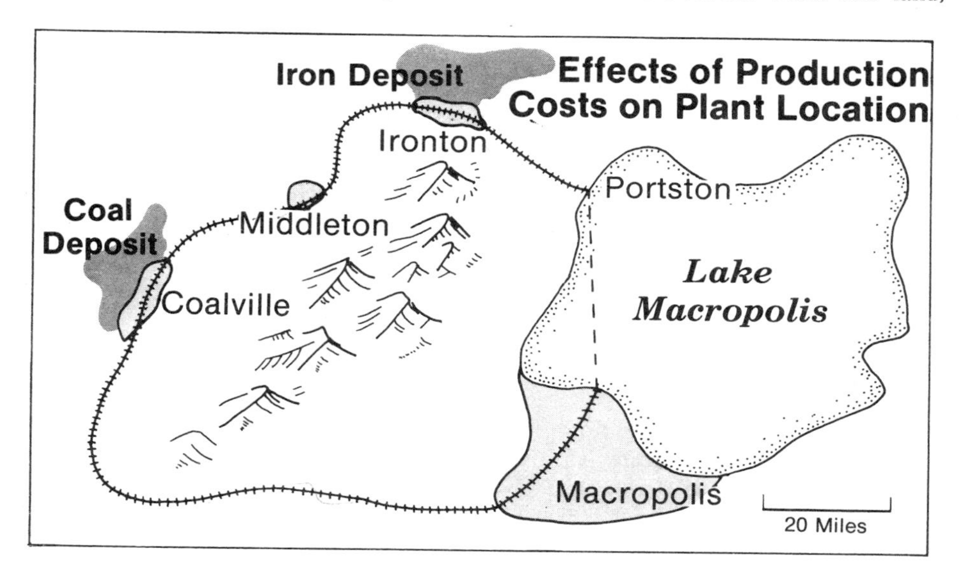

Figure 11–5 Effects of production costs on plant location. Assume 20 km between Ironton, Middleton, Coalville, and Portston, 30 km between Portston and Macropolis, and 90 km by rail between Coalville and Macropolis.

and break-of-bulk considerations. Assume that Coalville, Middleton, Ironton, and Portston are all 20 km apart, that Portston is 30 km to Macropolis, and that the southern route from Coalville to Macropolis is 95 km. To keep the mental calculations simple, we shall assume that the weight-loss ratio is still 0.50, but that it takes $1\frac{1}{2}$ units of RM2 and $\frac{1}{2}$ unit of RM1 to produce 1 unit of finished product (FP). We shall further refine this by introducing actual transportation cost figures:

Transportation costs on land:	RM1	£2/unit km
	RM2	£3/unit km
	FP	£5/unit km
Transportation costs on water:	RM1	£2/unit km
	RM2	£2/unit km
	FP	£3/unit km
Break-of-bulk costs at Portston:	RM1	£10/unit
	RM2	£10/unit
	FP	£15/unit

At this stage, the location problem may seem rather complicated. However, using the principles discussed here, calculations can soon be made. The important technique, as with most geography problems, is to focus on the map (Figure 11–5).

Inasmuch as the total market is at Macropolis, the location problem can be approached in a simple and logical way. The transportation cost for finished products from a plant at Coalville amounts to £475 (1 unit at £5 per unit for 95 km) via the southern land route to Macropolis, whereas such cost for shipping the finished product would be only £205 (1 unit at £5 per unit for 20 km, plus £15 break-of-bulk at Portston, plus 1 unit at £3 per unit for 30 km) from a plant at Ironton via the northern route. Moreover, Coalville would not be chosen because the heavy mix of raw material (RM2) is at Ironton. Ironton is thus far favoured. For this same reason, the plant would not be located at the intermediate point of Middleton, which is ruled out because there is no way to find a shorter distance to Macropolis from Middleton than either Coalville or Ironton. Thus because of the transportation route system, there are no possibilities for intermediate strategies, which was the case with the more abstract example shown in Figure 11–4. The best location initially, from the standpoint of raw material mix requirements and general proximity to market, would thus be at Ironton.

However, the best location from a market standpoint would be at Macropolis inasmuch as the market is exclusively at that location. The second major question therefore is: which is cheaper—a plant at Ironton or a plant at Macropolis? By a little quick calculation it is found that Ironton has some £45 advantage (£245 versus £290 for a plant at Macropolis).[1]

But before too hastily concluding that the best plant location is at Ironton, take note of a possible intermediate point between Ironton and Macropolis, Portston. Portston, located between the major raw-material-oriented production site at Ironton and the exclusive market-oriented site at Macropolis, is in a position to reduce costs even further because of the savings in break-of-bulk charges. A plant at Portston would save all break-of-bulk charges inasmuch as raw materials could come in the 'back door' of the plant by land and go out the 'front door' of the plant by water. Hence if such savings in break-of-bulk amount to any less than a plant located at Ironton, Portston would be the least costly location. This is clearly the case inasmuch as such break-of-bulk costs on raw materials amount to some £20 (£15 from RM2 and £5 from RM1). A plant located at Portston would also save transportation costs in water transportation inasmuch as 1 unit of FP would be shipped 30 km rather than 2 units of raw materials—that is, 1×30 km × £3 = £90 versus 2×30×£2 = £120. This gives an additional cost advantage at Portston of £30. Total savings thus amount to £50. Hence the plant at Portston will operate at only £240 as compared with £245 for one at Ironton and £290 for one at Macropolis.

BASIC WEBERIAN ANALYSIS AND ITS DEVELOPMENT

In the first part of this chapter we introduced many of the several economic circumstances which affect the choice of location in very varied conditions and, in the ex-

[1] The costs are calculated as follows: First, transport ½ unit of RM1 to Macropolis. The cost is £95 by southern route (0.5×95 km × £2), or £60 by northern route (0.5×60 km × £2) plus £5 for break-of-bulk at Portston, and (0.5×30 km × £2) or £30 for water transportation. The cost of shipping 1½ units of RM2 to Macropolis via the northern route (the cheaper route) is £195 (£90 land transportation, £15 break-of-bulk at Portston, and £90 for water transportation), making a total of £290. For the plant at Ironton, move ½ unit of RM1 to Ironton at a cost of £40 (0.5×40 km × £2), then ship 1 unit of finished product to market by the northern route at a cost of £205: (1×20 km × £5), £100 for land transportation, plus break-of-bulk at Portston at a cost of £15 and water transportation amounting to £90 (1×30 km × £3).

amples used immediately above, we have demonstrated the simplest forms of Weberian transportation cost analysis. The gulf between the real world and Weber's analysis may seem considerable. Yet we must remind ourselves that Weber was employing a model within which certain principles could be demonstrated. Given his basic assumptions, he was able to determine where manufacturing or processing should take place in that model.

Broadly, we can see that the concepts so far discussed do provide a framework by which the various locational factors for industry can be identified and particular spatial patterns can be anticipated. In summary, two types were emphasized: raw-material-oriented and market-oriented. Raw-material-oriented firms were characterized as those whose products (1) have a high weight-loss ratio, (2) contain a highly perishable raw material, (3) are less perishable than the raw material, or (4) are less bulky or breakable than the raw material. By contrast, firms whose products show the opposite characteristics were commonly market-oriented. However, as will be demonstrated, these abstract concepts become difficult to apply to any particular manufacturing type. In fact, most types of manufacturing are likely to have several spatial patterns: some plants will be located near their raw material, some at the market, and others at various intermediate locations. Thus an examination of the distribution of any particular industry leaves one with the feeling that something is wrong; the pattern is not pure, but a mixture of various locational positions.

Part of our problem, of course, relates to what we decide to 'feed-in' to our analysis. In trying to answer the question: 'What causes an industry to move from place to place?' Weber relaxed his assumption of fixed and uniform labour costs. Incorporating different labour costs from place to place enabled him to plot new spatial variations away from the optimum (least cost) transportation location. And just as Weber relaxed this and other assumptions, so may we with the exercise above related to Figure 11–5. The reader may wish, for example, to specify different labour costs at various production points as a reflection of different on-site processing costs. If he or she wishes to create a more realistic problem it is also possible to add other raw materials, other markets, and new transportation rates. Still another problem remains, however, if we are aware that like industries commonly do gather together. Weber himself, having conducted the experiment of combining the effects of transportation and differing labour costs, turned to the problem of how further 'deviations' in location may be caused by the tendency of firms to agglomerate. He saw that a firm might increase the concentration of production by enlargement at its chosen location, thereby obtaining savings by a larger scale of operations. Furthermore, he observed how it might benefit by selection of a location alongside other firms, thereby gaining benefits from shared services, a larger, trained pool of labour and large-scale purchasing and marketing.

Such circumstances were present in Weber's Europe in the early years of this century, and his detailed model analyses clearly provided important insights into the industrialist's locational behaviour. It is not surprising, then, that his work inspired a whole 'Weberian School' (especially after its translation into English in 1929). His basic ideas have been successively updated, developed and embellished by many scholars (including Hoover, 1948, Isard, 1956, and Smith, 1966; see Bibliography), while full acknowledgement of his role as pioneer in the study of industrial location theory is made in many textbooks. Most recent students tend to criticize the incompleteness rather than the inadequacy of what is now known as 'traditional industrial location theory'—incomplete because it is centred on transportation cost analysis in a world which offers new complexities to unravel in a period of great innovation and change. Other influences are brought to bear and the location pattern of any function is developed over a period of time.

As a result, some firms are found in places that are no longer optimum while others are located in areas that hold promise for the future. Tabulation of the majority of firms in a particular locational situation may merely reflect the major growth period in that industry. At any rate, there is always a lag effect inasmuch as firms often cannot make an immediate and detailed response to new or short-term locational opportunities. As opportunities for manufacturing continually change, there is always some residual effect whereby some industries remain in suboptimal locations. Once investment is made in a particular plant, this plant will often remain in operation long after its full usefulness and perhaps even its geographic rationale have departed. In fact, the investment in buildings and other nontransferable items may be so great that it prevents any spatial response whatsoever, even though another location may now be optimum. Any given distribution will usually contain plants developed at different times under different technological, monetary, and competitive conditions. Thus, for these and many other reasons, as we turn to patterns of industrial location in Chapter 12, we shall be able to examine some industries which appear to behave locationally much as Weber predicted and others that appear to react locationally in response to other influences.

BIBLIOGRAPHY

Alexander, J. W. 'Location of manufacturing: methods of measurement'. *Annals of the Association of American Geographers*, March 1958, pp. 20–26.

Alexandersson, G. *Geography of Manufacturing*. Englewood Cliffs, N.J.: Prentice-Hall, 1967.

Chisholm, M. *Geography and Economics*. London: Bell, 1966.

Estall, R. C., and Buchanan, R. O. *Industrial Activity and Economic Geography*. London: Hutchinson, 3rd ed., 1973.

Fulton, M., and Hoch, L. C. 'Transportation factors affecting location decisions'. *Economic Geography*, 35, 1959, pp. 51–59.

Hamilton, F. E. I. 'Models of industrial location'. In *Models in Geography*. Ed. Chorley, R. J., and Haggett, P. London: Methuen, 1967, pp. 361–424.

Harris, C. D. 'The market as a factor in the location of industry in the United States'. *Annals of the American Association of Geographers*, December 1954, pp. 315–348.

Hoover, E. M. *The Location of Economic Activity*. New York: McGraw-Hill, 1948.

Isard, W. *Location and the Space Economy*. Cambridge, Mass: M.I.T. Press, 1956.

Karaska, G. J., and Bramhall, D. F. *Locational Analysis for Manufacturing: A Selection of Readings*. Cambridge, Mass: M.I.T. Press, 1969.

Krumme, G. 'Towards a geography of enterprise'. *Economic Geography*, 45, 1969, pp. 30–40.

Lösch, A. *The Economics of Location*. New Haven, Conn: Yale University Press, 1954.

Marcus, M. 'Agglomeration economies: a suggested approach'. *Land Economies*, 41, 1965, pp. 279–284.

Miller, E. W. *A Geography of Industrial Location*. Dubuque, Ia.: W. C. Brown, 1970.

Moses, L. N. 'Location and the theory of production'. *Quarterly Journal of Economics*, 72, 1958, pp. 259–272.

Penrose, E. *The Theory of the Growth of the Firm*. Oxford: Basil Blackwell, 1959.

Pred, A. 'Behaviour and location. Pt. I'. *Lund Studies in Geography*, Series B. 27, 1967.

Rawstron, E. M. 'Three principles of industrial location'. *Transactions of the Institute of British Geographers*, 25, 1958, pp. 35–142.

Riley, R. C. *Industrial Geography*. London: Chatto & Windus, 1973.

Smith, D. M. 'A theoretical framework for geographical studies of industrial location'. *Economic Geography*, April 1966, pp. 95–113.

Smith, W. 'The location of industry'. *Transactions of the Institute of British Geographers*, 21, 1955, pp. 1–18.

Tornquist, G. 'Transport costs as a location factor for manufacturing industry'. *Lund Studies in Geography*, Series C, 1962.

Townroe, P. M. 'Locational choice and the individual firm'. *Regional Studies*, 3, 1969, pp. 15–24.

Weber, A. *Theory of the Location of Industries*, trans. Friedrich, C. J., Chicago: University of Chicago Press, 1929.

Wolpert, J. 'The decision process in spatial context'. *Annals of the Association of American Geographers*, 54, 1964, pp. 537–558.

Zimmerman, E. W. *World Resources and Industry*. New York: Harper & Row, 1951.

12

Industrial Location Patterns

The substitutions that occur among the factors of production (that is, procurement, processing, and distribution costs) make the analysis of any given type of manufacturing difficult. Detailed information as to the nature of the firm is generally required in order to make any good assessment of location decisions. Even with detailed data and intimate knowledge of the firm in question, the task is arduous and the results are not always correct. Peculiarities of regions, for example, their distribution of raw materials and transportation routes, cause deviations to occur among the factors of production even for identical firms located in different areas. Hence good territorial knowledge, as well as manufacturing-firm knowledge, is a requirement for sound understanding of locational variables. Of utmost importance, however, is the identification of the cost characteristics in the more recent locations so that trends can be ascertained. Fortunately, an approximation of the importance of the various factors of production can be made by careful analysis of the various patterns. The focus of this chapter is to make the nature of such pattern interpretation explicit.

Another major pitfall in interpretation of the factor of production costs results from the fact that one cannot merely add up the components of each of the three major production costs (that is, procurement, processing, and distribution) to ascertain which of the three is most critical in determining location. In general, if a firm is located near raw materials, then the raw procurement costs may be very low relative to finished product costs. Thus one might erroneously conclude that such a firm is there because of otherwise heavy raw-material transfer costs. But nontransfer cost considerations may be the dominating factor. In fact, on-site costs (labour, taxes, power, water, and other processing features) may outweigh all transfer cost considerations. Nonetheless, such firms will still try to use as little transportation as possible and will seek out strategic places, where transportation costs are minimal, whenever possible. On the other hand, a particular industry may be located at the market, but its location there may have little or nothing to do with distribution cost considerations. Rather, the determining factor may be that such a market area is the key labour area for that industry. Thus, unlike many raw-material-oriented and market-oriented patterns, industries dominated by such processing costs often do not reveal the real reasons for their location.

A further contaminating agent is differences in institutions and economic social systems from place to place. This is especially the case for those industries that are distributed, as in the Soviet Union, on the basis of variations in on-site features. Labour can be moved easily to particular localities. In the United States, taxes are often made a major inducement or deterrent. Differences in power costs, a function often controlled in the public sector, can also cause decided distortions in some industrial location patterns. Inasmuch as on-site costs are the prime locational determinant for many industries, power cost differentials make for considerable variation in industrial pattern for a single industry type. Even so, for many types of industry the locational components discussed above are so strong that institutional considerations cannot easily override them. Thus, when we look at an industry that is typically raw-material-oriented or market-oriented on a global scale, we will note little basic difference in pattern. Communism, socialism, or capitalism, do not have a major effect upon the location of types of industries tightly controlled by transfer cost considerations.

Finally, the models in Chapter 11 are somewhat limited in that they are based on the assumption that each firm's location is an independent entity. In fact, this is rarely the case. Branch plants of major corporations are the rule today. Thus the viability of any particular location can often be fully understood only in light of the total distribution of plants in any particular corporate structure. This is why the later part of this chapter is devoted to the multi-plant firm and its location decision-making, and why finally we examine two world-important industries which are dominated by a relatively small number of giant 'multinational corporations'.

MARKET-ORIENTED OR RAW-MATERIAL-ORIENTED?

A simple but operationally meaningful way to understand better the locational characteristics of a particular type of manufacturing is to study its broad regional pattern. Despite certain pitfalls caused by peculiarities of time and place, the examination of a particular industry's pattern of location is often highly fruitful and yields accurate data. Admittedly, it would be helpful also to have detailed cost information for the various components of production, but such information is often difficult to obtain, even in countries like Britain and the United States.

In the first part of the chapter, two basic pattern types will be examined in some detail: (1) market-oriented patterns, and (2) raw-material-oriented patterns. In most examinations of industrial location, the distribution seen on a map provides the primary input for locational understanding. The following discussion, therefore, should considerably aid such pattern diagnosis in two ways. First, it will air the positive identification of patterns by pointing out the critical features that need to be identified in any particular pattern before any conclusions are reached. Second, it will aid (in a somewhat negative manner) by demonstrating certain common fallacies in pattern analysis. Nonetheless, these operational devices are by no means foolproof. Only after long and patient studies of the trial-and-error type and great knowledge of particular activities will students become comfortable and sure in their diagnoses.

Market-oriented industries commonly include both ubiquitous and city-serving firms. A ubiquitous firm is not one that is found everywhere but one engaged in an activity that is found in about the same distribution pattern as the market, which in most cases means the urban population distribution. Thus a map of Europe or the world that shows the distribution of cities by population size and a map that shows the distribution of a market-oriented activity, such as soft drink bottling, would look very similar. Nevertheless, ubiquitous firms are not found in any city of a particular size until the threshold for a suitable market has been reached, much as is the case with retailing establishments. To support a viable plant, some manufacturing firms, such as soft drink bottling plants, need a city of only a few thousand people, whereas a brewery requires a city with a much greater population. Part of the reason is that not everyone drinks beer, whereas people of almost all ages consume soft drinks. These city-serving manufacturing types generally include printing and publishing, construction, and many food industries.

In contrast, raw-material-oriented industries are usually sporadically distributed and are called *city-forming*. Sporadically located manufacturing firms are simply located unevenly with respect to the market. In most instances, they are raw-material-oriented. A major difference between city-serving and city-forming industries is that the product of sporadically located firms is usually sold to consumers outside the urban area in which they are found. Thus the city-forming types are important for urban growth. Workers in city-forming industrial firms are hence called *basic* workers, whereas workers in the ubiquitous pursuits are called *service*, or *nonbasic*, workers. Inasmuch as a good deal of industry is of the sporadic type, it is not difficult to see why chambers of commerce and similar organizations the world over try to bring in more industry. It is postulated that the workers in such industries indirectly support other people in the community and hence create new jobs and new population growth for the urban centres. This idea is a crucial part of the *economic base concept* and hence of theories concerning the growth of cities.

Another dual classification in use distinguishes between durable and nondurable manufacturing types. Durable manufacturing includes most of the sporadic industries such as iron and steel, the metal industries, machinery, chemicals, and mining. Strangely, even the industries with such truly temporary products as apparel, paper, tobacco, fur, leather, and textiles are also classified as durable manufacturing types. The nondurable types are mostly market-oriented and include printing and publishing, construction, and most kinds of food processing such as meat, bakeries, soft drink bottling, and the manufacture of dairy products. Nevertheless, there are a number of exceptions, to this generalization inasmuch as such nondurable manufacturing types like meat packing, flour milling, and canning are raw-material-oriented and are hence sporadically distributed in conformance with their raw materials.

The major industrial categories are associated with urban or rural areas. For example, manufacturing types least associated with large urban places include mining, textiles, and furniture making, These are characteristically found in small and isolated areas and are clearly raw-material- or resource-oriented. By contrast, several manufacturing types like primary metals, fabricated metals, machinery, motor vehicles, apparel, and printing have a higher percentage of their work force in urban places than does the hallmark of urban employment itself, retail sales. In fact, a few industries of this class are highly concentrated in the largest metropolitan areas; these include particularly metals, machinery, apparel, printing, and chemicals.

MARKET-ORIENTED PATTERNS

Overall, market locations are becoming more common in industrial location patterns. This occurs primarily because considerable improvement has been made in the movement of raw materials, but relatively few improvements have been made in the handling of finished products, containerization being an exception. Because of bulk shipments and special handling procedures, the costs of moving raw materials have been greatly reduced relative to finished products over the past several decades.

A market-oriented manufacturing pattern is to be expected if one or more of the following conditions are met: (1) The raw material inputs and finished product outputs are similar in bulk and weight. This is the case because raw materials are generally easier and less costly to transport than finished products (see Chapter 6). (2) The finished product is more bulky, more perishable, or more fragile than the raw material.

Such manufacturing types include most kinds of bakery items, confectionery, and glass-making. (3) The product is augmented with a rather commonly found raw material during the manufacturing process. One common example is water, used, for example, in making beer and soft drinks.

A market location is particularly prominent in the manufacture of products that actually gain weight as processing occurs, in beer production and soft drink bottling, for example, water is added to the imported raw material. Inasmuch as water is a commonly available commodity, it makes little sense to pay transportation-to-market charges on it in the finished product. However, where water of a different quality is important in the manufacture of the product, as is claimed in the case of the beer brand that advertises, 'It's the water', the weight-gain factor may exert considerably less influence in favour of choosing a market location than it ordinarily would.

Products that become breakable or more bulky in the manufacturing process are likewise often made near markets. Examples include dishes and glasses, pots and pans, metal fabrication, furniture manufacturing and the assembly (construction) of machinery. However, because of other locational factors, not all of these activities are found at the market. For example, transportation costs to markets of these more bulky items can be reduced through 'breakdown' shipment. Thus furniture is shipped in a breakdown fashion and simply reassembled at the site of retail purchase. In this case, one might say that the manufacturing process is bifurcated: finished elements are made at one site and assembled at another. This is increasingly the procedure used in the manufacture of automobiles, whereby many regional assembly plants have developed.

Nonetheless, two types are most prominent: (1) type A, in which distribution coincides almost perfectly with the general urban population, and (2) type B, in which only higher order centres contain the firm, or in which a point of minimum aggregate transportation costs for the finished product is the chosen location. In this latter case, the plant may be located in a small community or even in rural territory, yet its main locational force is still proximity to markets.

The type A market-oriented pattern, one which coincides almost perfectly with the distribution of urban centres, is shown diagrammatically in Figure 12–1. Thus a map

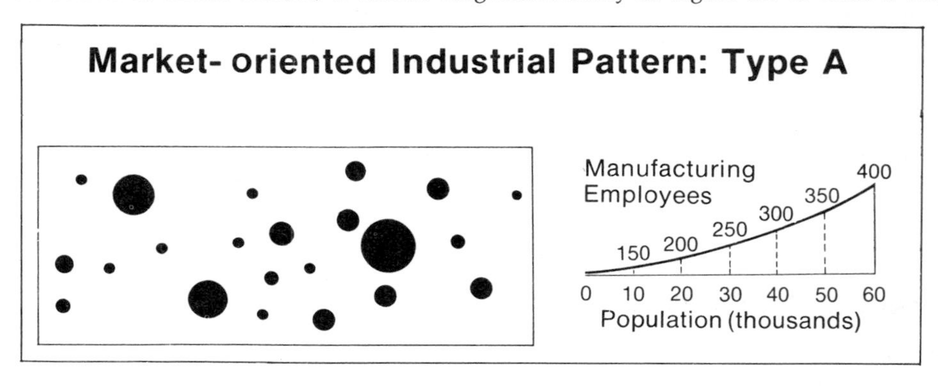

Figure 12–1 A market-oriented industrial pattern, type A. Note that all markets contain industrial employment in conformance with their relative populations.

showing the relative size of urban centres is almost identical with one showing the general extent of this manufacturing activity in those centres.

Various quantitative measures of deviation from the base population distribution can be developed. One method is shown in Figure 12–2. City population is plotted along the x-axis and some measure of manufacturing importance in that city is plotted along the y-axis. Thus, given the population of city X and the number of employees in manufacturing type Y, a direct relationship is obtained if this kind of market location pattern is prominent.

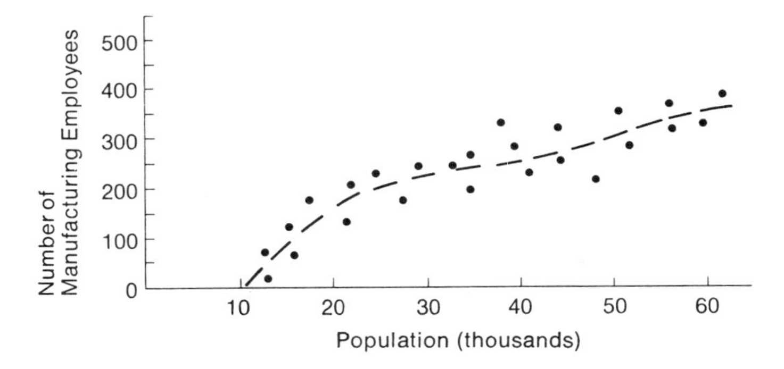

Figure 12–2 Graphic representation of a type A industrial location pattern (see Figure 12–1). Note that there is a direct, albeit not linear, relationship between the amount of population a place contains and the number of employees in manufacturing.

In the example shown, note that there is a population below which a firm cannot operate because of market-size limitations; thus cities below 10 000 population will have no manufacturing of this type. Such centres will then be supplied with the product from larger nearby centres, causing the latter to have a larger number of employees than their population would indicate.

Note also that there is rarely an equal increase in industrial employment with increasing population. For example, in this case a city with 20 000 population is expected to have 200 employees in a particular market-oriented industry. But a city twice as large, at 40 000 population, is not expected to have 400 employees (twice as many as the centre half as large) but only about 300 employees. The reason for this exponential relationship relates to economies of scale as well as other on-site efficiencies.

Given a higher threshold level for various manufacturing activities, only the larger city or cities will contain the industrial function (Figure 12–3). Examples in the United

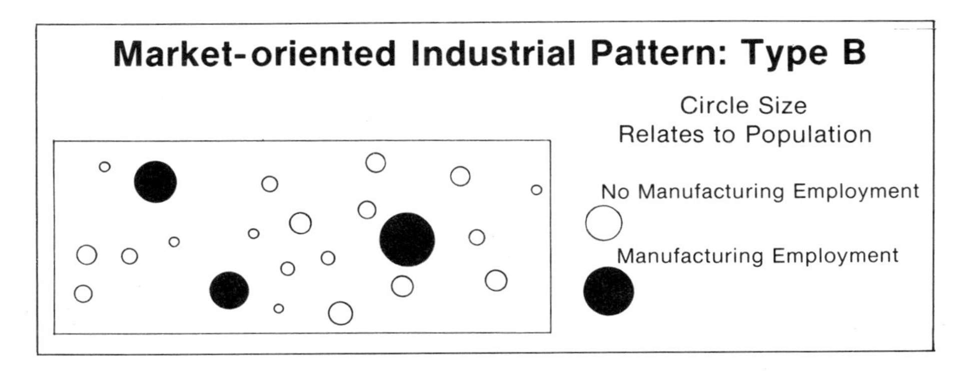

Figure 12–3 A market-oriented industrial pattern, type B—note that only the larger markets contain industrial employment.

States are regional capitals such as Denver, Atlanta, St Louis, Los Angeles, Seattle, Minneapolis-St. Paul, and Chicago. These regional centres develop because of scale requirements that enable only one plant or one location to best serve the entire area. In most cases, this plant will be found at the largest city because the latter will contain the largest single market. This large market centre also usually has the best access to all other centres in the plant's distribution area. Thus any function that is primarily restricted to New York or Chicago might qualify as a market-oriented activity.

This relationship is generally shown by the spatial covariation between the urban

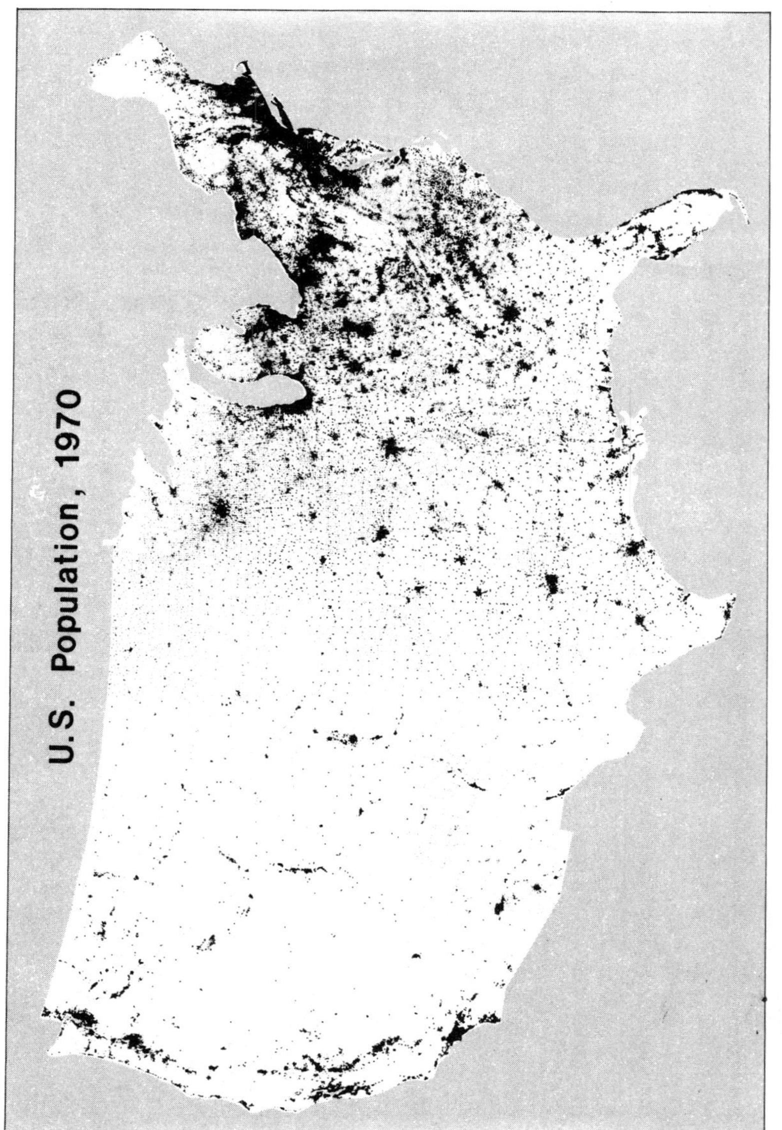

Figure 12–4 The population distribution of the United States. This distribution can be taken as the 'market' distribution for various patterns of manufacturing. A type A manufacturing pattern (see Figure 12–2) will appear to be identical to this one. In a type B pattern (see Figure 12–3) only larger cities will contain manufacturing employment. (From US Census.)

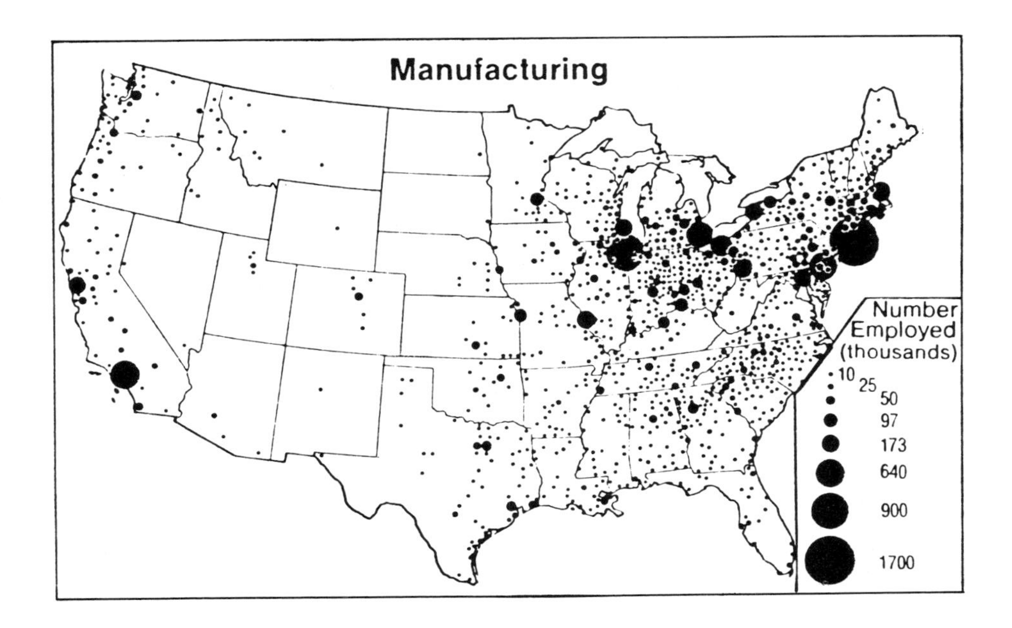

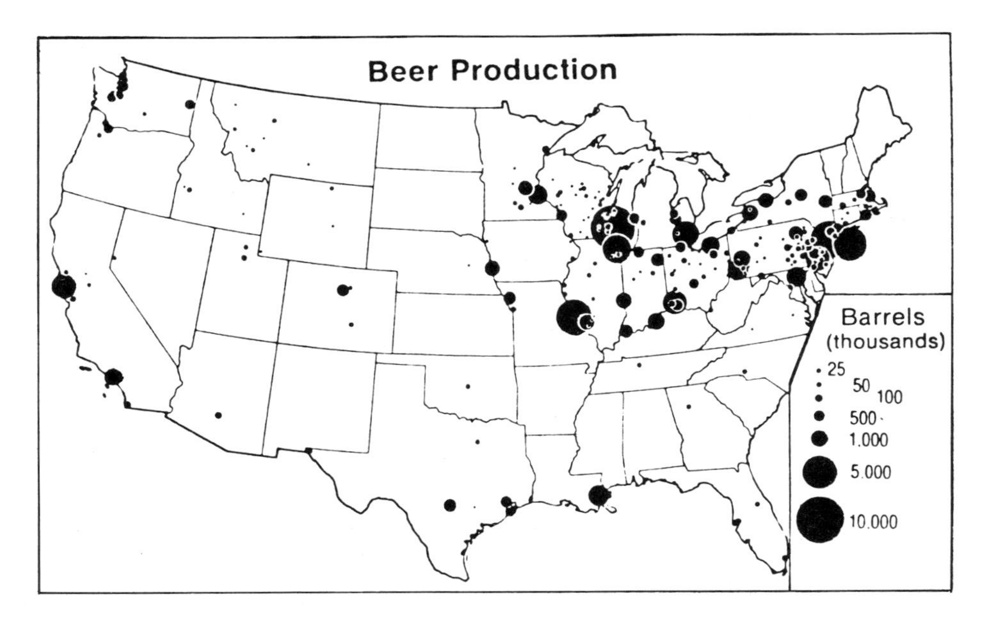

Figure 12–5 (a) Total manufacturing employment distribution in the United States: note the general conformance with Figure 12–4. A type A market pattern is suggested here. (b) Beer production by barrels: note the importance of the larger centres in any given region for such production. (c) Employment in printing: note the evidence of a type B market pattern. (From Alexander, J. W. *Economic Geography*, Englewood Cliffs, New Jersey: Prentice-Hall, 1963, p. 308, p. 330, p. 405. Reprinted with permission.)

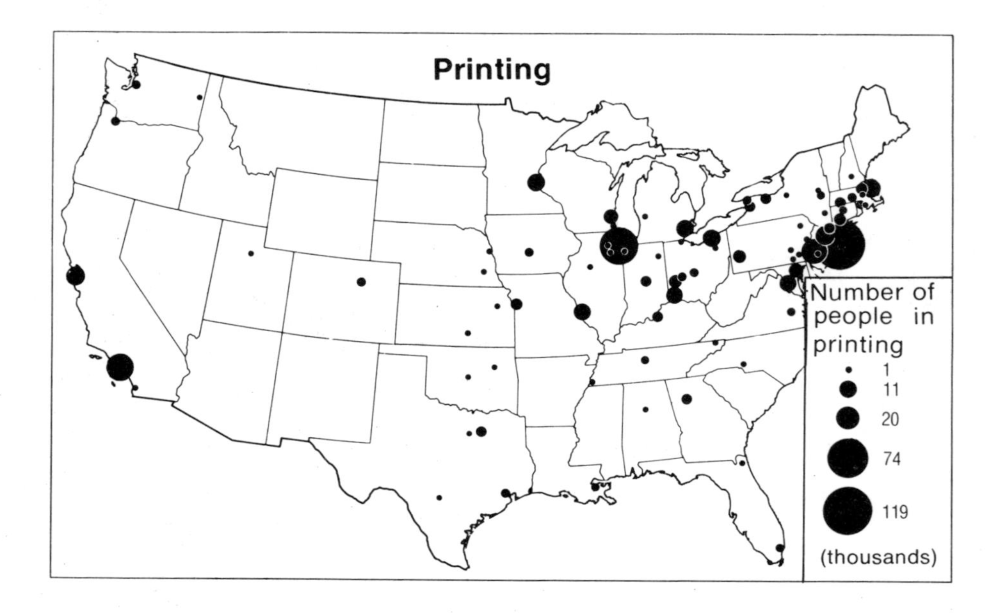

population distribution in the United States and the distribution of employees in all kinds of manufacturing (Figures 12–4 and 12–5a). The population distribution is almost identical in relative terms to the number of manufacturing workers in each centre. Note, as suggested, that the progression in the number of manufacturing employees does not increase in the same ratio as population size. Instead, it follows an exponential relationship similar to that shown in Figure 12–2. Note furthermore that this demonstrates the general market-oriented pattern of manufacturing as a whole. Thus the sporadic and raw-material-oriented pattern appears to be the exception to the rule. On the other hand, an analysis of each industry reveals a tendency for raw-material-oriented industries to set the initial settlement pattern of cities, a pattern which, in turn, is reinforced by market-oriented industries. From such a crude distribution it is difficult to answer the question of whether industry is the result of cities or cities are the result of industry. In fact, depending on the type of industry, both are correct.

As a method of pattern analysis, compare the circles representing manufacturing employment with those representing urban population (Figure 12–4). In this manner, both positive and negative deviations will become apparent and, once apparent, will suggest meaningful questions as to why such deviations occur. For example, note that there is an overabundance of manufacturing employees in relation to urban population in the industrial belt, but a relative scarcity of them in Florida. We wonder what kind of industry it is that is most prominent in the industrial belt; indeed, what kind of industrial employment is dominant in any one of the regions examined. Examination of the distribution of basic raw materials (for example, timber, vegetables, fruits, and mine products) gives us a first approximation of this answer. In order to gain good insights, however, it will be necessary to examine patterns of individual types of industry.

Three Levels of Type A Patterns. The pattern of beer production provides a good example of a type A market-oriented manufacturing pattern (Figure 12–5b). When this pattern is compared with the basic distribution of population, which is used here as a rough approximation of the market (Figure 12–4), it is at once apparent that while there is a general spatial coincidence, there are also some major exceptions. It is notable that some areas of the United States have almost no beer production. This is in some cases the result of state regulations that prohibit the making of beer. In other cases, perhaps the population is simply too small to support a brewery, a function that appears to require a fairly large population for entry. Nonetheless, there are many small towns in Wisconsin and New York where considerable production occurs. A common explanation for this is the large proportion of population of German stock with both a taste for beer and the know-how to produce it. Milwaukee is one of the outstanding centres of production, given its population. St Louis is also an extremely large producer. In the Rocky Mountain West, populations having breweries are smaller than are similar populations in other parts of the United States. Perhaps distance between centres is a major explanatory factor here. Nonetheless, the overall pattern is heavily market-oriented and is largely geared to major regional capitals, except for much of the South.

Figure 12–5c shows the distribution of employees in printing. The pattern is still a type A pattern, but the smaller centres have dropped off as compared with the overall pattern of beer production. Most metropolises having over 1 million population have over 1000 persons employed in printing—a good deal more than would be necessary merely to print newspapers. Nonetheless, there are some small cities, and a few large ones, that are outstanding relative to their population in this respect. Among these, New York is the paramount centre. New York and vicinity accounts for almost four-fifths of all the books published in the United States. Chicago and Boston are also outstanding relative to their population. In proportionate terms such medium-sized centres as Indianapolis, Indiana, Columbus, Ohio, Des Moines, Iowa, Rochester, New York, and Grand Rapids, Michigan, have much more printing employment than their populations, as compared with other centres, would indicate. In contrast, such cities as Phoenix, Arizona, New Orleans, Louisiana, and Miama, Florida, have comparatively little such employment. Thus, even though the pattern very nearly coincides with the market pattern, beyond a certain size market level there are many exceptions. One of the reasons for these exceptions is the mixed nature of the employment category. If it were limited to newspaper publication or to publication of certain types of books (for example, general, academic, or religious) a clearer market pattern might emerge.

The Type B Market Location. Type B patterns are far more difficult to evaluate (Figure 12–6). Here industry may be found in a small city or even in a rural area. Even so, if the location of such a firm is the result of an attempt to minimize transportation costs to market, its location is market-oriented. Thus, in the United States a central location is often found in Chicago or in other parts of Illinois, in Michigan, Indiana, Ohio, Missouri, or even Iowa. Transportation networks tend to make such locations highly

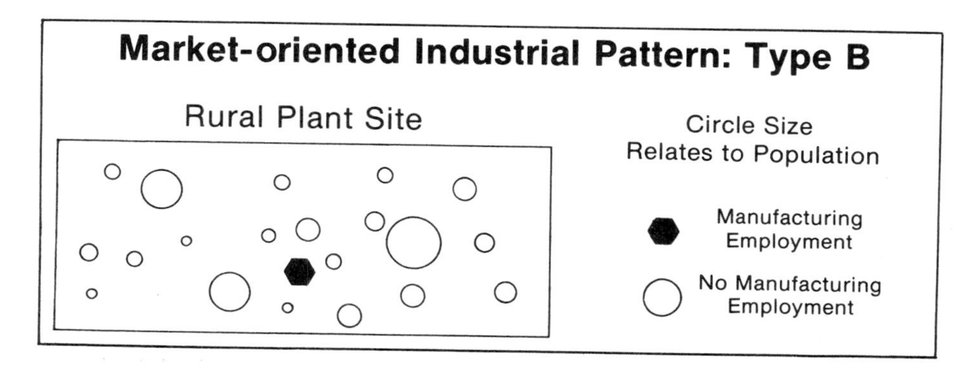

Figure 12–6 A market-oriented pattern, type B. Compare this pattern with the type B pattern shown in Figure 12–3. In this case the plant is still market-oriented inasmuch as it is centrally situated to serve all markets; however, it is not situated in any market but in a rural area. Certain on-site cost saving may be the cause.

advantageous with regard to distribution of a product to the entire population of the United States.

Despite the simplicity of its first appearance, a particular pattern requires careful assessment to determine whether it is market-oriented. This is especially the case with type B patterns. For example, vegetable canneries are raw-material-oriented. Nonetheless, container-making plants, which are found near canneries and which have a pattern almost identical to theirs, are market-oriented. They are market-oriented because they locate next to their market, the canneries. In determining whether a particular industry is market- or raw-material-oriented, specific information about the location and nature of its raw materials and about the nature and location of its customers is required.

Figure 12–7 shows this type of market-oriented pattern. In fact, two sub-patterns

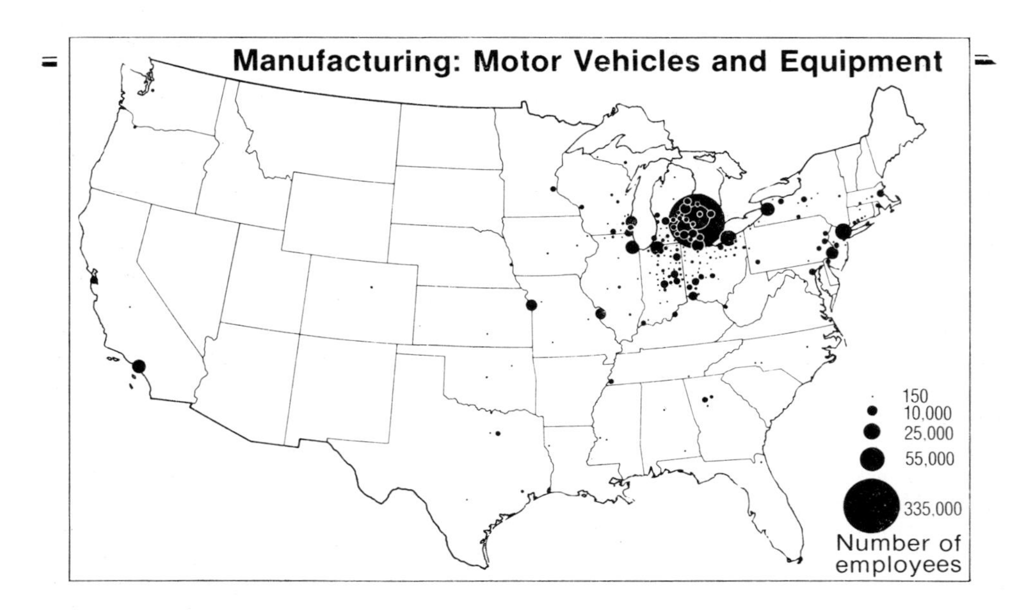

Figure 12–7 Employment in motor vehicles and equipment manufacturing. Note the tremendous concentration in Detroit and vicinity. Actually, considerable manufacturing occurs in very small towns and rural areas which are centrally situated in the American industrial belt. (From Alexander, J. W. *Economic Geography*, Englewood Cliffs, New Jersey: Prentice-Hall, 1963, p. 373. Reprinted with permission.)

are evident. In the first instance, the industry, concentrated in and around Detroit, is highly market-oriented and relegated to one general area of production, an area somewhat centrally situated within the total market area of the United States. The large number of employees in Ohio, Indiana, and Michigan may work in small towns or even in rural areas. Nonetheless, these firms, which produce motor vehicle equipment such as tyres, carburettors, and batteries, are market-oriented in that they are clustered about the major motor vehicle assembly plants. Recently, as markets have risen throughout the United States and population thresholds have been reached, branch assembly plants have been developed in major regional centres such as St Louis, Kansas City, Dallas, Los Angeles, San Francisco, and Atlanta. Thus a newly developing market-oriented pattern is being superimposed over an old market-oriented pattern. This example provides a useful introduction to the survey of the world-wide automotive industry given later in this chapter.

RAW-MATERIAL-ORIENTED PATTERNS

Just as generalizations can be made about the types of firms that are typically market-oriented, so can they also be applied profitably to types that are raw-material-oriented. Such firms are usually peculiar in one of several respects: (1) the raw material input may contain a major impurity so that there is a high weight loss between it and the finished product; (2) the raw material is generally perishable or subject to difficulties in handling compared with the finished product; or (3) the raw material sources are diversified or are highly scattered in their location. In this latter regard, an assembly problem is involved. Therefore, the activity will locate at strategic places whereby the costs of assembling the raw material might be minimized.

Nevertheless, there are several exceptions to these rules. One exception concerns the frozen food industry, an industry that generally occurs at or very near the source of raw materials. In such cases the finished product requires special shipment facilities (refrigeration); given proper handling, however, it is much less perishable than are its raw materials. Perhaps the critical factor to be considered is that without some preservation of the raw materials, the finished product could not satisfactorily reach the market at all. Thus foods are frozen at points accessible to farms for much the same reason that foods are canned there. Cattle slaughtering is another case in point. Historically, the finished product (fresh meat) was more perishable than the live raw materials. Therefore packing formerly occurred at markets, to which the live animals were brought. Today, with refrigeration, the slaughtering industry is moving closer to farm production areas, or most often, production occurs at intermediate loca ions at strategic break-of-bulk points on the side toward major markets.

Further distinction should be made at this stage as to what is meant by raw-material-oriented. For example, if the leather industry is parasitically tied to the meat-packing industry, it is raw-material-oriented, regardless of what pattern the meat-packing industry assumes. In this regard, locational understanding within the manufacturing sector is greatly aided when one thinks within a highly linked raw material and market system. Most manufacturing is bound within this chainlike system. Thus we might break down further the secondary manufacturing sector.

A useful framework for thinking about the many types of manufacturing within such a linked system is the connection between the primary sector and the tertiary sector. In this regard, particular attention should be given to the *primary manufacturing* firms as commodities move from the primary sector into the secondary sector. This path should be followed until the terminal portion is reached, that is, the point at which products are prepared for final markets. Primary manufacturing is that type of manufacturing whose raw materials are commodities, that is, raw materials obtained directly from the primary sector. This includes most food and agricultural processing, canning, and packing plants as well as most firms that change ores into pure metal and mineral form. *Secondary manufacturing* receives its raw materials as products from primary manufacturing firms. Thus the raw materials of secondary manufacturing firms are the finished products of primary manufacturing. The locational tendency is for primary manufacturing firms to be raw-material-oriented and for secondary manufacturing firms to be market-oriented. Even so, many secondary manufacturing firms are raw-material-oriented inasmuch as they are located at the source of their raw materials, the primary manufacturing plant.

The simplest type of raw-material-oriented pattern is one that covaries closely with the locational pattern of the raw material used in the manufacturing process (Figure 12–8). Thus this pattern, type Y, is highly analogous to the type A market-oriented

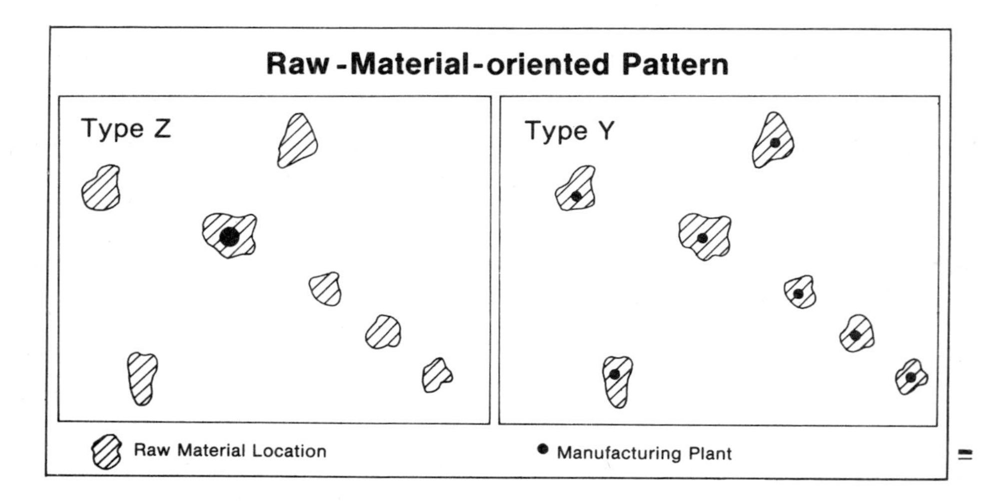

Figure 12–8 A raw-material-oriented pattern, type 2. This pattern is the type A market-oriented pattern (Figure 12–1). In this case each source of raw material contains a manufacturing plant relative to the size of raw material.

pattern discussed above. If the pattern of raw materials is known, it can be fairly easily determined whether or not the pattern of production coincides with it. For example, if a firm uses lumber as its basic raw material and if similar firms are always found in lumber-producing areas, then one would assume that such a firm is raw-material-oriented. If a firm's raw material is orange rinds and if similar firms are always found near the source of orange rinds, it would likewise be considered raw-material-oriented.

The simple correlation type of raw-material-oriented pattern is seen by the covariation of the distribution of forest land and the production of wood pulp in the United States (Figure 12–9a and b). Again, there are some exceptions, such as the relative absence of wood pulp production in the Rocky Mountain states, but this is easily understood given the type of tree required for wood pulp production and the relative inaccessibility to markets. Coastal locations, it will be seen, are highly favoured.

A more complicated pattern is demonstrated by flour milling (Figure 12–9c). In this case, several subpatterns are evident: (1) a general coincidence of milling areas with grain-growing areas, in the Great Plains, Montana, and much of the West, (2) the extreme dominance of major assembly centres such as Minneapolis-St Paul, St Louis, and Buffalo, and (3) the historically older pattern in much of the eastern Piedmont and the Old West. In the first instance, milling centres are centrally situated in the producing areas as based on certain scale requirements. In the second instance, major regional centres dominate large areas. An important exception is Buffalo, which is a major break-of-bulk point for both American and Canadian wheat to international markets. Buffalo behaves in much the same way as export ports like Seattle, Portland, and Los Angeles, which are some distance from producing areas.

Yet grain milling might be expected to be market-oriented rather than raw-material-oriented. Certainly milled grain is more perishable than grain kernels. The solution is twofold. First, most of the grain exported is in the whole kernel stage and thus does not show up in the pattern. Second, although highly perishable relative to the raw material, flour is nonetheless much less perishable than bakery products. Finally, it must be remembered that there is a fair weight loss between raw wheat and finished white flour. On the average, it takes 100 pounds of wheat to produce 72 pounds of white

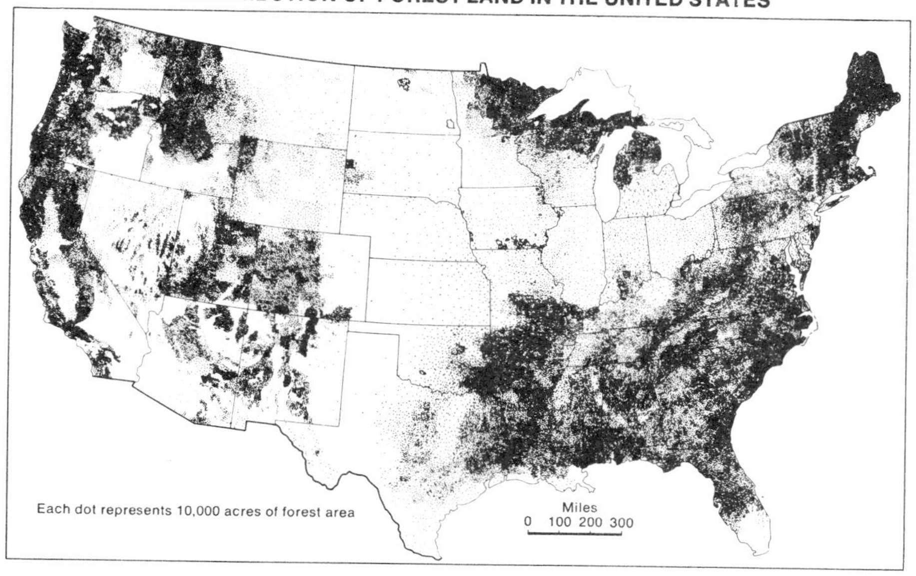

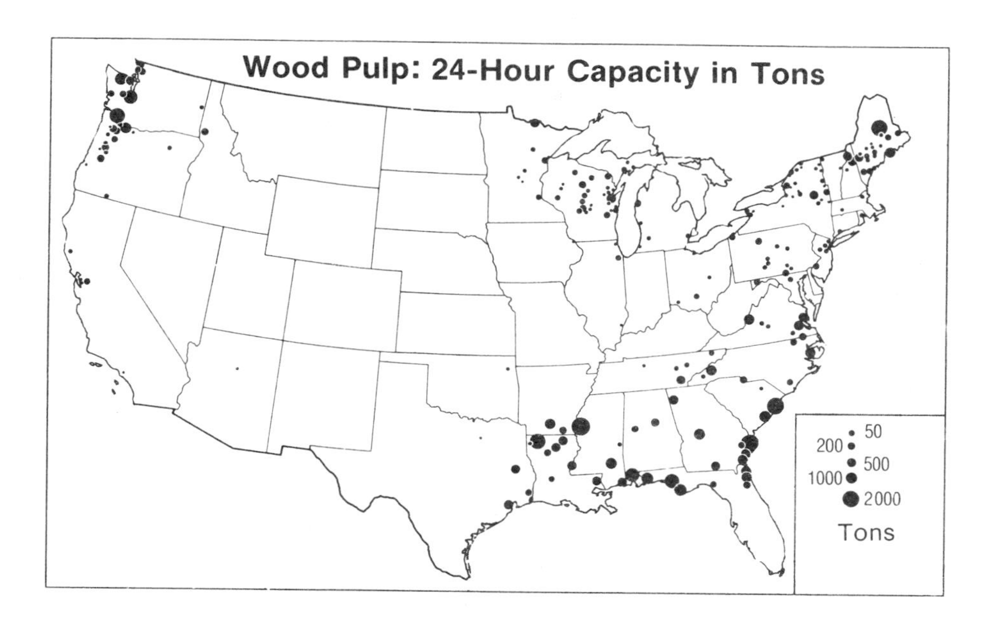

Figure 12–9 (a) Distribution of forest land in the United States. (b) Production of wood pulp per 24 hour tonnage capacity; note the general correspondence of this manufacturing type with the pattern of forest land (raw materials) as shown in Figure 12–9a. (c) Flour milling production per daily cwt capacity, a pattern which is heavily raw-material-oriented and follows generally the type Y pattern above (see Figure 12–8). (From Alexander, J. W. *Economic Geography*, Englewood Cliffs, New Jersey: Prentice-Hall, 1963, p. 303, p. 327. Reprinted with permission.)

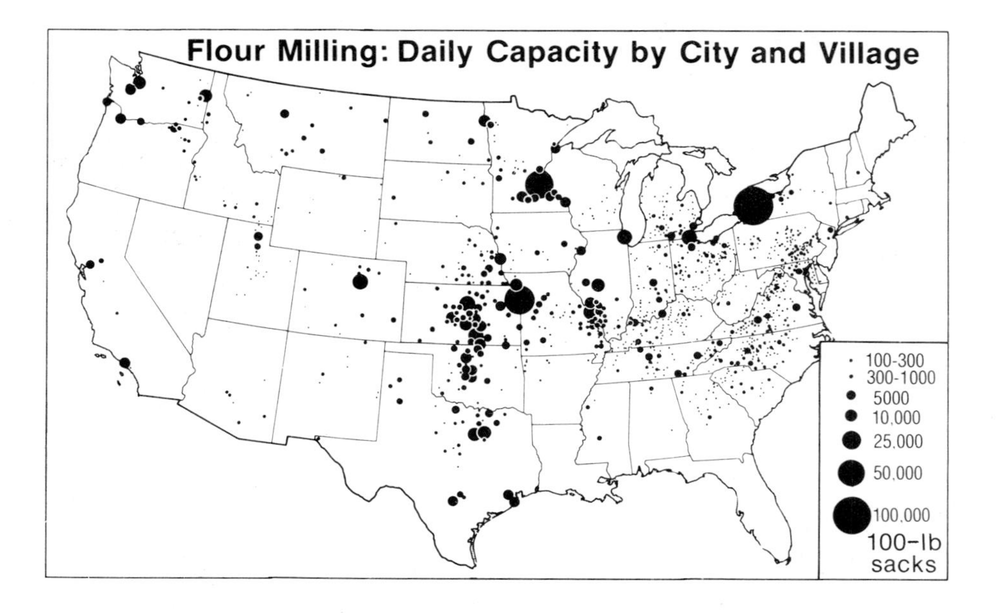

flour. [The weight-loss ratio is therefore 28 (100−72/100).] Moreover, if 28 pounds were of no value, there would undoubtedly be much more milling in remote rural areas than is the case. In fact, the 28 pounds are *shorts* and bran, which are used mostly as cattle feed. Thus there is a considerable incentive to mill at raw-material-oriented sites near cattle-feeding areas; and vice versa, there is some incentive to feed cattle near grain-milling sites.

For these reasons any given manufacturing type would not be expected to have just one locational pattern but a group of patterns. At any one time these firms would be located in a vast number of different kinds of places. One key to decoding such groups is in terms of the size and age of firms in any particular pattern type. As a general rule, the large firms are the newer ones. Thus these might be indicative of the trends taking place in the industry and, more importantly, reflect the future pattern. On the other hand, such firms may be mavericks doomed to extinction. Surely the safest course of interpretation is to discuss the predominant pattern type. Another key to spatial understanding involves the regionalization of the pattern into meaningful territories.

With most types of manufacturing, however, there are many sources of raw materials (type Z pattern). In such cases the determination as to whether or not a firm type is raw-material-oriented becomes much more complex. Obviously, where one firm uses raw materials from several sources, it must make a choice among raw-material locations. As noted in Figure 12–8, the firm may be at one raw-material site or at some point in between. Usually the plant will be located at the largest raw-material input source. In other cases, a raw-material-oriented firm might locate at some place between the various raw-material sources. In fact, a firm could be anywhere within the dashed area shown in Figure 12–10 and still be raw-material-oriented.

The pitfalls in the interpretation of the type Z pattern are demonstrated in Figure 12–10. Note that a manufacturing activity is shown to be distributed over an area in a fairly even pattern, which coincides perfectly with towns in the area. Our first impression is that such an industrial activity is market-oriented. This would be a logical assumption if we did not know what activity was being discussed. Paradoxically, this is also a classic pattern for type Z raw-material-oriented plants. Such manufacturing types include flour milling, food canning, food processing, and beet sugar plants. As can be seen by Figure 12–10, such plants are centrally located with reference to the farms surrounding each town. The town thus represents a logical place in which to assemble

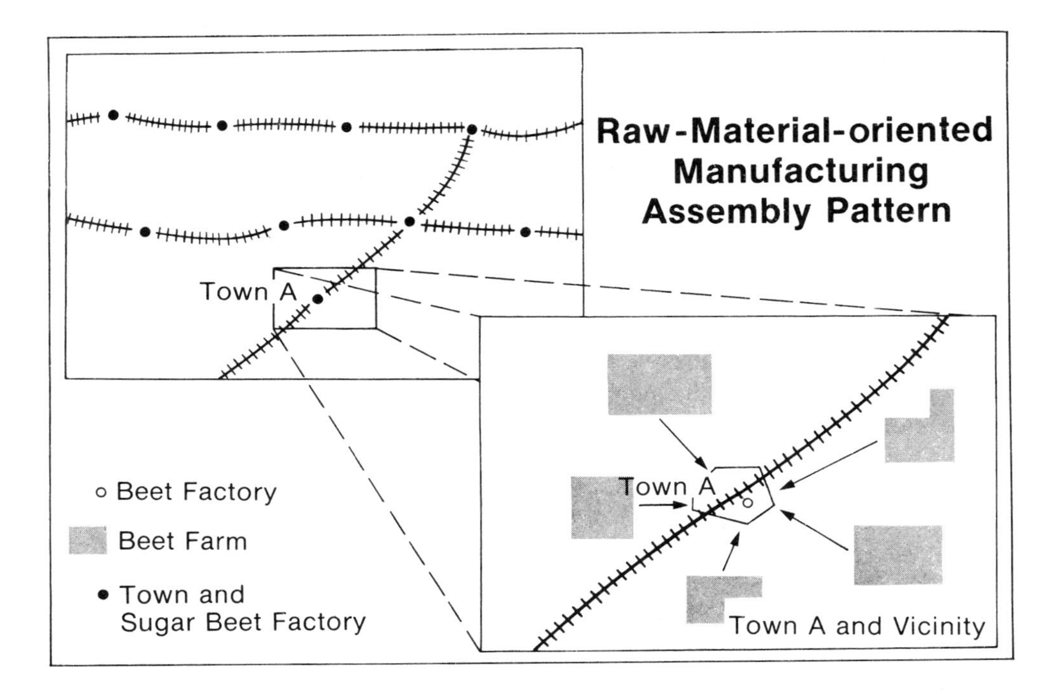

Figure 12–10 A raw-material-oriented manufacturing assembly pattern. In this case the plant is located so as to minimize its total raw material procurement costs. Such a pattern commonly results when several different raw materials are required for production or where raw materials must be gathered from a number of different areas.

the raw material for processing and shipment to markets. (Here a change in map scale is most helpful.)

In another respect, however, the town localization might be considered as a distorted form of raw-material orientation whereby the town, through its labour force and special transportation facilities, provides a good trans-shipment point. Thus one might say that the location pattern is skewed toward the most convenient break-of-bulk point.

It should be apparent that each particular type of manufacturing has its own locational peculiarities. Often the pattern is not the pure one that is discussed here but is blatantly multifaceted and distorted. In this limited study we cannot examine the many types of locational patterns for manufacturing, nor would this suit our purpose, which is to understand the general nature of the industrial location problem.

It is hoped, however, that the student will carefully investigate other industrial types as they exemplify these locational patterns and principles. Nonetheless, anyone who hopes to understand all types of manufacturing and their patterns has undertaken an almost impossible task. For example, there is a different manufacturing pattern for almost every product, agricultural crop, mineral, and every commodity of forest and sea.

THE IRON AND STEEL INDUSTRY

Examination of the present distribution of iron and steel production (based on plants with major furnaces such as the blast, Bessemer, and open hearth) reveals some of the complexities and pitfalls involved in pattern interpretation (Figure 12–11). Some production locations are clearly raw-material-situated in that they are near sources of coal or iron ore. Major world centres of production at sources of coal include the valley of the Ruhr in West Germany, Pittsburgh, Pennsylvania, Karaganda in the USSR, Youngstown, Ohio, and the older steel-producing towns in England, South Wales and Central Scotland. However, on the basis of prominence, one might believe that the source of iron ore is the dominant factor affecting production location. It will be noted that this is primarily the case in the USSR, France, and Brazil, but is not the case in the United States and Japan. In fact, major iron ore source areas like Labrador, Venezuela, Chile, and Australia have very little or no steel production. In these countries, production appears to be at the market. Thus, as based on the present world pattern of iron and steel production, production is found at just about every conceivable combination.

A glance at Table 12–1 and at an atlas might cause us to assume that there is a play-off between a location at either coal or iron ore, depending on which is closer to the market. If the iron ore is closest to the market, then coal will be shipped to the iron ore (as at Magnitogorsk). If the coal is closest to the market, then iron ore will be shipped to the coal (as at Pittsburgh). Production occurs at intermediate locations, such as Cleveland, because these locations are classic break-of-bulk points. Production at the outer extremities of the raw-material source is due simply to backhaul possibilities. This all makes sense until it is realized that some of our largest steel centres are located neither at nor between raw materials sources, but are clearly located at markets.

This is easily seen within the United States (Figure 12–12). Note the major production along the eastern seaboard. Coal is obtained from West Virginia and iron ore from eastern Canada, Venezuela, and Peru. Similarly, on the West Coast, the plant at Fontana, California, in the Los Angeles metropolitan area, receives its iron ore from the southwestern part of Utah and its coal from various parts of the Rocky Mountains, especially Colorado. Other centres in the West are likewise not truly raw-material-oriented. In fact, the blast furnaces at the Geneva Plant (near Provo, Utah) and at Pueblo, Colorado, were developed during World War II largely for national defence purposes.

Table 12–1 Location of iron and coal production.

At source of iron ore	At source of coal	At market	None
Duluth, Minn.	Pittsburgh, Pa.	Japan?	Japan?
Krivoy Rog, USSR	The Ruhr, West Germany	East Coast of US (Philadelphia, Baltimore)	Cleveland
Magnitogorsk, USSR	Karaganda, USSR		
Lorraine area, France	Yorkshire South Wales	Los Angeles	Detroit
Belo Horizonte, Brazil	Bethlehem, Pa.	Chicago-Gary	Buffalo
An-shan, China	Sambre-Meuse area, Belgium		Geneva Steel Plant, Utah

This extremely diverse pattern raises questions as to the general importance of the various locational forces. In some areas markets appear to be most important; in others, coal; and still others, iron ore. In some places the plant is located neither near raw materials nor in major markets. Thus, careful examination of the pattern leaves us with an enigma.

Examination of actual cost data also reveals contradictions on every hand. Data from some plants indicate that it costs more to transport coal than iron, while data

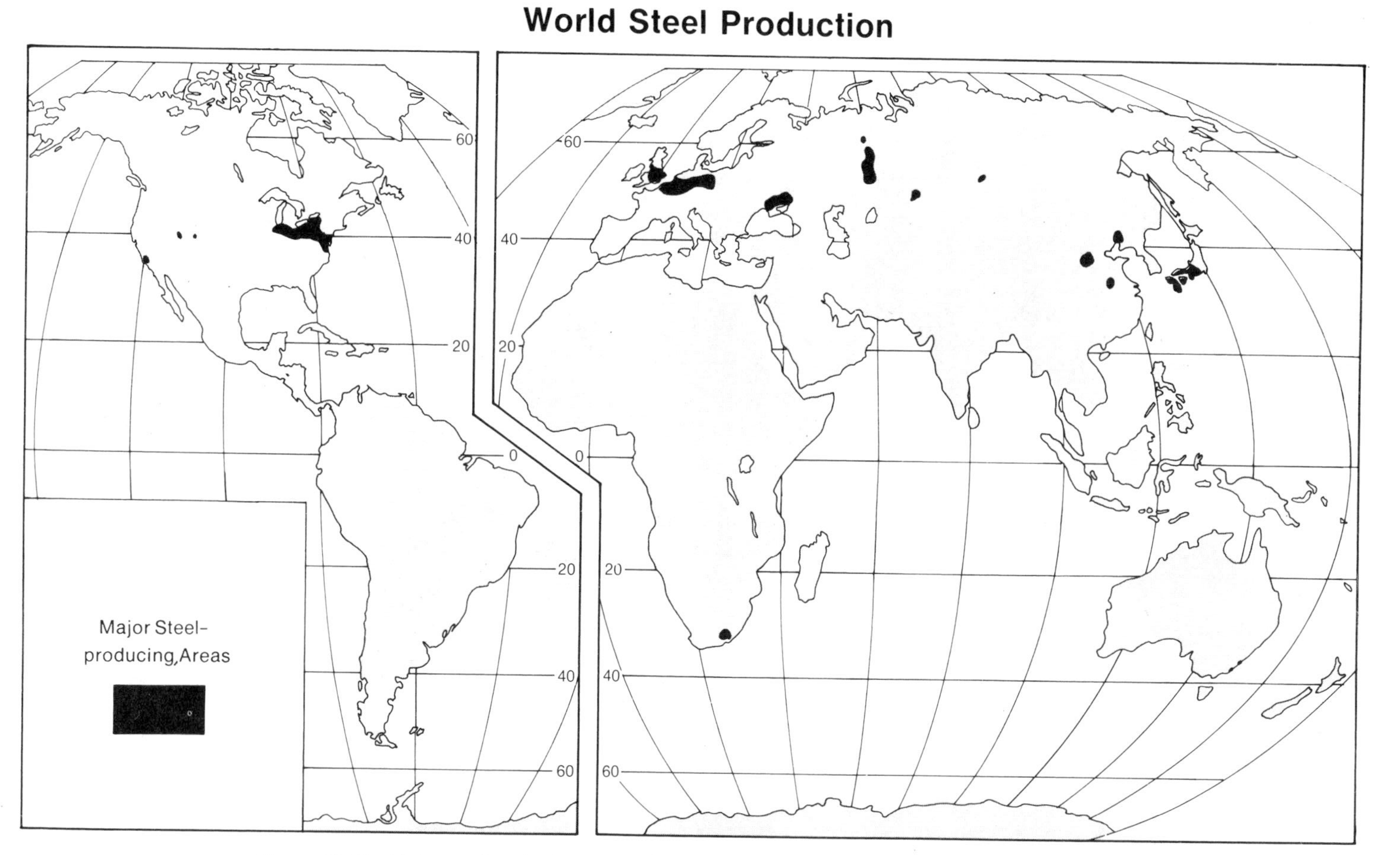

Figure 12–11 Major steel-producing areas of the world. The USA, EEC, Japan, and the USSR are the major producers. Third World countries, notably China, Brazil, and India, produced 10 per cent of world production in 1977 and aim to raise this to 30 per cent by the year 2000.

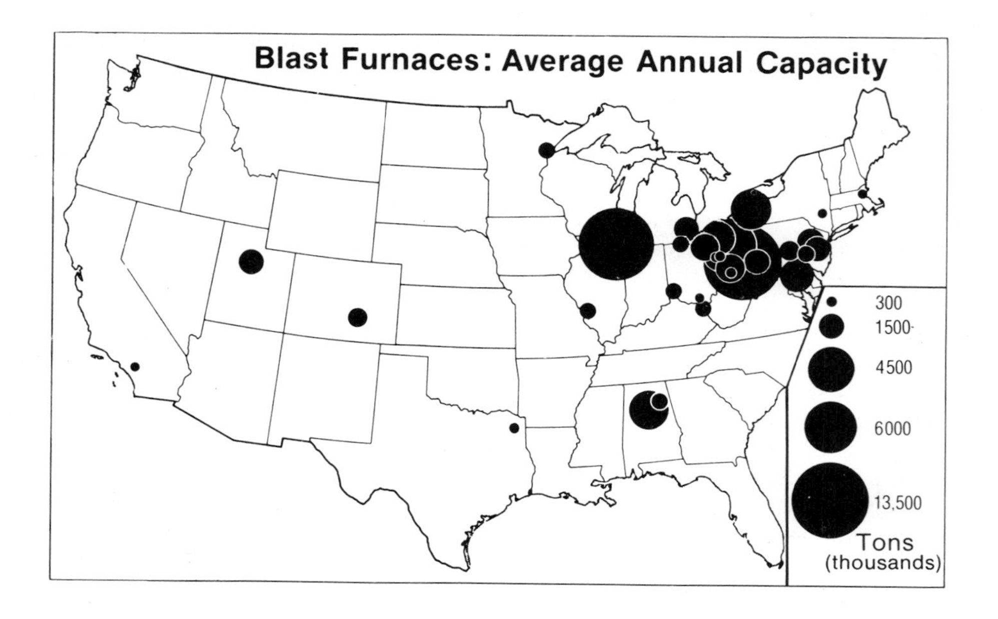

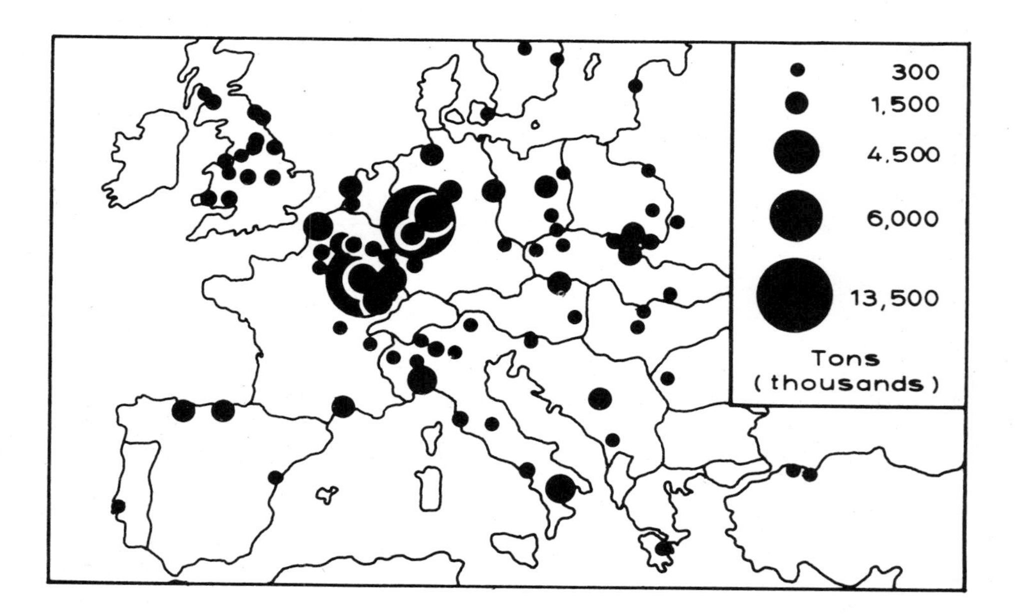

Figure 12–12 (a) Blast furnace capacity in the United States. Note the tremendous concentration of furnaces in the American manufacturing belt. (b) Blast furnace capacity in Western Europe and the Mediterranean. Western Europe shows a similar concentration within the manufacturing belt, but new capacity has been built on the shores of the Mediterranean in several countries.

from other plants indicate the opposite. Some plants maintain that the costs of delivering the finished product to market are less per tonne-km than the transportation costs of raw materials, while other plants maintain that the opposite is true. How can this be?

The sharp geographer has already reached an answer. It is evident that the peculiarities of the resource and market location, and the nature of the transportation available in any given area, will lead to different locational responses for any given steel plant. For example, if water transportation is available, as in the Great Lakes, then that raw material farther from the market (in this case, iron ore) is likely to be transported toward the raw material nearer the market (coal) and toward the market itself. Likewise, raw materials near ocean transportation might be expected to be transported to plants near markets.

As we all know, we cannot make a safe generalization from only a few examples. Based on a case or two, we could prove what we wanted to prove about the locational pattern of the steel industry. The real question, however, is: what is the predominant locational pattern? Moreover, by this is meant, what is the predominant locational pattern *today*? Thus it is important to make distinctions appropriate to the age of the steel complexes involved.

In order to answer such questions, the general trends in location over time need to be examined. Historically, the location of iron and steel production has witnessed at least three critical stages: (1) a fuel-oriented stage, (2) an iron-ore-oriented stage, and (3) a market-oriented stage.

During most of the eighteenth century, iron making (the predecessor of steel making, which requires use of ferro-alloys and more advanced production methods) was widely scattered in small plants within forested areas. The source of fuel was charcoal from forests. Less than 20 tonnes of pig iron could be produced from one hectare of woodland, and since the largest furnaces produced up to 30 tonnes of pig iron per week, they consumed hundreds of hectares each year. Charcoal could not be economically transported more than about 24–30 kilometres. The relative input was so great that this was the factor determining location. At this stage, therefore, local concentrations of the industry were difficult to sustain, so that in both Western Europe and in the settled parts of North America iron production had a very diffused locational pattern. Not surprisingly, since they both possessed abundant iron ore and great forests, Sweden and Russia were important producers.

By the end of the nineteenth century, coal had replaced charcoal as the dominant fuel, and iron and steel production had shifted to coal fields. Notable in this regard were the Ruhr Valley in West Germany, the Midlands of England, and Pittsburgh, Pennsylvania. As late as 1939 it took about 1½ tonnes of coal to produce 1 tonne of iron. This technological condition was perhaps further strengthened geographically, inasmuch as in most industrializing nations, iron ore was more distant from the major markets than was coal. Thus iron ore was commonly the major raw material to be transported. It was one of the important early advantages in Britain that in the mid-nineteenth century 95 per cent of the ore used was obtained from coal measure sources. As this supply waned, overseas sources became important both to Britain and Germany; first from Spain and later from Sweden. Many West European iron ore fields, like that in Lorraine, closer to the steel-producing areas could not be used until the Bessemer process (1879) enabled them to be used in France, Belgium, Luxembourg and Germany, for the most part without the need to change the location of the steel industry. One notable exception to the iron ore to coal movement, was in the Soviet Union, where iron was abundant in the Urals (home of the earlier charcoal-based industry) but coal was much farther away. In fact, suitable coal beds could not be found for at least 1600 km farther east in the vicinity of Karaganda in central Asia. Therefore, in order to achieve steel production in the Urals (the resource closest to the market), the USSR undertook the longest massive land movement of coal ever to occur anywhere.

Recently electric furnaces and other technological advances have permitted the effective utilization of scrap iron as a raw material in conjunction with iron ore. The use of scrap requires less coal for fuel and also, of course, less iron ore. Because scrap is located (as might be expected, given its source) at major consumption points, particularly at most large metropolitan centres, production would be expected to occur there also. Scrap is particularly important to the steel industry in the Midlands and in Sheffield, for example. Nevertheless, scrap is insufficient in most areas. Iron ore is in greater demand, and the British steel industry relies on many overseas sources for it, besides employing the remaining iron ore fields in eastern England, particularly in Northamptonshire and Lincolnshire. These gave rise to new steelworks in Scunthorpe and Corby.

Today, in Britain, the United States, Western Europe, and Japan, some of the largest integrated steel mills are located close to deep water near major markets. In fact, Japan is a major importer of both coal and iron ore, and also a major exporter of steel. It has one of the largest and most up-to-date operations in the world and is now one of the two or three largest producers. Undoubtedly, steel production is rapidly becoming market-oriented and highly dependent on imported raw materials from underdeveloped nations.

A factor that contributes heavily to the mixed pattern of iron and steel production is the heavy investment involved in developing a major production centre. Once production begins in an area, it does not readily shift to new areas despite changes in production-cost components. In fact, the predominant pattern of steel production still reflects decisions made a century or so ago. This discrepancy would be even more noticeable were it not for the fact that markets themselves are likewise quite fixed because of the heavy investments in the infrastructures of cities (see Chapter 4).

Thus the distribution of most manufacturing types contains crops of the past and plantings for the future. The general age of industrial plants is therefore critical in pattern interpretation for those activities that are experiencing major locational shifts. When viewed within a time sequence and with an appropriate appreciation for local transportation variations, the basic locational conditions and forces usually become evident: steel production is largely market-oriented.

SCALE, AGGLOMERATION AND LINKAGE

There are many ways in which we can build on the insights so far gained in relation to industrial location patterns. As a reminder, so far we have emphasized the importance of the various costs of procurement, processing and distribution, and we have introduced transportation-based location models based on Weberian principles (Chapter 11). So far in Chapter 12, we have pursued market-oriented and raw-material-oriented patterns which are evident in many manufacturing activities, and—still in pursuance of the same objectives—we have looked closely at successive influences which have been important in determining the location of the iron and steel industry. One way forward is to look more closely at some of the agglomeration economies which influence industrial locations, and especially to examine what may be regarded as an important aspect of modern industries, their inter-linkage with other industries.

We may return briefly to the iron and steel industry to suggest the fundamental importance of linkage. One point that emerges from this case study is that while in the past the iron and steel industry was highly mobile (as raw material sources, markets, and technical characteristics changed), it is today an industry highly market-oriented. Very many iron and steel plants remain where they were sixty years or more ago. Some authors would suggest that they persist in such locations long after these localities have lost their initial cost advantages: they display *inertia* because they fail to relocate where costs are lower. But there may be good reasons why the iron and steel plant remains where it is, reasons which lie beyond considerations of locational response to raw material procurement. In the modern integrated steel plant there may be many internal economies to be won by keeping all the processes close together, but other economies may be locationally significant. As coal is converted to coke for the blast furnace, tar ammonia may be produced as a by-product and supplied to the chemical industry. The direct products of the blast furnace may serve both the forging industry with wrought iron and the foundry trades with castings necessary for many types of engineering. The steel converter, too, supplies ingots (via a cogging mill) to forging presses which may serve factories which make such items as wheels and axles, while the steel finishing mill provides plate, sheet, strip, and tube to engineering, transportation and construction companies.

To point out such things may well be said to emphasize only the importance of the market. But we are also addressing ourselves to linkage of a particular kind, for if a range of dependent industries and trades is established in relation to an iron and steel producing plant or plants, they become *mutually* dependent. Indeed, each sector of the whole iron and steel-based activity may not only 'feed off' the next, but also gain a number of other external economies which, separately or together, may perpetuate location.

There is a variety of such external or agglomeration economies, but perhaps the most important to emphasize in the present context are those usually termed localization economies. These arise from the clustering of individual plants all engaged in similar or related activities in a restricted geographical area. They include: an established reputation acquired by goods produced in a given locality, such as chinawares in Staffordshire (the Potteries) or cutlery and fine steels in Sheffield; the creation of a pool of skilled labour from which many specialist manufacturers can draw; local utilities and services which have become adapted to the needs of a particular group of industries and trades, such as banks, insurance agencies or freight-haulage concerns suitably equipped for special loads and deliveries. Auxiliary specialists may arise in either the manufacturing or service sectors who can provide many firms with their exact needs better, and more cheaply, than they can themselves. Inasmuch as many firms associate in these ways, each can benefit, too, by quick deliveries, low transportation costs and the concomitant advantage of low stockholding costs.

Localization of this kind, which has both seen and unseen advantages, may add up to an 'industrial atmosphere' which pervades the whole cluster and so acts as a powerful magnet upon those wishing to set up new firms in its allied trades. The best example in Britain, and possibly in the whole of Europe, is the Birmingham and Black Country area where constellations of the metal-working trades are heavily localized in constituent parts of the conurbation. Here, although the various branches almost defy classification, we can recognize the intense localization of (1) a large number of firms performing common metal processes—the founders, forgers, stampers, welders and galvanizers. Their basic products serve (2) a very large number of firms making nuts, bolts, screws, tools, springs and wires, and (3) an even greater range and number of smaller, workshop-type trades which make patterns, hand-tools, machine-tools and much more necessary to the manufacture of metal goods. These three main types of metal-industrial activity support the fabricators and assemblers of all kinds of hardware. These, also highly localized, used to specialize in such items as nail- and chain-making, wrought iron domestic wares, ornamental brassware, riding gear, swords and flintlocks, clock and lock mechanisms and jewellery. Such trades have evolved and modernized, their inter-linkages growing with time rather than diminishing in importance. In an era which demands new kinds of metal products, guns are still made in Birmingham, locks in Wolverhampton, and remarkably, the old jewellery-making quarter still survives in its original locality in an industrial area which for many miles around is now dominated by mills, factories and workshops which today focus on the production of a multitude of components for the motor vehicle.

INDUSTRIAL LINKAGE: DEFINITIONS

Following the reasoning above it may be useful to think about other industries which reveal linkage and how this linkage conditions the location of constituent or related parts: in forestry—pulp, paper, printing, lumber, furniture; in fishing—fish freezing, canning, pet-food canning, fish-based fertilizers; in petroleum—refining, petrochemicals; and so on. Careful study of linkages in a spatial context will enable us to gain new insights into how places and things are held together to form a regional whole.

Such an exercise, however, will also make us very wary of reaching quick conclusions concerning the locational forces responsible for any given pattern. Indeed, if industrial linkage is to be a useful term and concept, we must give it more precision. A conventional definition is provided by Townroe (see Bibliography) who noted that it involved:

1. *Process:* movement of goods between different firms as stages in the manufacturing process, including subcontracting;

2. *Service:* supply of machinery and equipment and of ancillary parts such as tools and dies, as well as repair and maintenance requirements when supplied by different firms;

3. *Marketing:* ties with other firms that aid in the selling and distribution of goods;

4. *Financial and commercial:* ties with financial and advisory services.

However, as we examine a given industry closely, we will find the relative importance of material and non-material connections difficult to assess. Secondly, we must try to identify the strong connections which have overriding importance in making locational decisions, separating them from the many other weaker contacts that are merely part of the daily operations of industrial organization. In addition to these, we may need to distinguish between single patterns of connection with relatively few linkages and complex patterns with many linkages.

At this stage, we may find that the term 'linkage' demands even further clarification. So far in this text we have used the terms 'forward linkage' and 'backward linkage', but others have been proposed as more appropriate to the study of localization economies.Thus P. Sargent Florence (1948; see Bibliography) distinguished between vertical, diagonal, lateral and common service linkages. It is appropriate here to illustrate their meaning from the metal trades of the West Midlands (both because we have noted these already and because Sargent Florence was himself located in their midst during his career). *Vertical linkages*, he proposed, are a series of processes linked to each other in succession which contribute to the gradual transformation of raw materials into finished products. These are the links described above, the in-steel plant processes from molten iron ore to metal sheet, etcetera. *Diagonal linkages* are characterized by the trades which serve a number of different industries and need contact with them all to maintain viability. These are trades like engraving, plating, polishing as undertaken in jewellery making and in machine-tool finishing. *Lateral linkages* involve all those industries which produce parts, accessories, or services that feed different stages of the assembly processes. And finally, *common service linkages* relate to the common processes and skills available in a given local area, upon which, in the metal trades example, each branch may draw.

INDUSTRIAL LINKAGE: VARIETIES

So far in our discussion of linkage, we can see that a broad view of operational connections is necessary for understanding the impact of linkage upon industrial location; but at the same time we need to define the geographical scale most relevant to our examination. In a perceptive review, P. A. Wood put the whole problem succinctly:

> '*Obviously, most plants have transactions over a wide range of distances and to describe these is to do little more than define the scope of manufacturing. If "linkage" is worth consideration, it must provide an explanation for distinctive industrial patterns or important locational trends in manufacturing. Essentially, strong or complex linkage ties are assumed to operate only over limited distances, so that the problem becomes one of firstly separating the effect of these from wider exchanges or communications, then discovering how much of the remaining "regional" pattern can be attributed to operational linkage. On a national scale the systematic ties of a plant to others have locational significance broadly in relation to raw material sources or major market areas. The former may be indigenous, as on iron ore fields, or foreign, available at ports, while the latter gives patterns of distribution, for example that are weighted towards southern and central Britain. In general, however, plants located primarily in relation to raw materials or markets form a relatively small proportion of total industrial activity. On a local scale, connections to adjacent or nearby plants undoubtedly exist, but such connections do not account for the concentration of massed industrial areas that have been formed during the twentieth century. Areas such as Park Royal in North West London or the Team Valley Estate in Durham have been generated by forces and inertia of land ownership and rent, labour supply and quality or planning regulations, rather than by positive operational linkages. Only in older, specialized metropolitan industrial areas, such as East London (Hall 1962, Martin 1966) or the jewellery quarter of Birmingham (Wise 1949) may local agglomeration be explained in terms of this type of linkage and, in many senses, such areas are survivals from an earlier period of industrial history.*' (P. A. Wood, 'Industrial Location and Linkage', *Area*, 2, 1969, pp. 32–39.)

To Wood, therefore, 'linkage' and agglomeration implied something more than the existence of heavy industries near raw material sources or the occurrence of industrial zones and estates. The chief clues to understanding were wrapped up in the variety and complexity of materials and information linkage as part of the general system of spatial communications: 'These must surely be essential elements of any theory of industrial location that hopes to explain the paradox of many modern locational trends; the apparent need for regional proximity to other plants in spite of the vast improvements that have been made in communications and transport technology.'

Wood carefully devised a typology of linkage which helps us to sort out the full range of links amenable for study (Figure 12–13). He lists:

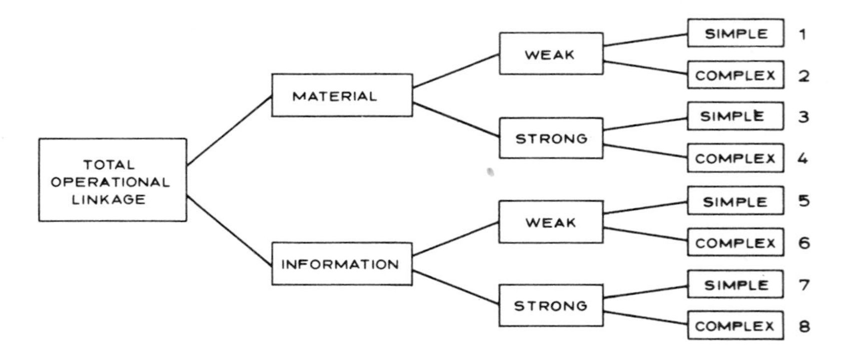

Figure 12–13 A typology of industrial linkages (after Wood; see Bibliography).

1. Material linkages few in number and of little importance in deciding about location. These are found in traditional 'footloose' light industries and may have been exaggerated in importance because of neglect of categories (6) to (8) below.

2. Material linkages that are individually weak but which form a complex organizational arrangement which must be maintained for successful operation. Would include examples found in large scale assembly industries such as motor vehicle production.

3. Material linkages that are simple but dominated by strong individual ties to material sources or markets. A category found in heavy industries with high transport costs most closely fitting the conditions postulated in Weberian location theory.

4. Material linkages in which a few dominate as part of a complex pattern of supply and marketing. Occurring in firms operating as part of horizontally integrated industrial agglomerations, such as in the Lancashire textile or West Midland metal industries.

5. Information exchanges that are few and of little importance in deciding about location. A state generally assumed as normal in studies of industrial location, and which applied presumably in cases where strong material connections are undoubtedly of overriding importance (as in (3) above).

6. Information exchanges individually weak but which together form a complex system that must be maintained. A situation common in manufacturing industry and may have important implications for locational change. Often ignored, however, because of difficulties of measurement in relation to other, more tangible elements of linkage. Much of the attraction of manufacturing to metropolitan areas and conurbations may be explained in reference to this type of linkage.

7. Information exchange simple in structure but dominated by strong individual ties to other plants. The clearest case occurs in the subsidiary or branch plant, whose operation may be related to instructions from a head office or parent factory, but it also applies in industries where monopoly suppliers or buyers of goods are dominant.

8. Information exchanges that are complex but dominated by a few strong ties. These

would include non-manufacturing commercial and service activities as well as perhaps manufacturing organizations that are closely tied to the service sector, through sales to a consumer market (for example, clothing trades near metropolitan centres).

Traditional location theory, Wood points out, seems capable of embracing material exchanges under types (1), (3), and (4), with the assumptions of (5) widely accepted for information flows. Categories (2), (6), (7), and (8) were too frequently ignored because of the difficulty of measurement.

THE MULTI-PLANT FIRM AND ITS LOCATION DECISIONS

Indirectly, Wood's typology of linkages leads to a new area of consideration in our quest for understanding industrial location patterns. In (7) above, he refers to the importance of instructions from a head office or parent factory. We have earlier made the point that the branch plant is the rule today. Indeed, in a recent study of manufacturing employment in Inner Merseyside, it was found that multi-plant firms employed over four-fifths of the area's workforce, and that nearly 88 per cent of the area's multi-plant jobs were controlled by firms headquartered outside Merseyside County. Here, as elsewhere in our modern industrial economy, the organizational decision-making may be made from afar. In such circumstances, how are location decisions made, and how do such decisions vary in what goes into them? How far are the many cost considerations and localizing influences perceived by entrepreneurs and business corporations?

ADOPTIVE AND ADAPTIVE LOCATIONAL BEHAVIOUR

Berry, Conkling and Ray (see Bibliography) draw attention to the contribution made by two American economists which illuminates these problems. These economists distinguished between *adoptive* and *adaptive* locational behaviour, circumstances which lead to the apparently rational locational patterns 'about which people think they can theorize'. The adoptive stance acknowledges that the locational choices of individual firms may appear quite random if viewed case by case, but it also argues that this random behaviour takes place within a competitive economic environment. In this environment, individual locations do matter in the long run, because some locations are by their nature more profitable than others. An important point in this argument is that initial or individual location decisions may turn out to be irrelevant to the emergence of spatial regularities, for those in weak locations will fade or shut down. Such differential industrial growth leads to apparently rational locational patterns of plants within an industry, made up, of course, of the successful survivors. By contrast, the adaptive viewpoint focuses on rational business decision-making and, from this, rational locational choice. 'Adaptivists' would stress the systematic way in which the large firm invests in any of its activities: in product research, labour availability, market research and much more. With their systematic processes of internal decision-making they continuously analyse and monitor the economic environment in which they operate, and so can locate themselves in the best places to achieve their corporate goals.

The adaptive strategy, it may be argued, is one that commends itself to the large company which wishes to expand into new markets. Rees developed a decision-making model with such a circumstance in mind (Rees, J., 'The Industrial Corporation and Location Decision Analysis', *Area*, 4, 1972, pp. 199–205). Such a corporation might consider whether to relocate existing plant, acquire a plant by purchase, or construct a new branch plant. Assuming the third alternative fulfils company requirements better than the first or second, it works from the whole to the part. That is, it selects a potential sales region within which market prospects are expected to be good enough to warrant a branch plant; within that region, it then selects a community most likely to satisfy the company's profit goals; and thirdly, it selects a site for the plant within the chosen community.

It is at the level of the site that the casual enquirer into the reasons for its location may be puzzled by the locally-given answers. For, as Rees observes, the branch plant site may well be based on social, rather than economic grounds. Indeed, a common reason offered to the student enquirer is that it 'was chosen by the Managing Director's wife!'—a reply that may make more sense than it seems, for it may imply company perception of where it is thought the probable plant manager would like to live, or his employees would like to shop. Good worker relations may be important, of course, to the long-run economic success of the firm, and they may be especially critical at a new site. They will, however, be but one of many circumstances reviewed by the parent organization, and whatever may be learned from the evaluations of the branch plant's performance is fed back to the central decision-making group. The sequence of rational location decision-making which establishes the branch plant in untried territories may be, of course, equally brought to bear to close it.

In bringing ourselves back to the single plant location, we have gone full circle and are on the threshold of a more complete spatial understanding. We are suitably armed to tackle the locationally-significant aspects of a variety of industrial patterns. The particular real world pattern to which we now turn is that adopted by the automotive industry, whose world-wide ramifications are both multi-plant and multi-national. Indeed, it has been called 'The Industry of Industries'.

THE WORLD AUTOMOTIVE INDUSTRY: INDUSTRY OF INDUSTRIES?

By employment and value of output, the automotive industry is one of the world's largest manufacturing industries. In 1973, the twenty-six largest motor vehicle makers alone employed well over three million workers world wide. In many industrial countries vehicle manufacture occupies a major role. Both in Britain and the US, for example, it accounts for over 6 per cent of the national manufacturing employment and over 5 per cent of gross manufacturing output. In these same countries, personal expenditure on new or second-hand vehicle purchase (alone) absorbs over 10 per cent of total consumer spending. Indeed, the production and consumption of the road vehicle has become something of a 'barometer' test of the world's developing economy and so great is our dependence on it that it is difficult to see how we might cope without it. In a major study of the industry, which is obligatory reading for serious students of the topic, Gerald Bloomfield rightly suggests that the motor vehicle has made a very substantial contribution to the human geography of the twentieth century:

> *'By providing more flexible transport for goods, it has affected the detailed location of commerce and industry. As a new mode of passenger transport, it has modified urban and rural areas, allowing new dimensions of personal mobility*

> *for commuting and recreation. Motor vehicle manufacturing has directly shaped the geography of only a few areas, although indirectly it has had a significant influence in most industrial regions. Unlike many of the great industries of the nineteenth century, automotive production has scarcely created a distinct industrial landscape; though the vehicle, as a piece of portable sculpture, has become an integral part of any landscape, while the supporting infrastructure has been a major force shaping urban and rural places.'* (Bloomfield, G. *The World Automotive Industry*. Newton Abbot: David & Charles, 1977, p. 23).

Thus the automotive industry is of world-wide importance today, but almost from its very beginnings it has been international in its technology and organizational structure. It is an industry presently characterized by oligopoly, for a small number of large corporations dominate, each of which manages every aspect of design, marshals the manufacture of components, streamlines the routine assembly of vehicles and organizes extensive chains of retailing. It is also characteristic of the world's big-name manufacturers that they do not only build cars, trucks and buses. As examples, Ford makes agricultural tractors, Chrysler produces outboard marine engines, and Honda makes motor-cycles. British manufacturers are similarly diversified, for British Leyland manufactures fork-lift trucks and military vehicles and the famous name of Rolls-Royce is carried over into locomotives, tanks, jet engines for aircraft and industrial power plants. Finally, apart from direct links like these, through the long period of its growth, the vehicle industry has stimulated demand for an improved range of products in many other sectors of manufacturing, like metals, rubber, plastics and, of course, petroleum fuels.

In defining the automotive industry proper, we can follow Bloomfield in recognizing three closely related sectors. The core of the industry, and the larger sector, is the manufacture and assembly of complete motor vehicles by the large corporations. The second sector is the making of bodies and trailers, mainly for buses and trucks. And thirdly, the whole range of motor vehicle parts and accessories, including engines and parts, clutches, gears, transmissions, brakes, frames, body fittings and wheels. In recent years, the large motor corporations have acquired many firms in this last sector and thus become increasingly integrated. But beyond this classification there is a wide periphery of ancillary industries, including tyres, glass, electrical equipment, fabrics and plastics. And there is, too, the enormous 'aftermarket' of spares, accessories and vehicle service and repair.

THE MOTOR GIANTS AND THEIR ACTIVITIES

It is well-known that the Ford Motor Company achieved an early dominance of automobile mass production in the USA. In 1921 it produced 58 per cent of the world output of vehicles. Joined by General Motors and Chrysler, the 'big three' in the USA accounted for over 74 per cent of the world annual production in 1935 and had begun to develop minor assembly plants in other parts of the world. But the world position of the US corporations has been challenged in recent decades by the rise of new international manufacturers in Europe and Japan. By the early 1970s Volkswagen produced 6 per cent of world output, followed by Toyota (5.9 per cent), Nissan (5.2 per cent), Fiat (4.3 per cent), Renault (3.7 per cent), British Leyland (2.6 per cent) and Citroën and Puegeot (each 2.0 per cent). At the same time, the big three in North America had expanded assembly and manufacturing plants in many countries so that one-third of total Chrysler and Ford output and one-fifth of G.M. production was achieved out-

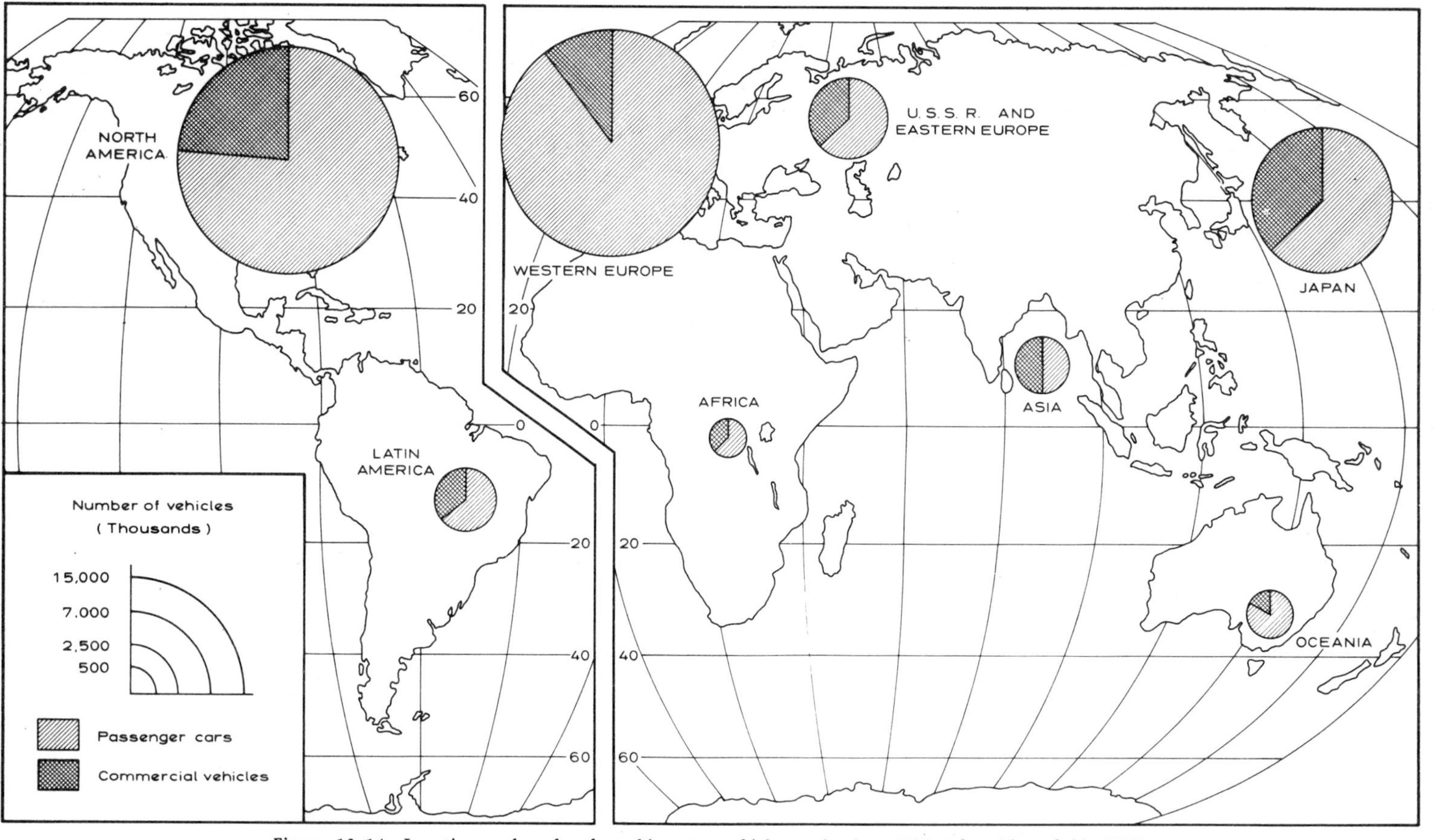

Figure 12–14 Location and scale of world motor vehicle production, 1973 (after Bloomfield, 1977). Within each country, 'product mix' tends to reflect national demand and export trade. In the USA, it is approx. 76 per cent cars, 24 per cent commercial vehicles. In the USSR and India commercial vehicles amount to over 40 per cent, while in Italy, France, and West Germany the output of cars is over 90 per cent. In all, over 70 countries are engaged in some aspect of vehicle manufacture or assembly; they range from the millions produced in the USA down to the few hundred of Malawi.

side the USA and Canada. World motor vehicle output, by continent, is shown in Figure 12–14.

The expansion of the giants and the development of the new rivals on the international scene was facilitated by the massive growth of the market. In 1955 only fourteen companies had an output of over 100 000 units; by 1975 there were thirty-two. Since the international oil crisis of 1973, corporate mergers (always a feature of the industry) accelerated, so that the size difference between the largest US corporations and the largest European and Japanese producers has narrowed. By this time, the sales revenue of the largest corporation, General Motors, equalled the national income of Belgium, while Chrysler's was about that of Venezuela and British Leyland's equivalent to Morocco's.

Despite the size of the typical motor corporation, the very large plant is rather rare. The Wolfsburg Volkswagen factory in Germany employs 65 000 workers and produces about a million units per year, and Ford at Dagenham employs 28 000 workers with an output of about 350 000 units. But both of these examples are large integrated plants at which many stages of manufacture and assembly take place in situ. The average size of plant in the US is about 225 000 for cars and 80 000 for commercial vehicles, and most car plants across the world are much smaller than this. Size of labour force varies with the nature of production and the degree of automation incorporated into the many processes. The most recently planned and constructed plants, notably those in Japan, have remarkably small work forces for their high volume output.

All the large assembly plants, however, are centres of a constellation of component manufacture and assembly plants devoted to feeding the 'track' (final assembly line; Figure 12–15). Indeed, the need to keep the track continuously running necessitates rigorous control of component parts, so that it is not uncommon for the major component factories to be wholly or partly owned divisions of the parent corporation. It is unusual, however, for the motor company to be ownership-linked backwards as far as basic raw materials. Ford and Fiat (see later) are two exceptions, for they are both major steel producers, and Ford (USA) also manufactures its own glass. Nearly all the major corporations have extensive iron foundries for casting engine blocks, casings, brakes and transmissions. Making the engine and transmission is the most costly single material element in car-making, but the cost of labour is high. Labour disruption, which appears to be all too common an occurrence in both the North American and European car industry, can have a disproportionately serious effect on the volume and quality of output. Some manufacturers, like Volvo and Saab in Sweden, have spent heavily on new systems of production to obviate labour instability which is related to monotonous, repetitive, final assembly-line activities.

ECONOMIES OF SCALE

From an early stage, economies of scale have been significant in shaping the structure of the motor vehicle industry. These economies relate to marketing as much as manufacturing, and they also apply to the development of new models and to the raising of new development capital.

Scale economies in the passenger production process are illustrated in Figure 12–15. The graph, however, does not indicate that 100 000 units per year is the most desirable level of car production, since various studies suggest the best economies of scale with engines at 400–500 000 per year. Inasmuch as viability in the internationally competitive situation that now prevails is important, it has been suggested that a produc-

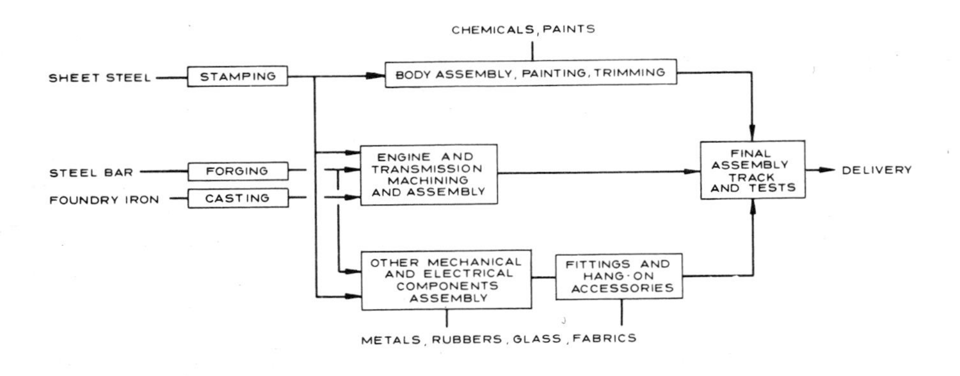

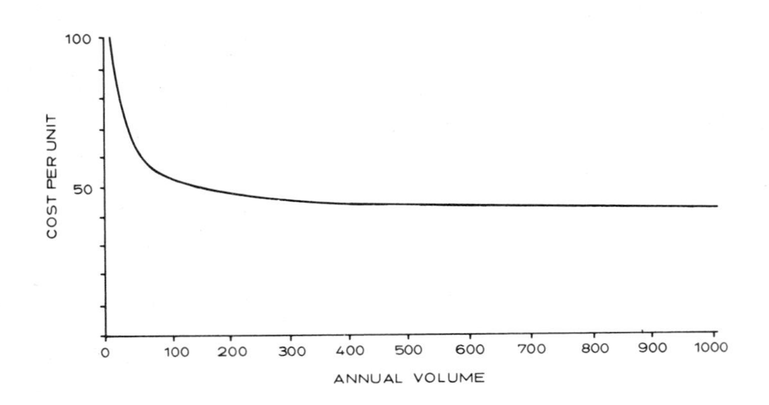

Figure 12–15 **(Above) The motor vehicle production process (generalized) and (Below) technical economies of scale. The graph indicates a sharp decline of costs up to 100 000 units per year, but economies of scale in other operations like stamping and engine-building are only achieved at much higher levels of production.**

tion of 400–800 000 cars is necessary. Much depends, of course, on cost-controls which will vary between each major producer, but it is interesting that among British car makers in the mid 1970s only British Leyland and Ford exceed 250 000 units per year. All other British based firms were placed at a serious disadvantage because of their failure to obtain the necessary technical economies of scale. The search for these is incessant among the 'majors' and it has become a compelling reason for new international linkages.

Scale economies, of course, can be beneficial either in the whole motor corporation or at the level of the individual plant. In both cases, however, there can be problems. Long production runs of a single model can maximize benefits (Ford's 'Model T'; Volkswagen's 'Beetle'), but difficulties of readjustment face the corporation that tries to diversify its range of models. Also, very large firms and plants are awkward to manage and labour relations appear to be more difficult in large units (BLMC). With heavy investment necessary for plant redevelopment and retooling, the financial penalties of under-utilization of capital and productive capacity become severe.

THE MOTOR CORPORATION: CONCENTRATION AND DIFFUSION

In the early 1970s, the twelve largest companies (each producing over 700 000 units) accounted for by far the greater proportion of the world output of cars (85 per cent) and commercial vehicles (68 per cent). Concentration is a long-standing feature of the industry, as old as the process of mass production. Ford captured over half the US market by 1914, and Morris Motors' share of the British market reached 42 per cent in the 1920s. Citroën in France did likewise, while in Italy Fiat held 80 per cent of the national market by 1925.

Once a few producers achieved these commanding levels, other companies either had to emulate them with similar high-volume/low price cars or specialize in low-volume/high price cars for a limited clientele. The larger European companies all adopted Morris's high volume methods, either by going it alone or by mergers. Rootes and Auto Union were created in this way and while Ford entered the European market directly, GM entered it through the purchase of Vauxhall and Opel. Other European car companies, however, could not meet the challenge. Thus in Britain the total number of manufacturers declined from 90 in 1920 to 41 in 1929 and to 33 in 1939.

Some new mass producers emerged in Western Europe after World War II encouraged by the growing market for personal transportation (including Simca in France, Lancia and Alfa Romeo in Italy); but there were few entrants to the industry, Volkswagen in West Germany being a grand exception. However, by the 1960s several Japanese firms emerged, first behind a protected national market and then on an international scale. Fears for their viability in both national and world markets had already brought new protective mergers, such as BMC in Britain (from Austin and Morris, 1952) and American Motors in the United States (from Nash and Hudson, 1954). The largest merger united BMH with the Leyland Motor Corporation (plus several specialist vehicle builders) to form British Leyland (BLMC) in 1968. By the early 1970s, the number of British car manufacturers producing over 1000 vehicles had fallen from 14 to 7 and in Germany from 18 to 10. The financial weakness of some European firms enabled the Chrysler Corporation to expand into Europe by acquiring Rootes in Britain and Simca in France. Citroën and Peugeot merged in 1975. More recently both Renault in France and BLMC in Britain have been injected with Government capital to strengthen their competitive positions.

Concentration in the vehicle industry has reduced consumer choice and strengthened the oligopolistic power of the large surviving manufacturers. But unless tariff walls are high, the international nature of the modern car industry also means that the consumer may gain through the rapid growth of the imported car segment of the market (20 per cent in the USA, 40 per cent in the United Kingdom). A most important point, however, is that a small number of large-scale producers in Detroit, Tokyo, Toyota City, Paris, Torino, and Wolfsburg have become the effective decision-makers.

Their decision-making includes, of course, location decision making. When, where, and how to enlarge their motor empires is a corporate matter undertaken after exhaustive inquiries into commodity supply, labour availability, and market research. Volkswagen, for example, had to calculate carefully the risks before its development in Brazil and it made several investigations of the North American situation before deciding (in 1976) to assemble cars in the United States. Diffusion, however, is as marked a feature of the volume production vehicle industry as is concentration. Indeed, the great concentration of productive might in the core areas has enabled the car giants to expand into the peripheral areas of the world (Figure 12–16). It is a special feature of the motor vehicle industry that this expansion is still under the con-

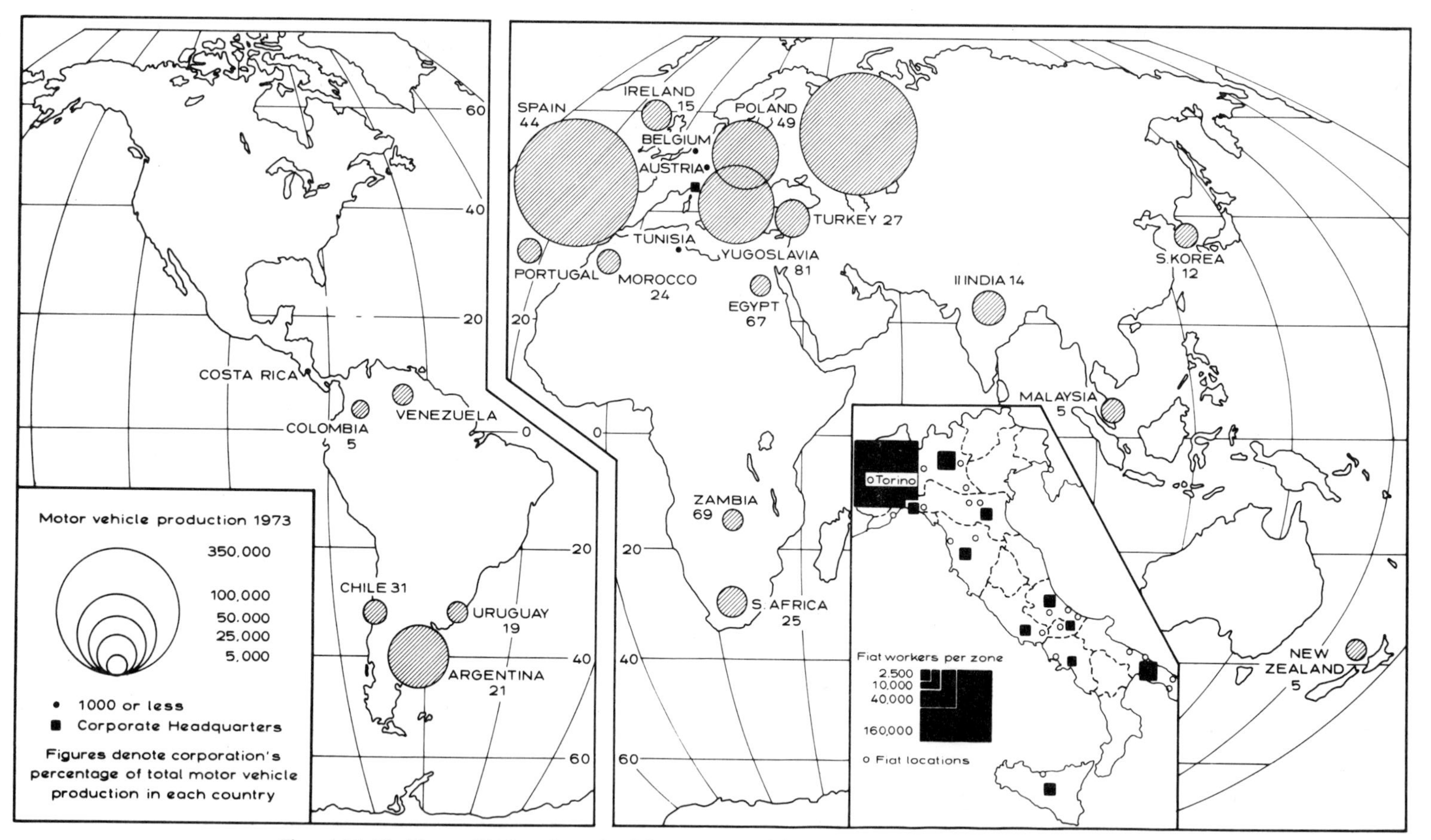

Figure 12–16 The multi-plant 'car giant' becomes a multi-national operation. Fiat's scale of operations in the 1970s and their world-wide locations (from data in Bloomfield, 1977; see Bibliography).

trol of the major American and European manufacturers everywhere outside Japan and Eastern Europe. Unlike the cotton textile industry noted earlier in this book the original manufacturers, if not the original countries, have not been undercut by products made elsewhere. One of the reasons for this is that the major motor vehicle market remains concentrated in the core regions (North America, Western Europe and Japan still account for 84 per cent of the world total of vehicles in use).

In expanding into new markets abroad, Bloomfield recognizes several stages:

1. Import of completely built-up (CBU) vehicles by local retailers. In this stage, the ready-for-the-road vehicle is dispatched to a peripheral country after branch sales offices are established. Back-up spares and supplies are also supplied from the home parent plant. Body building for basic trades and buses may be the only market-area production in the peripheral location.

2. Assembly of completely knocked-down (CKD) vehicles imported from the home plants of world manufacturers. Undertaken this way to save on ocean freight costs and avoid damage in transit. Also to allow vehicle modifications to suit local market conditions. This stage is only an overseas extension of existing branch plant operations at home. It persists if high tariffs on CBU imports are established by developing countries anxious to create employment by fostering local assembly lines. In the 1970s, Bloomfield reports that 37 countries were involved in CKD operations.

3. Assembly of CKD vehicles with increasingly locally-made content. A new stage brought about by the increased scale of adoption of motorized transportation consequent upon stages 1 and 2. The importing country steers this development by selection tariffs which encourage the deletion of 'less sophisticated' parts from imported CKD packs. New Zealand reached this stage in the 1950s.

4. Full manufacture of motor vehicles. The manufacture of vehicle bodies is the first step to this achievement, but the various economies of scale already noted make it difficult to bring to fruition. It is a stage, however, reached in Australia, Brazil, Argentina, Mexico, South Africa and several other countries with a large population and growing market.

Today, a reverse flow of vehicles and components is beginning to develop from the 'periphery to the core'. For example, Brazilian VW engine plants now supply the parent factories in Germany, and a small Ford plant in New Zealand makes gearbox parts fitted to all 'Cortina' production wherever it occurs. Perhaps a full and final stage may be reached when world-wide motor vehicle production is characterized by full 'complementation'—that is, an international integration through specialization in engines in one country, transmissions in another, body pressings elsewhere, with final assembly in each country. Such a development has already been pioneered by Ford and General Motors, principally to overcome some of the scale diseconomies of small manufacturing operations.

LOCATION FACTORS AND LOCATION THEORY

Most students of the early automobile industry are agreed that the first entrepreneurs relied on intuitive knowledge in their locational decision making. If they succeeded in their location, like Ford and Olds in Michigan, their location was 'right' and others followed. William Morris's choice of the unlikely location of Oxford is rather similar, and his acumen (as buyer, assembler and salesman) was much more important than his location. But the pioneering companies became large corporations, gathering expertise in marketing, transportation and real estate, all important for locational decisions. Thus the desirability of expanding in situ, nearby, at a distance from the parent plant, or even to relocate in a new region or country is carefully reviewed. After Bloomfield, we can summarize the most significant location-influencing factors:

(*a*) *Production-orientated*

Materials. While the value and volume of material supplies are substantial, they are also in great variety since the motor vehicle consists of well over 2000 separate components derived from a multitude of materials/processes. Generalizations are difficult, therefore, although the major centres of manufacture are generally within the material-producing and engineering regions (Figure 12–18).

Labour. While the industry depends on skilled designers, testers and prototype builders, the great proportion of the work-force is semi-skilled. The proportions in the UK are 25 per cent skilled/75 per cent semi-skilled. Increasing mass-production/automation decreases the need for skilled labour and may allow much greater locational choice for the larger-scale manufacturer. Regional variations in wage rates were important in the early development of the industry; wage differentials attracted US manufacturers to expand in Europe, and they are still an attractive influence in the development of new locations in developing countries. But availability and pro-

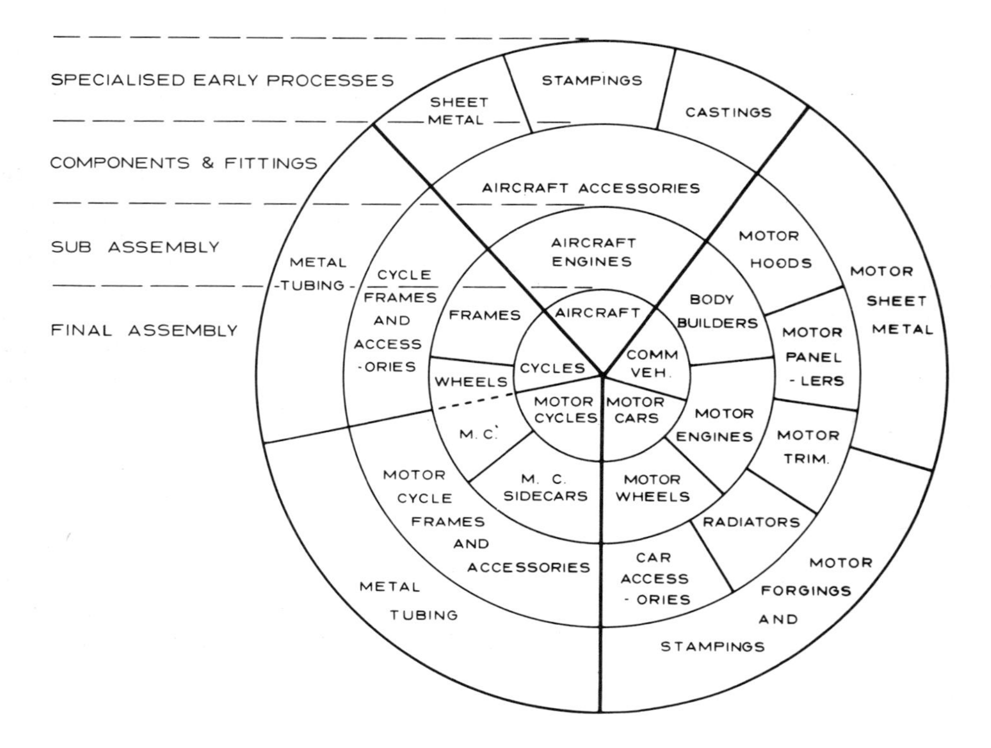

Figure 12–17 Ramifications of the motor industry in Coventry, 1940. The presence of the cycle, motor cycle, motor vehicle, and aircraft industries suggests strong agglomeration economies and hints at various kinds of industrial linkage.

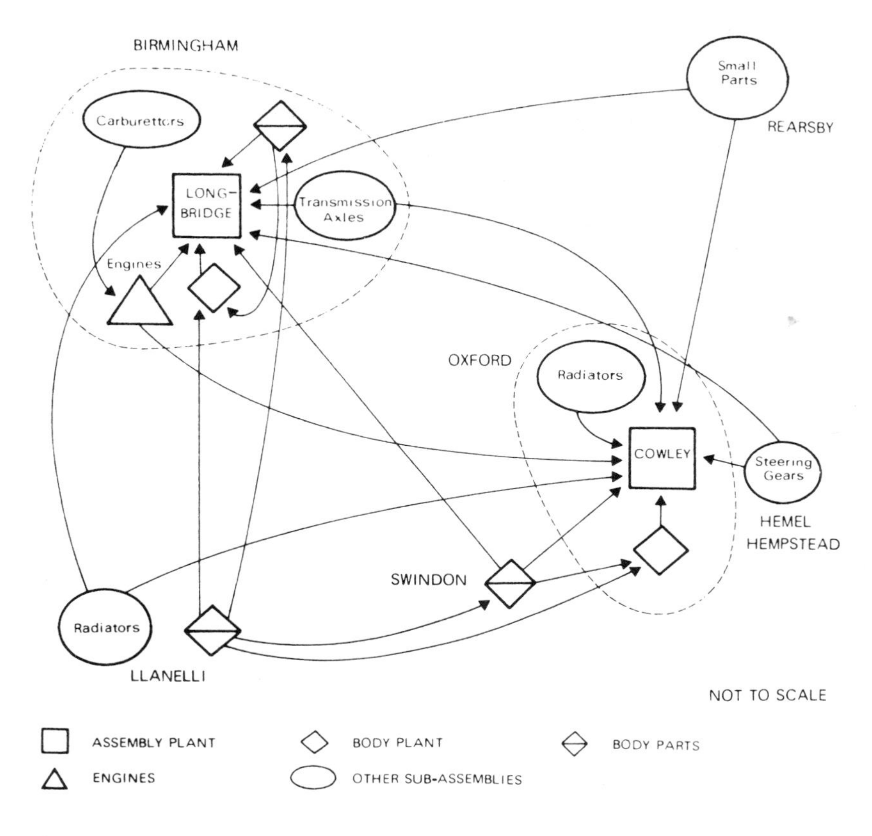

Figure 12–18 Inter-plant linkages between component suppliers and British Leyland car assembly plants in the Midlands (from Bloomfield, 1977; see Bibliography).

ductivity of labour is a more vital consideration. Labour demand was met in earlier decades by migration into Detroit, Birmingham, Coventry and Torino, while today it is often met by decentralizing into areas of high unemployment. Productivity is difficult to measure, but in vehicles per employee per year there is great variation (USA and Japan between 12–15; France and Britain 5–7).

Capital. Historically, there was greater ease in obtaining risk capital for motor manufacturing in some already successful locations. After the early years, however, capital availability within national areas did not appear to be of major locational significance, except as a barrier to new entrants.

Agglomeration. As the motor industry grew in some centres, the significance of external economies tended to increase. In places like Coventry, Detroit, and Paris, where an infrastructure of engineering and metal-working business existed (Figure 12–17), capital could be saved by the extensive, but local, sub-contracting of component production. Such areas also offered a pool of skilled and semi-skilled labour, and some of them also offered 'open-shop' (i.e. non-union) policies of local employers' associations.

(*b*) *Market-orientated*

Market access. Although the large metropolitan city regions in the USA and Western Europe have always been the principal market centres, comparatively few such places became early major production centres. Paris was a notable exception—and today Greater New York accounts for 8 per cent of US output and holds the US headquarters of eleven major European and Japanese car importers. We can generalize that large regional markets draw assembly plants away from the core areas, as in the USA where the majors established themselves in the Far West, the South and the East Coast. Ford, too, considered the farm market sufficiently large to build a series of regional assembly plants in Iowa, South Dakota, and Oklahoma.

Transportation distribution costs. Savings in transportation costs were a major element in the decision to establish regional assembly plants in the USA. By their development, manufacturers not only made freight savings on CKD shipments but also made profits on the practice of charging 'phantom freight' (charging of assembled car freight rate from Detroit to the customer's retailer, irrespective of the place of actual assembly). In recent times, the cost of railway delivery has been reduced (as a proportion of total costs) by the adoption of the unit train with roll-on tri-deck wagons. In the US, this innovation has led to the closure of some branch plants in California and Texas in favour of supply from the 'Motoropolises' in Michigan. At sea, too, the arrival of the 'drive-on' bulk carrier has enabled the large-scale and cheap delivery of European vehicles to the USA and Japanese vehicles to Western Europe. This phenomenon, too, has brought specialized 'autoports' into being at such places as Long Beach, California, Seattle, Washington, Halifax, Nova Scotia, and Liverpool, England.

In setting out production and market-oriented influences, and offering examples, it is possible only to see the complexities of the location patterns in the world's vehicle industries. There is little doubt that Weber would have been drawn to the analysis of the location problems they pose, for it has been rightly pointed out that the automobile industry stands for modern industry all over the globe. It is to the twentieth century what the Lancashire cotton mills were to the early nineteenth century: the industry of industries. But as we have seen, classical theories of industrial location for the most part attempt to determine the location of the single plant, and in the motor industry we have suggested that the initial factors in plant location appear to have been idiosyncratic. Today, the industry has core areas of production, but its plants in such locations are inter-linked with many more that have been spawned across the world. Concentration and dispersion have been mainly inspired by the profit motive, but Government involvement, whether in core or periphery, has also steered the industry into yet other new locations.

If the list of locational influences is long, we can see that the modern vehicle industry is neither tied to material sources nor limited to distinct market areas. We have previously referred to the term 'footloose' in relation to industries which were neither resource nor market-oriented, and some authors have applied this term to the motor industry. But it is also true that for more than fifty years the industry has been capital-intensive, with its high fixed costs of manufacturing plant giving a strong tendency towards geographical inertia. To Hoover, labour and management were critically-important factors in explanation:

> *'When an industry is young and its problems unfamiliar, it prospers best in those few places which provide the combination of appropriate basic skills (generally developed in pre-existent similar industries), together with experienced managers and some venturesome enterprisers and financial backers. The product is then perfected and standardized, the best methods of cheap large-scale manufacturing are worked out in those places, and the economies of mass production and geographical concentration assert themselves.'* (Hoover, E. M., *The Location of Economic Activity*, New York: McGraw-Hill, 1948.)

Hoover's comment, however, dates from a period before the great motor corporations embarked fully on their modern course of dispersed manufacture almost worldwide. On the broad spatial dimensions of the industry, more recent theorists emphasize a 'zone of locational tolerance', while on the detailed scale of site selection the new behavioural school of location theorists (Pred and others; see Bibliography) emphasize the role of the entrepreneur, the scale of output, and change over time.

Finally in this case study, we may note Bloomfield's view that the industry today provides almost every type of location pattern and, significantly, he advances the view that change over time is the key to locational understanding. He promotes this by means of an 'idealized sequence of phases in location of automobile manufacturing' (Figure 12–19). Like other models of this kind, it is an empirically-based stage-model, the main aim being to identify important locational elements:

Phase I Experimental. Widespread entry of considerable numbers of vehicle makers, few with more than a small output and even fewer succeeding with a genuine commercial production. Workshops dispersed and deliberate site selection not very important because entrepreneurs most concerned with technical success. Some geographical effects of early business leadership and success reflected in beginnings of agglomeration where founders located.

As France, Britain and the United States, 1890–1901

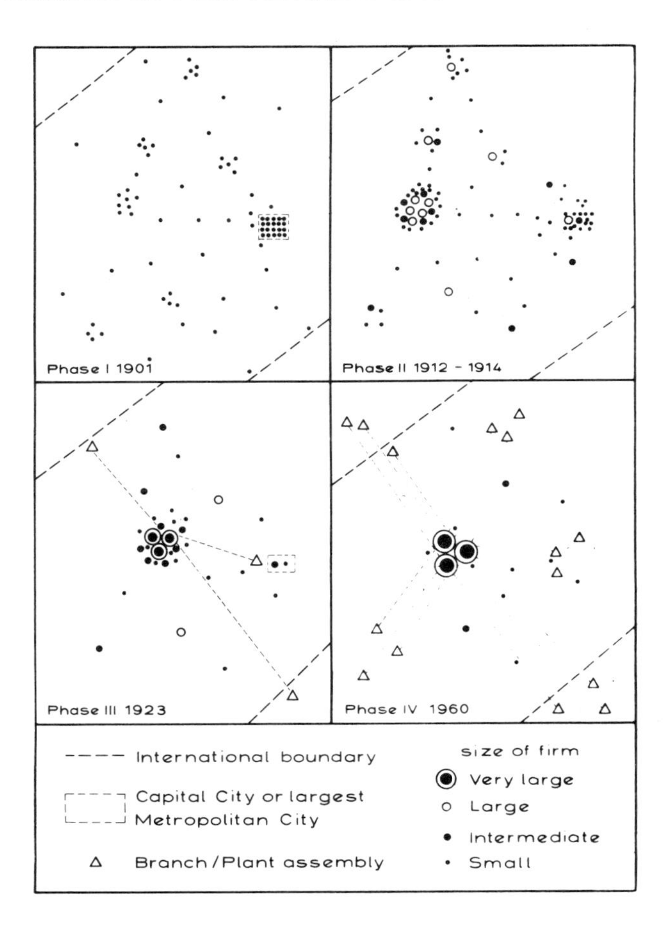

Figure 12–19 Idealized sequence of phases in the location of automotive manufacturing. (Redrawn from Bloomfield, G. *The World Automotive Industry*, Newton Abbot: David & Charles, 1977.)

Phase II Small-Scale Engineering Production. New firms continue to enter, despite growth in size of successful firms. Skilled labour emphasis as factory production developed all characteristics of an established precision-engineering industry. Some degree of localization becomes apparent, since for some firms there are advantages in agglomeration. Degree of geographical concentration varying from country to country (Britain and Germany more dispersed patterns than France or Italy). Motorcar still luxury product, so wide spatial margin of profitability. Many firms anchored in original location, but some new and established firms deliberately select new sites. Some locational characteristics of this phase persist into the next in low-volume car and specialist commercial vehicle sectors.

As USA from 1900–1914, extended to 1920s in Western Europe.

Phase III Mass Production. Developed on foundations of Phase II. Limited number of firms converted because of high capital cost of production plant. Industry more concentrated in both structure and location. Increased mechanization emphasizes semi-skilled and unskilled labour. Mass production and mass marketing allows continuation of leading agglomerations (Detroit, Paris, Torino), the survival of independent locations (Peugeot at Sochaux, Doubs, France) and the rise of new centres (Oxford). Concentration of production in main agglomerations varying considerably between majors, and partly dependent on location of market leaders. Some deliberate site selection of new integrated plants (Ford), but in general, mass producers expanding on existing sites selected in early phase (Austin, Longbridge, Birmingham).

USA in this phase 1913–1917, Western Europe in early 1920s and Japan in 1950s.

Phase IV Industrial Maturity and Dispersion. Product design and methods standardized by beginning of this phase. Market expansion continues, but at slower pace than before. Reduced retail price (in real terms) of finished vehicles makes regional and world scale distribution cost a more significant element; thus dispersion of assembly facilities for mass market. Movement of both assembly and complete manufacture from major agglomerations, related with problems of labour supply, management, obsolescence, and technical change (this phase in USA almost parallel with phase III, but West European producers not entering it until 1945). Continued growth of mass market encourages dispersion of production throughout major areas of the world (1955–1975). Most plant site selection in this phase deliberately located by large corporations, with increased government involvement in the location process.

BIBLIOGRAPHY

Aaronovitch, S., and Sawyer, M. C. 'The concentration of British manufacturing'. *Lloyds Bank Review*, 1974, pp. 14–23.

Alexandersson, G. 'Changes in the location pattern of the Anglo-American steel industry, 1948–1959'. *Economic Geography*, April 1961, pp. 95–114.

Bannock, G. *The Juggernauts: The Age of the Big Corporation*. Harmondsworth: Penguin Books, 1971.

Beacham, A., and Osborn, W. T. 'The movement of manufacturing industry'. *Regional Studies*, 4, 1970, pp. 41–47.

Berry, B. J. L., Conkling, E. C., and Ray, D. M. *The Geography of Economic Systems*. Englewood Cliffs, N.J.: Prentice-Hall, 1976.

Beesley, M. 'Birth and death of industrial establishments: experience in the West Midlands conurbation'. *Journal of Industrial Economics*, 4, 1955, pp. 45–61.

Birch, B. P. 'Locational trends in the American car industry'. *Geography*, 51, November 1966, pp. 372–375.

Bloomfield, G. *The World Automotive Industry*. Newton Abbot: David & Charles, 1978.

Britton, J. N. H. 'A geographical approach to the examination of industrial linkages'. *Canadian Geographer*, 13, 1969, pp. 185–198.
Chapman, K. 'Agglomeration and linkage in the United Kingdom petrochemical industry'. *Transactions of the Institute of British Geographers*, 60, 1973, pp. 33–38.
Chisholm, M. 'In search of a basis for location theory'. *Progress in Geography*, 2, 1971, pp. 111–133.
Dicken, P. 'Some aspects of the decision-making behaviour of business organizations'. *Economic Geography*, 47, 1971, pp. 426–437.
Dunning, J., ed. *The Multinational Enterprise*. London: Allen and Unwin, 1971.
Estall, R. C. *New England: A Study in Industrial Adjustment*. London: Bell, 1966.
Fleming, D. K. 'Coastal steelworks in the common market countries'. *Geographical Review*, January, 1967, pp. 48–72.
Florence, P. S. *Investment, Location and Size of Plant*. Cambridge: CUP, 1948.
Gardner, C. M. *The Economics of Soviet Steel*. Cambridge, Mass: Harvard University Press, 1956.
Goodwin, W. 'The structure and position of the British motor vehicle industry'. *Tijdschrift Voor Economische En Sociale Geografie*, 61, 1970, pp. 148–159.
Gwynne, R. N. 'The Latin American motor vehicle industry'. *Bank of London and South American Review*, September 1978, pp. 2–11.
Hall, P. *The Industries of London*. London: Hutchinson, 1962.
Hamilton, F. E. I., ed. *Spatial Perspectives on Industrial Organization and Decision Making*. London: Wiley, 1974.
Hamilton, F. E. I., ed. *Contemporary Industrialization: Spatial Analysis and Regional Development*. Harlow: Longman, 1978.
Hayter, R. 'Corporate strategies and industrial change in the Canadian forest product industries'. *Geographical Review*, 66, 1976, pp. 209–228.
Heal, D. W. *The Steel Industry in Post-War Britain*. Newton Abbot: David & Charles, 1974.
Hoover, E. M. *The Location of Economic Activity*. New York: McGraw-Hill, 1948.
Hurley, N. P. 'The automotive industry: a study in industrial location'. *Land Economics*, February 1959, pp. 1–14.
Keeble, D. E. *Industrial Location and Planning in the United Kingdom*. London: Methuen, 1976.
Keeble, D. E., and McDermott, P., eds. 'Organization and industrial location in the United Kingdom'. *Regional Studies*, 12, No. 2, 1978, pp. 139–265.
Kennelly, R. A. 'The location of the Mexican steel industry'. In *Readings in Economic Geography*, Ed. Smith, R. H. T., Taaffe, E. J., and King, L. J. pp. 126–157. Skokie, Illinois: Rand McNally, 1954.
Krumme, G. 'The inter-regional corporation and the region'. *Tijdschrift Voor Economische En Sociale Geografie*, 61, 1970, pp. 318–333.
Luttrell, W. F. *Factory Location and Industrial Movement*. London: National Institute of Economic and Social Research, 1966.
Martin, J. E. *Greater London: An Industrial Geography*. London: Bell, 1966.
Massey, D. 'Towards a critique of industrial location theory'. *Antipode* 5, No. 3, 1973, pp. 33–39. (Reprinted in Peet, R., ed. *Radical Geography: Alternative Viewpoints on Contemporary Social Issues*. Methuen: London, 1977, pp. 181–197.)
McNee, R. B. 'Functional geography of the firm, with an illustrated case from the petroleum industry'. *Economic Geography*, 34, 1958, pp. 321–327.
Parsons, G. 'The giant manufacturing corporation and balanced regional economic growth in Britain'. *Area*, 4, 1972, pp. 99–103.
Pred, A. 'Toward a typology of manufacturing flows'. *Geographical Review*, 1964, pp. 65–84.
Pred, A. 'The concentration of high value added manufacturing'. *Economic Geography*, 41, 1965, pp. 108–132.
Rees, J. 'The industrial corporation and location decision analysis'. *Area*, 4, 1972, pp. 199–205.
Riley, R. C. 'Changes in the supply of coking coal in Belgium since 1945'. *Economic Geography*, 1967, pp. 261–270.
Rogers, A. L. 'Industrial inertia: a major factor in the location of the steel industry in the US'. *Geographical Review*, 42, 1952, pp. 56–66.
Rogers, H. B. 'The changing geography of the Lancashire cotton industry'. *Economic Geography*, 38, 1962, pp. 299–314.
Smith, D. M. *Industrial Location*. New York: Wiley, 1971.
Taylcr, M. J. 'Location decisions of small firms'. *Area*, 2, 1970, pp. 51–54.
Taylor, M. J., and Wood, P. A. 'Industrial linkage and local agglomeration in the West Midlands metal industries'. *Transactions of the Institute of British Geographers*, 59, 1973, pp. 127–155.
Townroe, P. M. 'Industrial linkage, agglomeration and external economies'. *Journal of the Town Planning Institute*, 56, 1970, pp. 18–20.
Turnock, D. *Eastern Europe: Studies in Industrial Geography*. Westview: Dawson, 1978.
Wallace, I. 'The relationship between freight transport organization and industrial linkage in Britain'. *Transactions of the Institute of British Geographers*, 62, 1974, pp. 25–43.
Warren, K. *The American Steel Industry, 1850–1970: A Geographical Interpretation*. Oxford: Clarendon Press, 1973.

Watts, H. D. 'The location of aluminium reduction plants in the UK'. *Tijdschrift Voor Economische En Sociale Geografie*, 61, 1970, pp. 148–159.
Watts, H. D. 'Spatial rationalization in multi-plant enterprises'. *Geoforum*, 17, 1974, pp. 69–76.
Wise, M. J. 'On the evolution of the jewellery and gun quarters in Birmingham'. *Transactions of the Institute of British Geographers*, 15, 1949, pp. 57–72.
Wood, P. A. 'Industrial location and linkage'. *Area*, 2, 1969, pp. 32–39.

Part Four

ECONOMIC ASPECTS OF URBAN LOCATION

'Cities are the focal points in the occupation and utilization of the earth by man. Both a product of and an influence on surrounding regions, they develop in definite patterns in response to economic and social needs.'

Harris, C. D., and Ullman, E. L. 'The nature of cities', *Annals of the American Academy of Political and Social Science*, November, 1945, p. 7.

13

Bases of Urban Location: Interurban and Intraurban

It is the purpose of this chapter to provide exploratory insights into the nature, function, and locational characteristics of cities. The most fundamental reason for this focus is that the location of cities over the earth is a critical and paramount base from which other activities can be understood. In fact, with cities as a 'given', the distribution of many other activities can, in crude fashion, be ascertained. Hence in earlier chapters the locations of cities was treated much as transportation routes, physical features, and various national and institutional features. For example, the location of agriculture was partly explained by the necessary proximity of certain crops to markets. Thus when accessibility to markets is under discussion, what is really being discussed is the spatial relationship of particular activities to cities, the places where most of the people are found. Cities are both a cause and an effect of manufacturing inasmuch as they are the places that have the available labour force and the channels of transportation so necessary for the acquisition of raw materials and the distribution of finished products to various other markets. In most cases, it is difficult to ascertain which is the cause and which is the effect. In practice, both are simultaneously interrelated. For purposes of graduated understanding, however, we may treat each activity separately and as if it were locationally flexible, while treating its associated activities as if they were spatially fixed.

THE IMPORTANCE OF CITIES

The importance of cities for spatial understanding was evident in the earlier chapters, where cities were treated as the prime controlling force in the placement of any given activity. Much of the discussion of the extractive activities, agriculture, and industry used cities as the fixed locational element in providing explanation of patterns.

Cities are also important as leaders for the degree of specialization over the earth's surface. They can occur only in areas of territorial and occupational specialization. Thus the more territorially specialized an area, the greater will be the urban response. In this regard, cities offer the best reflection of technological and economic progress in an area. Those areas with no cities (discussed in Chapter 7), which are therefore the least economically advanced places on earth, have not met the basic prerequisites for cities.

Cities are also important because of their control over the territory around them, that is, their tributary areas. They are important out of all proportion to the areas they occupy. First, they contain over one-third of the world's population (generally from 60 to 85 per cent in industrially advanced countries). But the land actually occupied by cities accounts for less than ·1 per cent of the earth's surface. The sheer concentration of population in such limited space, in itself, makes cities important places of geographic interest. But most important, cities contain those decision-making functions that set the stage for the use of the earth. It is from them that government and law radiate; they contain the commodity and stock exchanges, the financial institutions and capital reserves, the major concentrations of manufacturing and industrialization; they also contain the major libraries and institutions of higher learning and the arts; they are the centres of innovation, brainpower, and communication. In short, they include the controlling elements for the use of earth space.

The above comments are not meant to imply that cities can operate independently of rural areas. To the contrary, it is the activities in these areas that give cities their *raison d'être*. Because cities fundamentally are important only in conjunction with such territory, however, does not mean that they are in an inferior position. Here is a case where logic might suggest that the country should control the city, but in practice it is the other way around. In a geographical context, however, the country and the city are intertwined in one complementary system; both are critical parts of a highly complicated whole.

Unfortunately, there are no hard-and-fast rules as to just what is rural and what is urban territory. In fact, such criteria vary widely among specialists in urban studies and must be modified to fit various areas of the world. Several criteria are commonly used to differentiate urban from rural land. First, there must be a certain amount of incorporated population. In the United States, such places must contain at least 2500 persons. In many European countries, an incorporated place is not considered to be urban unless it has 5000 or more population. Second, population is an almost universal measure for urbanization. In the United States, the general breaking point between rural and urban is 800 persons per square kilometre.

However, some of the so called farm villages in China and elsewhere in Asia often have populations of 10 000 or more and have population densities exceeding 2500 persons per square kilometre. Nonetheless, this hardly strikes a westerner as meeting the essential criteria of a city since most functions characteristic of cities are missing. A third measure often concerns the occupation of the inhabitants: for an area to be considered urban, the majority of its nucleated population, must be engaged in nonagricultural pursuits. This is an arbitrary criterion, but it does solve the knotty problem of Chinese farm villages. Finally, a city's functions and road pattern, as well as the amount of land covered by its structures, will be discernibly different from such features in the surrounding rural area.

But perhaps the easiest way to visualize and delimit the kinds of urban areas, or cities, we will be discussing is simply to examine an aerial photograph of an urban place. The outer edge will be ragged and somewhat indefinite, but the general outline of such urban territory will be clear. The formal municipal boundaries have very little effect on the territorial extent of urban development in an urban area. We might thus term this built-up area and urbanized complex the true or geographic city, as distinct from a municipality. In this book the term city will be used as a synonym for geographic city.

In the United States the nearest formal delimitation of a geographic city is the designation *urbanized area* used by the 1970 US Census. Such an area includes all municipalities with 50 000 or more population, plus surrounding territory that meets certain political, demographic, and density criteria.

TYPES OF CITY PATTERNS: INTERURBAN APPROACHES

Most satisfying to the geographer are definitions of cities that include an explanation of their spatial pattern. Why are cities placed where they are? Why are some larger than others? Why are some growing and others declining? How is it that some parts of the world have many cities and other parts none?

The first approach to this is to categorize city types within such a framework. In one sense, there are hundreds of different city types. Some specialize in making aeroplanes, some in making steel; others are state capitals, university towns, resorts, ports, and so forth. From a geographical viewpoint, however, there are four major types that make spatial sense: (1) those that respond to transportation, (2) those that respond to territory, (3) those that respond to resources on site, and (4) those that respond to the institutional needs of people. Each of these types has a particular locational pattern that is readily identifiable on a map.

It should be evident also from the above that cities do not just happen, nor are their locations and particular functions accidental. Where a city is located and what it does are directly affected by the demand for such functions as transportation, territory, or resources, and by the demands of people. Thus in order to understand the location of cities, five major items are considered: (1) the distribution of rural population, (2) the location of man's extensive space-using occupations such as agriculture, forestry, and mining, (3) the location of transportation routes, (4) the industrial structure of government needs as reflected in geographical space, and (5) the functions that cities generally perform. However, it must be quickly pointed out that these things are also, in turn, much contaminated by the location of cities.

PATTERN WITHIN CITIES: INTRAURBAN APPROACHES

The second part of this chapter will attempt to provide a basis for understanding the placing of people and functions within cities. Attention will be given not only to *how* the land within cities is used, but to *why* it is so used. It will be demonstrated that the land use patterns of cities are not only quite understandable, but are spatially orderly. These patterns can be understood by using concepts and theories much as they were used in the preceding analysis. In fact, such principles possibly provide more powerful locational explanations for intraurban activities than they did for the activities previously studied.

Intraurban spatial study also includes understanding urban problems like decline of the central business district and the central city, marginal business districts, decentralization, and sprawl. Solutions to these problems and improvements in the quality of the urban environment and the urban fabric must surely stem from a solid base of intraurban spatial appreciation and understanding.

Our approach to the city is therefore two-fold, and we first tackle its locational and economic characteristics from the outside before looking closely within. Both approaches, we will find, can make extensive use of model-building related to many of the theories, concepts and principles so far used in this book, together with the introduction of new ones.

Interurban Considerations

CITIES THAT RESPOND TO TRANSPORTATION

Cities affect the location of transportation routes possibly more than any other single feature of the earth. In fact, they create the demands for movement. The major things a city needs are contact, communication, and connection with the outside world. A city without transportation cannot exist; but cities also respond to transportation conditions. For example, cities are needed where breaks occur along a transportation route, the most spectacular of which is the break between land and water where a change of transportation mode is necessary. Workers are required for such a transfer, and this further creates opportunities for adding value, through manufacturing, to the commodities being moved. Thus, all port cities are, in part, transportation cities of this type. As such, port cities are also strategic transfer points that serve the land area behind them (their hinterland) and bring in or ship out produce to and from areas in front of them (their forelands). Inasmuch as ocean water transportation opens up numerous trading possibilities in different parts of the world, port cities are possibly the most numerous; they are clearly the most numerous large cities (type A in Figure 13–1).

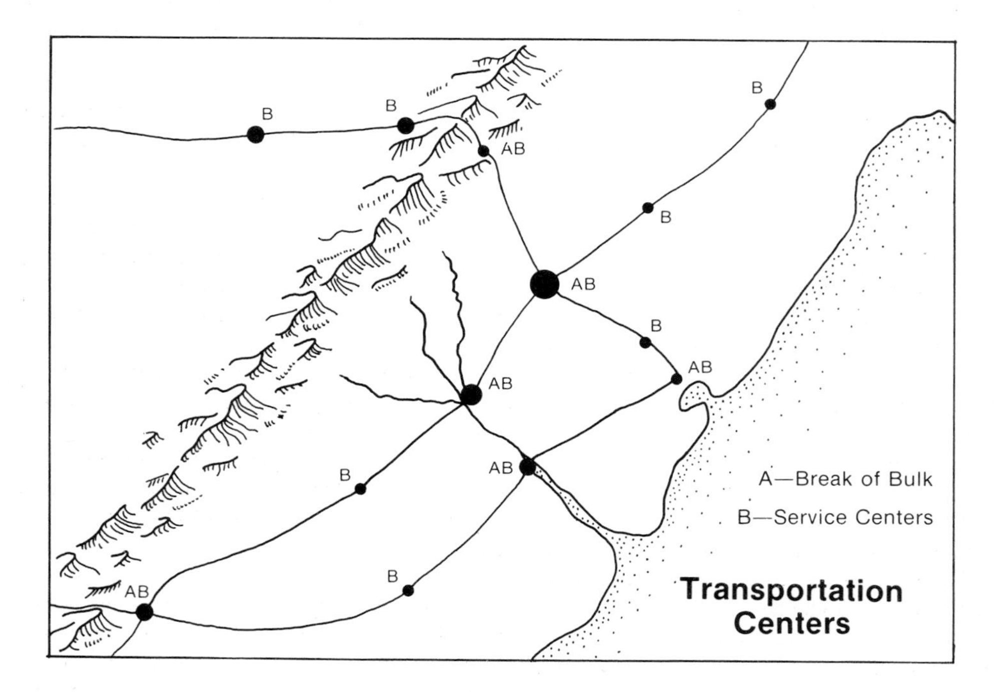

Figure 13–1 A pattern of transportation type cities. Type A cities are break-of-bulk 'scores', and type B cities represent service centres designed to aid the flow of movement along the routes.

But break-of-bulk also occurs along transportation routes on smaller scales, as between a plains area and a mountainous region, or where land routes cross major rivers. Such topographically based breaks may create new kinds of transportation responses. Thus cities have developed in many of these places to serve such transfer needs. River towns like those on the Rhine are an extremely common example of this type. Where river transportation is the dominant route, cities develop at barriers along this route (for example, the waterfalls at Cincinnati) or at strategic places (for example, where a change from large craft to small craft is made, as at Duisburg, or where an ice-free river becomes a periodically frozen river, as at St Louis). Changes in topography also result in transfer adjustments. Perhaps extra diesel engines will be required before a mountainous area is traversed, or people will wish to rest before engaging in a difficult portion of a route. Consequently, cities often locate at such topographical breaks.

Finally, cities respond to transportation-induced opportunities at strategic points created by routes. The best known is the crossroads or junction position, a break-of-bulk opportunity. Here transfer of people and freight from one route to the other might be required. Consequently, this is a good place for railroad consolidation yards or like functions. Such centres are really transportation hubs where a number of transportation routes converge. Atlanta, Salt Lake City, and in fact most large cities in the interior of the USA, have some activities based on these needs.

Cities also respond in a periodically spaced fashion along routes that have no physically based differences (type B in Figure 13–1). At intervals, stops are required for rest, food, repairs, or other services. The spacing required varies from one mode to the other, but generally increases as one proceeds from highway to railway to air. Spacing has also tended to increase over time as improvements are made among the modes. Thus coal-burning railroad engines needed many watering stops (whistle stops) as compared to diesel engines. Road transportation needed far more stops with horse powered than with gasoline-powered vehicles. It needs even fewer today in countries that have developed extensive motorway systems. Consequently, a number of former transportation service centres located like beads along major transportation lines, have been abandoned.

Other questions are also raised. Why, for example, should cities be on one side of the river rather than on the other? Generally, the reasons for such site features are multiple and are based on the direction of the traffic and the direction from which the route was built, on physical factors such as a high and low side of the river, or on institutional factors such as formerly-important political boundaries. The precise location of the port and the routes themselves are also the result of a combination of factors such as the time at which the routes were constructed (that is, sequence in construction), the physical conditions affecting route alignments, the best place to cross streams and mountains, the best place to anchor a ship, and the external regional requirements that dictate the position of the routes within the study area.

If such places are developed initially in relation to such circumstances, other functions also find it advantageous to locate there. Some manufacturing plants can reap considerable benefit by locating at break-of-bulk points, and others may find that locating at strategic crossroads positions gives them real advantage in assembling raw materials and distributing products to a wide market area. Thus manufacturing of these types (discussed in Chapter 12) might be thought of as transportation-induced manufacturing as distinct from other manufacturing types.

Note in Figure 13–1 that even among transportation cities there is an overlap of different kinds of such cities in various places. Thus type A and B cities are sometimes found separately, sometimes together. They may also be duplicated. The larger the city, generally the higher the score of particular urban functions. Note for example that the port city shown has five scores: it is a break-of-bulk between land and water for two different routes (two); it is a periodic service centre for two different routes (two) and it is a crossroads point for three routes: two land and one water (one). Therefore its symbol of importance might be symbolized AAABB. On the basis of transportation response alone, we would expect cities of different sizes to develop.

The overall pattern is one of alignment and linearity along routes. The general feature to look for in such a pattern, aside from periodically spaced centres, is the general absence of clustering of cities in any one location.

CITIES THAT RESPOND TO TERRITORY: CENTRAL PLACES

Another common characteristic of cities is their relatively evenly spaced pattern, generally most prominent in agricultural areas. Cities in this kind of area locate centrally to this rural market in order to provide the goods and services needed by the people in the surrounding area. This centralizing location principle is one of the more general features in the placement of all activities. It is particularly critical with regard to the site of certain kinds of cities, namely, those formally labelled *central place cities*.

The general characteristics of agriculturally based central places are fairly easy to visualize. Farmers and their families need various goods and services they cannot provide for themselves, including food and clothing, and legal, educational, governmental, and social services. Second, farmers need a market; thus grain elevators, warehouses, and the like are often adjunct features of such central places. Moreover, when farm commodities are perishable or otherwise difficult to ship, opportunities often exist in the city for manufacturing. Thus processing and other nondurable manufacturing are also often characteristic 'services' of a central place city. With the money farmers get for their crops, they not only buy goods and services, but require machinery, seed, and other things critical to the operation of their farms. Thus a typical central place is not simply a location for providing retail and service items, but is a complex reflection of and response to a multiplicity of demands in a surrounding area. As such it is a reflection of the opportunities for economic activity that the production of farm commodities affords. These things all combine to create jobs in the central place.

Finally, the people who are directly engaged in central place activities within a city also require goods and services. Thus a second round of demand functions occurs. In some cases, these people merely augment the number of customers otherwise available for an establishment and hence increase profits; in other cases, the number of customers is raised sufficiently to justify new stores that would not otherwise be possible with only the rural area customers.

As might be expected, there is a formal theory that attempts to explain the number of cities, their distribution, hierarchy of sizes, and the territory controlled by each place in the hierarchy, given such a dispersed market. This is called the *central place theory* and was originally formulated by Walter Christaller and published in German in 1933 (see Christaller, W., *Central Places in Southern Germany*, Englewood Cliffs, N.J.: Prentice-Hall, 1966. This is a translation by Carlisle W. Baskin of Christaller's *Die Zentralen Orte in Suddeutschland*). The theory is based on three primary principles: (1) centralization as a principle order, (2) economic distance and the range of a good or service, and (3) the nature of complementary regions, that is, trade areas of different order, or size, of places.

Given an even spatial distribution of customers and given the inability to locate a business adjacent to each customer, the best locational strategy is to locate centrally with regard to those customers who accumulatively will constitute a viable market. Cities will therefore locate at the place that minimizes transportation costs and time to the potential customer market. If the people are scattered over an even surface and if transportation is equally free in any direction, then such business (and hence urban places) will be evenly spaced over the landscape, with each one serving an equal number of surrounding customers. The territory each serves cannot, of course, be a circular one, inasmuch as circles would create uncovered or overlapping spaces. Such space would therefore result in competitive zones. In exercising their options to minimize transportation to each market, the consumers would recreate zones of a hexagonal nature, the geometric form nearest to a circle, which work in such a fashion as to leave no unused spaces.

But what size is necessary to support any given function, such as a service station, a department store, a grocery store? Clearly, there is a different-sized minimum market necessary for each activity to carry on a viable business. The number of persons required will also depend on consumption patterns, incomes, and other variables, but we shall assume these to be homogeneous over the rural surface. In this case the minimum number of persons, the *threshold* level, for each function can be estimated. Thus there are establishments that can be maintained with very few customers, for example, service stations and grocery stores, and hence are called *lower order functions*, and establishments offering goods or services that require a large number of customers and hence are labelled *higher order functions*.

In terms of absolute number of customers higher order functions often require fewer total customers than a lower order function. Lower order functions are those that are participated in by a larger proportion of the general population than are higher order functions. Thus everyone eats and thereby has interaction with a food store or supermarket (lower order functions) but not everyone buys jewellery (higher order function); nor does the average person go to the jewellery store as often as to the grocery store.

According to this theory, there are about seven levels, a hierarchy of seven places (or cities) of different order, that result from these differing market sizes. By thinking about these functional levels in terms of such concrete things as hamlets, villages, towns, cities, and metropolises, we can better visualize this hierarchy. Such functions locate together in a central place as a result of certain linkages and aggregation economies mentioned in Chapter 4 and explained in detail in Chapter 14.

A threshold level is the approximate number of persons in a region necessary to support any particular establishment or function. Thus if there are groups of establishments, each having similar threshold levels, a hierarchy of places should occur over space. But the settlement pattern would still be one based on the original, even spacing of farmers and hamlets. What happens is that those higher order functions will locate at an already established lower order centre, again in a way central to the market, and the result will be the creation of the next higher order centre. For example, if town functions centralized at a former village (V) site, then that village would now move up the hierarchy to town (T) status. However, it will continue to contain all the functions of a village (although the number of establishments would be much greater) as well as those of the town.

At this point, the distinction between number of *functions* and number of *establishments* should be clarified. For example, a village may have two service stations, three food stores, two barber shops, and one drugstore. In this case, it has only four functions, but eight establishments. Thus, a town might have five service stations, four food stores, four barber shops, and three drugstores, but it has the same number of functions as the village above; it has more establishments, however, because of a greater market area. Hence a typical town has more establishments than a village.

A hypothetical distribution of central places is shown in Figure 13–2, for the same area used for the analysis of transportation influences. Several things are quite different from the transportation city pattern. First, the distribution of places is highly regular; an even spacing prevails, although disrupted by physical features. (Transportation routes for the moment are assumed to be different from those shown in Figure 13–1.) Second, there are many more central place nuclei than there were 'transportation cities'. Third, there is a definite order in the number of different-sized places. There are 57 villages,

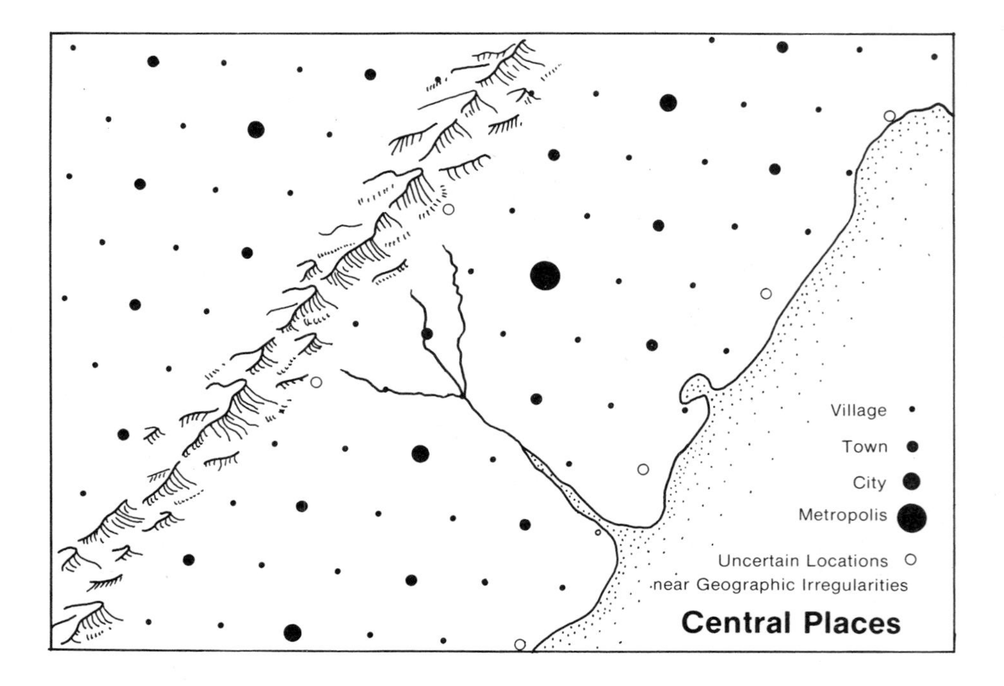

Figure 13–2 A possible pattern of central places. Note the even spacing in the 'plains' area.

20 towns, 4 cities, and only one metropolis, or 82 places in all. Even so, the physical factors are somewhat distorted from the ideal. If we had a completely uniform plane, there would be a progression in number of places by threes, for example, 1:3:9:27:81, whereby there would be 1 metropolis, 2 cities, 6 towns, 18 villages, and 54 hamlets, thus making 81 places in all. The reason this area has more cities and towns than expected is because we are not dealing with a single, uniform region. Some of the places are dominated by higher order centres outside the area shown on the map. Also, only a hierarchy of four different-sized places is shown. Nonetheless, this should suffice to demonstrate the manner in which territory is divided.

The second major spatial principle of this theory involves the range of a good or service. The notion of range is based on the assumption of threshold level, identical with the *inner range of a good*, and the central placement of various activities to serve this customer level. Thus inner range is the geographical equivalent of the marketing concept of threshold. Threshold provides information on the minimum market necessary to viably support a particular establishment or function, whereas the inner range is the mimimum territory around the establishment or function necessary to provide the minimum market (Figure 13–3). It is also based on the competition with other nearby centres of equal order of importance. Given this arrangement, the area from which each similarly sized centre will draw customers would be halfway between each (Figure 13–3). Note that in the example, customers to the left of the ideal range travel to town A in order to get the item at lowest cost; those to the right of the ideal range will travel to town B for the same reasons. Thus the breaking point for customers is precisely halfway between the two centres if both towns are identical in price of goods and in size. According to the pure central-place system, the breaking points will always be halfway between all similarly labelled places.

The three aspects of range are illustrated in Figure 13–4. Note that each central place has an *inner* and an *outer* range for each good or service. The inner range is the

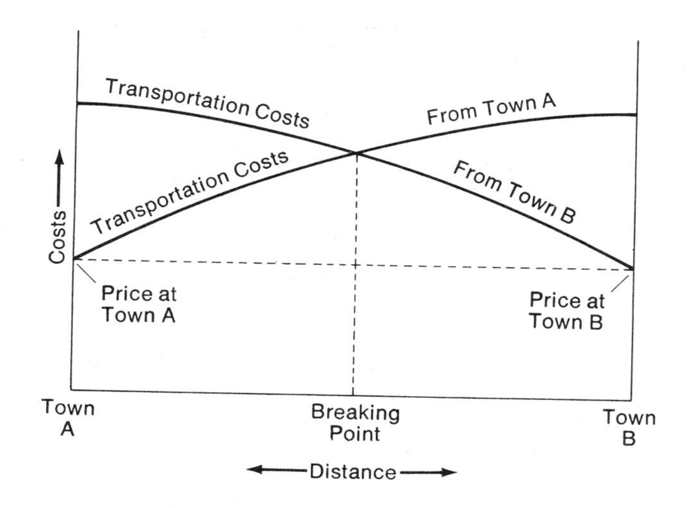

Figure 13–3 Breaking points between two towns. This assumes that customers will shop in the town that offers goods at least cost when the customer's travel costs are taken into consideration.

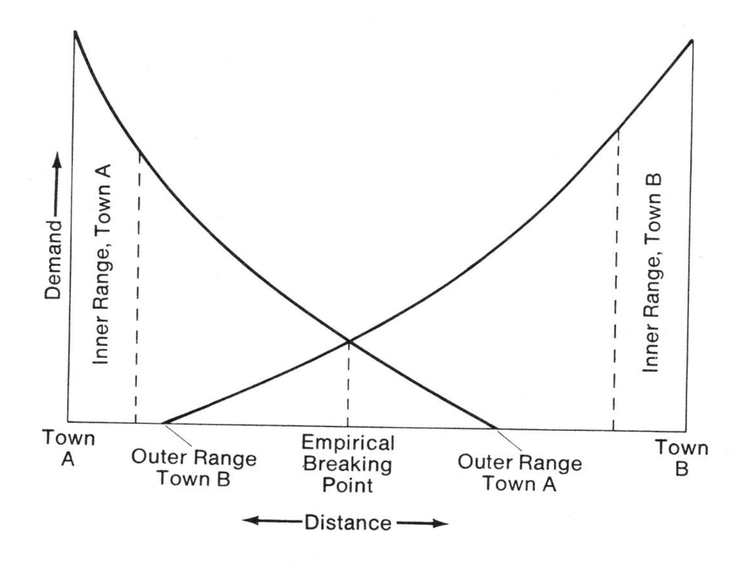

Figure 13–4 Inner and outer ranges of goods and services. Note that the outer range is the furthest distance people will travel to purchase an item in a community. The inner range is the furthest distance people would have to travel to constitute the minimum market population for a particular good or service.

ideal empirical range and is identical with the concept of threshold, or minimum customer requirements, just discussed. A smaller number of customers will keep a business from operating at a profit. The outer range is the limit of territory from which customers would theoretically come to any given centre. The empirical range is the actual range of customers as determined by competition among central places.

Careful examination of Figure 13–5, however, which illustrates town trade areas, indicates that all towns are not the same in terms of their potential trade areas. Many towns along the coast and near mountains have less trade area and hence fewer customers than towns further inland. This means that they are not able to achieve customer levels to the ideal limit (as shown in Figure 13–3), but perhaps only to a level slightly beyond the threshold. Thus in competing with full-status towns they are not going to fare very well. Other areas have 'no-man's lands' near them which may logically fall to them. For example, the breaking point between A and B, for instance, will not be the hypothetical dashed line of Figures 13–3 and 13–4 but will shift to the right, so that town A will become more important and town B will become less important. Both towns may have the same number of functions, but they will vary greatly in number of establishments and hence in territory controlled.

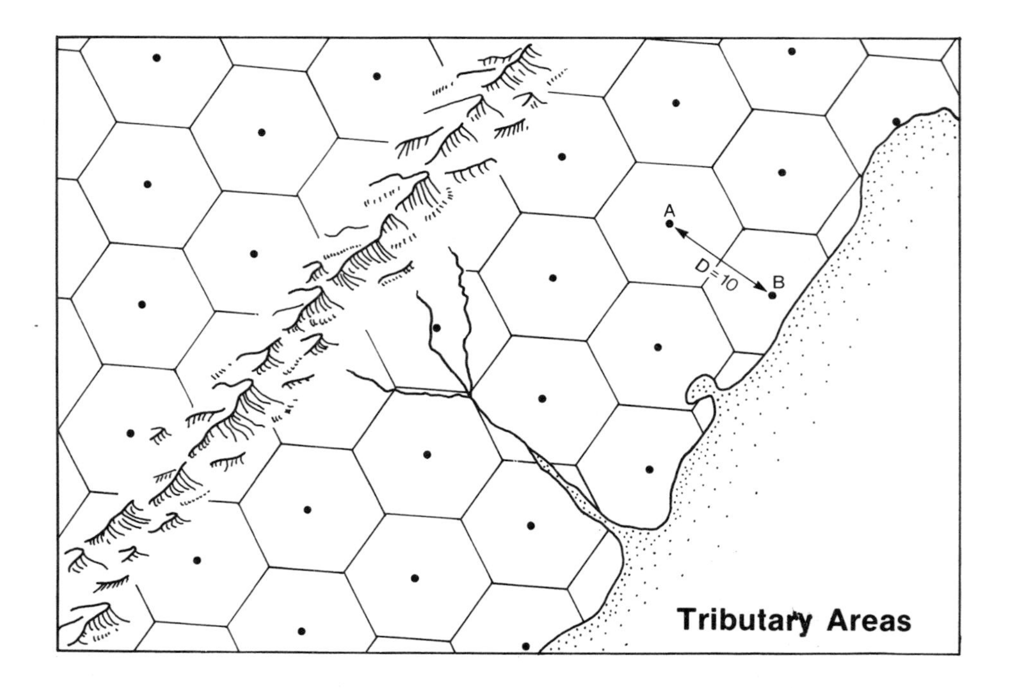

Figure 13–5 Breaking points between towns. Breaking points equidistant between towns result in a hexagonal tributary area network. Partially developed tributary areas, however, may not be of adequate size to support the full services of a town, and the breaking point might be expected to shift toward the smaller town.

One empirical way to compute the breaking point on the real earth is to use population of the town as a surrogate measure for number of functions and establishment importance. (Researchers also often use measures such as retail floor space.) A gravitational formula, often called *Reilly's Law*, is a very useful device in this regard. Only two pieces of information are needed: (1) the general size, or mass of any two places, and (2) the distance between them. From this information the breaking point (D) from the smaller centre (P_1) can be computed as follows:

$$D = \frac{d}{1 + \sqrt{\frac{P_2}{P_1}}}$$

where d is the distance between places, P_2 is the size of the larger place and P_1 is the size of the smaller place. Thus, for example, if place A in Figure 13–3 has a population of 20 000 persons and place B has a population of 5000, and the distance separating A and B is thirty km, then D, the breaking point from B, will not be 15 km as shown by the dashed line, but only 10 km. This formula works for all kinds of trade area estimations and is also of considerable value for computing the territorial domain of centres within cities.

In fact, the trade areas of cities are much more complicated than shown above. First, there are different threshold levels, and hence different range characteristics, for each good or service offered. This means that there are many different service areas of a central place. The trade area just discussed is merely a composite average (one kind of tributary area) for a number of similar types of goods and services. In practice, each establishment has its own competitive territory. Second, the trade areas will not be pure, as shown, but distorted to the extent that various no-man's lands are picked up. This makes even similarly ranked places slightly unequal in population and importance. Finally, transportation routes, not yet considered, will cause serious distortions in the actual position of the central places. The general effect is for centres to be located at accessible positions along transportation routes, that is, centrally in terms of time accessibility to customers but not in terms of the physical geometry. Such deviations from pure, even distributions cause still further deviations in the trade area patterns.

It must also be pointed out that even with the constraints mentioned above, there are many possible patterns of regularly spaced cities. Simply by starting at another place or by rotating the even pattern of city distribution, other patterns emerge. Clearly, time is involved in the development of any particular pattern. The fully developed pattern of central places does not occur simultaneously over an area. The places that first develop affect the location-opportunity of the ones that come later.

CITIES THAT RESPOND TO ON-SITE RESOURCES

Some cities develop at sites of, or very near to, a particular resource, for example, a mineral deposit, a physical environment suitable for recreational purposes, or some other equally distributed feature. For example, in order to make use of minerals, the mining activity must be at the source of minerals, and inasmuch as many minerals contain impurities, it is usually advantageous to remove the waste material at the site. In the process of removal it is also often most economical to carry on various other activities such as smelting and refining. Thus this activity becomes the main economic base of a town or city. The location of such cities is, of necessity, governed by the location of some mineral resource.

Resorts and recreational functions also must of necessity be placed where such resources are found. Consequently, places at a sandy beach or near good mountain recreational land are selected and urban settlement occurs.

CITIES AND INSTITUTIONALLY BASED FUNCTIONS

A fourth type of settlement is related to the functions that are located in particular places for a multitude of institutional purposes. County seats, state capitals, universities, prisons, military bases, and a wide variety of other government functions are examples. In many respects such functions would be located anywhere, insofar as they are not tied to resources, to surrounding areas, or to transportation; the kind of cities that characterize them are perhaps the most difficult to understand because of the multitude of factors and uncertainties involved. Political considerations, difficult to generalize about, also underlie the location of government installations and military bases. All this does not mean that there are no locational principles. In fact, this is the difficulty. There are so many specifics required to explain each case that one is plagued with trying to explain unique circumstances.

Figure 13–6 illustrates one possible arrangement for our study area. Note that a

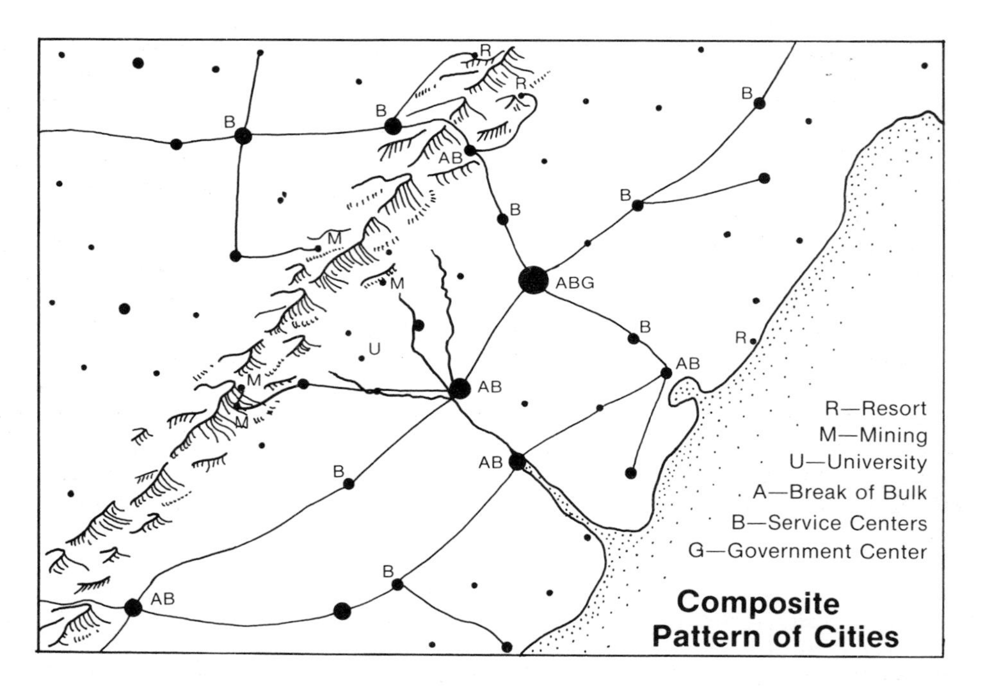

Figure 13–6 A composite pattern of cities. When the various scores of city types are combined, a composite pattern of cities results. Note that in this example it is assumed that the transportation routes and the urban responses came first in time and that the institutional cities came last. Thus the institutional city functions are located at the sites of formerly existing cities.

government centre (G) is placed near the centre of the land mass east of the mountains. It might be assumed that this is a regional or state capital. For many government activities, accessibility to the general population is a primary concern. Many state capitals in the United States are centrally situated within states, just as many county towns are found in the centre of English counties. A university (U) is assumed to be situated in an area somewhat remote from the major population centre. Nonetheless, depending upon the spatial philosophies of the time, such governmentally placed centres might be located almost anywhere.

INTERPRETATION OF A HYPOTHETICAL URBAN LANDSCAPE

Given the four patterns of city placement, one can begin to see the complexity of interpreting an actual landscape of cities. Rarely is one city formed in response to only one of these factors. Most likely a number of functions combine to produce a settlement. Hypothetically, the greater the number of overlapping functions, the larger any given place will be. Thus any given city might be a regional capital, transportation centre, and central place, and might have some local resource feature as well. The patterns that one looks at on the earth, therefore, are contaminated by this juxtaposition of various locational responses.

Patterns are also contaminated by time, or sequence, in settlement. If an area developed primarily as a central place pattern and later as a response to transportation routes, its pattern will be quite different from that of an area developed primarily on the basis of on-site resource responses.

Figure 13–6 shows one possible hypothetical pattern as based on the assemblage of all four types of cities in the study area. In this case the general sequence over time is (1) transportation, (2) central places, (3) resources, and (4) institutionally based patterns. In other words, Figure 13–6 represents the consolidation of patterns in Figures 13–1 and 13–2. Note that some smaller places that entered later in the sequence have been consolidated with previously existing centres. You can confirm this by making a composite map of all the previous patterns in this chapter and comparing this with the pattern shown in Figure 13–6.

The composite pattern, however, is composed primarily of scores of the various functions. In this regard, the largest city to emerge is that near the centre of the map. This city, labelled ABVTCMG, has seven scores: (1) as a periodic service centre along transportation routes (A); (2) as a major break-of-bulk point at the crossroads of two transportation routes (B); (3) as a central place village (V); (4) as a central place town (T); (5) as a central place city (C); (6) as a central place metropolis (M); and (7) as an institutional centre (G). These combined functions create employment, thus making this place larger than those around it. Not shown or anticipated on the composite map is the probability that manufacturing would also find such a large and strategically placed city to its advantage and thereby create another factor of growth in the city's economy.

TOWARD AN UNDERSTANDING OF ACTUAL PATTERNS OF CITIES

With the urban background so far provided, the reader might turn to an atlas to compare, for example, city size, location and distribution in Western Europe, North America or elsewhere on a national or international scale. Yet, as such an exercise will confirm, interpretation or understanding of any real pattern of cities is difficult indeed. First, each pattern is in process of change. Most important, it has evolved out of past conditions in a veneerlike fashion. City types that arrived on the scene first have affected the location of cities (and their functions) that followed, a circumstance that was fully revealed by Walter Christaller in his studies on Southern Germany. Thus the actual city pattern in any area is a composite of the sequence of city occupancy over time. On a world scale, understanding the present pattern of cities can best be approached from the standpoint of hearth areas for the initial development of cities as these spread to other areas. By far the most important factor in this spread has been the development of cities from hearth areas of Europe through the colonization process.

Second, the pattern of cities can be understood on a detailed basis only by careful attention to what cities do. For example, inasmuch as cities are focal points for the control of the earth, they may be located strategically. They are also portals of interchange and hence would be greatly affected by transportation routes. Third, cities reflect various features in the surrounding area. Some are service centres for agriculture and some are convenient places from which to assemble and distribute materials for manufacturing.

Finally, the pattern of cities reflects in their interconnection. There is a general hierarchy involved—few large places but many small ones. There may also be clustering of cities either for site reasons or because of the agglomerative economies to be derived in manufacturing or retailing. Full understanding would involve careful presentation of historical, functional, and theoretical information detailing the ways in which cities have responded and functioned at various times. What should now be clear, however, is that a full appreciation of city locations must be grounded on the four bases previously presented: transportation, territory (central places), on-site resources and institutionally based functions. Thus an understanding of these bases and their various combinations, interactions, and ramifications provides the key to appreciating what is perhaps people's greatest artifact—the city.

Intraurban Considerations

THE SPATIAL CHARACTER OF URBAN LAND USE

Having focused so far on interurban aspects of the city, we can now turn to intraurban aspects. Our leading purpose must be to explain the location of the many functions the city is called upon to provide, particularly those which are economic in character. We have so far demonstrated that cities are no simple locational problem: that they are, in fact, so complex that for the purpose of understanding other economic activities, we must use cities as 'givens'. But if cities are one of the many bases for understanding the distribution of many of man's other activities, we must also seek to understand the spatial character of the city. Indeed, since the United States and most other industrialized countries are truly nations of cities and city dwellers, understanding the distribution of people and their activities within cities ranks as one of the most important parts of spatial study. In the United States, Australia, Great Britain, and most other European countries, more than four-fifths of the population reside in cities. In such countries, far less than 10 per cent of the working force is engaged directly in agriculture. Moreover, the urban population is heavily concentrated in a few large centres. For example, in France, the United Kingdom, Argentina, and many other countries, over one-fifth of the total population is found in the largest metropolis. Thus in terms of the number and the concentration of people in cities alone, inquiry into the internal location of things within such centres appears to be fully justified.

Despite the potential for intraurban spatial understanding, surprisingly little attention generally has been given to the placement of people and things within cities. Such paucity of treatment in economic geography may be related to the small amount of territory occupied by cities. In some respects, however, there is an inverse correlation between the amount of space used by a function and its importance.

Cities, even in highly urbanized nations, occupy only one or two per cent of each country's total land area. Yet these microcosms contain over one-third of the world's population. Perhaps more importantly, these small areas are the control centres for much of the occupation and utilization of the earth. It is in these small focal points that much of the world's consumption, production, and services are centred, controlled, and administered.

But the importance of such small territory is even more dramatic than this. A mere few blocks of land within cities often comprise, for a particular activity, the control centre for a whole nation or a major part of the world. In New York City, for example, the Wall Street area is the financial centre; Broadway, the entertainment centre; and other small areas, the fashion, news, and advertising and publishing centres of the United States and much of the Western world. (New York is the world's supreme 'headquarters' city.) Chicago's La Salle Street area is a control centre for many of the world's commodity prices. London also has lilliputian areas that loom giantlike in commerce, government and world affairs: the financial 'City', the shopping 'West End', Westminster's government and bureaucracy, Fleet Street's newspapers and publishing, and so on.

The importance of such minute parcels of land is further demonstrated by urban land value relative to rural real estate. Land within cities sells for more than any other real estate on earth (certain mineral deposit areas are exceptions). In most nations the value of the urban land outweighs the total value of all other land in the country. Prime land within American cities commonly sells for well over $1 million an acre. Buildings, of course, add considerably more value. In fact, urban land is so costly that in business areas its price is commonly calculated by the front foot.

The study of the spatial character of urban land use has in many cases been deliberately neglected under the wishful premise that the location of things within cities is chaotic, random, or simply the result of a series of peculiar historical events. The romance of popular writers sometimes promotes this view, and even a well-known American planner-architect expressed this fallacious attitude when he proclaimed, 'Cities are nothing but a chaotic accident the summation of the haphazard, antagonistic whims of many self-centered, ill-advised individuals' (attributed to Clarence Stein by Jane Jacobs in *The Death and Life of Great American Cities*, New York: Vantage Books, 1963, p. 13). This assertion is the exact opposite of that expressed by urban land economist Richard Hurd: '. . . if cities grew at random, the problem of creation, distribution, and shifting land value would be insoluble. A cursory glance reveals similarities among cities, and further investigation demonstrates that their structural movements, complex and apparently irregular as they are, respond to definite principles' (Richard M. Hurd, *Principles of City Land Values*, New York: The Record and Guide, 1924, p. 13).

Still others argue that city land-use patterns are so orderly as to be almost foreordained. Such a simplistic approach to urban land use is illustrated by the too common assertion that the city is an organism. The architect Eliel Saarinen expressed this philosophy by stating, 'As long as . . . the expressive and correlative faculties are potent enough to maintain organic order, there is life and progress of life. Again, as soon as this ceases to be the case, and the expressive and correlative faculties are impotent to prevent disintegration of organic order, decline and death occurs. This is true, no matter whether it happens in the microscopic cell tissues of cell-structure where cancer causes disintegration, or in the hearts of our large cities of today where compactness and confusion cause slums to spread.' (Eliel Saarinen, *The City: Its*

Growth, Its Decay, Its Future, New York: Reinhold Publishing Co., 1943, p. 15.) Thus almost every part of the city is presumed to be analogous to some part of the human anatomy: the central business district becomes the heart, parks become lungs, telephone lines become the nervous system, and wharves, depots, and warehouses become the mouth through which the city is fed. Even residences become cells, and various other parts of the city are given physiological counterparts.

If such a simile is accepted, the desired outcome and policy is foreordained: the parts of the city, like the parts of the human body, must be kept healthy. Every function has a purpose, and it is to be kept in its right and proper place. Thus a changing city pattern might be viewed as serious; it might be assumed that the city is suffering from some kind of undiagnosed or perhaps incurable disease. Major surgery may even be needed. Clogged streets, or circulatory systems, are perhaps the result of something akin to arteriosclerosis, and cancerous growth and cell decay take place. If sales in the central business district are declining, then the heart of the city is sick and something must be done quickly if the urban body is to survive. Major surgery (urban renewal, for example) may be needed. Perhaps even a heart transplant would be considered.

This analogy is fascinating, entertaining, and dangerous. It ascribes far too much order to the spatial matrix and is thus highly deterministic. More seriously, it provides a ready excuse for not studying further the nature of the city.

Although the locality of things within cities is in some respects not nearly so simple as asserted above, the study of intraurban location is largely free of one variable that was critical in the study of earlier activities: the location of functions within cities is little affected by the intrinsic qualities of the land. Low land can be, and often is filled; hilly land can be levelled; swampy land can be drained. The primary value of urban land therefore is not derived from the quality of the soil but is based primarily on the competition for sites among functions. Simply put, land within cities is such a scarce commodity that it is often 'created' in a profitable location.

Nevertheless, in any particular city, topography does affect the general land use pattern. Hills may provide problems of construction, yet are often valuable as residential sites because of the view they afford. Flat land, although subject to flooding, may be held in high preference by transportation and industry. Thus topography does provide general shaping forces. Nonetheless, these forces are peculiar to particular cities and vary greatly in their importance to different functions and over time.

THE GENERAL LAND USE PATTERN WITHIN CITIES

Although the specific location of each land use varies from city to city, depending upon local historical and site conditions, the general arrangement of residential, commercial, and industrial quarters, with respect to each other, is relatively consistent. Although each city is somewhat unique in its pattern of land use, the degree of similarity among cities far outweighs local idiosyncrasies. Moreover, through such a generalized spatial framework, the truly unique aspects of any particular city can be more fully appreciated.

A number of generalizations can be made about the similarities. For example, each land use type is generally separated from others. Every city has differentiated business, residential, and industrial patterns. As one observes the arrangement of these land uses, a repetition of pattern from city to city is clearly discernible.

Why this should be so has long attracted the attention of geographers, many of whom would claim the overall spatial pattern of activities reflects a variety of adjustments to the factor of distance. These adjustments have been made by factory owners, office managers, wholesaling and retailing firms, householders and others, all of whom have attempted to minimize the frictional effects of travel and transfer. We have already seen (Chapter 12) how various industrial activities can take advantage of economies of scale by agglomeration, while other kinds of activity, both social and functional, may also gather together in clusters. All large cities in North America and Western Europe—and many other parts of the world where European and American influence has been paramount—tend to reveal the same elements: a high-intensity business centre or 'downtown' area; high-intensity residential areas, primarily in the inner zones of the city; and low-density, largely single-family residential areas in the outlying or suburban areas. Shopping centres and business districts are distributed in a fairly regular pattern and at strategic crossroads throughout the city. Moreover, a definite hierarchy of business centres is noted, ranging downward from the central business district through regional, community, and neighbourhood business centres. Most large cities have a recognizable sector, or sectors, devoted to industrial uses.

Recognition of these circumstances has produced a number of classic urban land-use theories, each of which can be briefly described and presented in model form (Figure 13–7). One of the earliest was the concentric or zonal model (Figure 13–7 (a) of E. W.

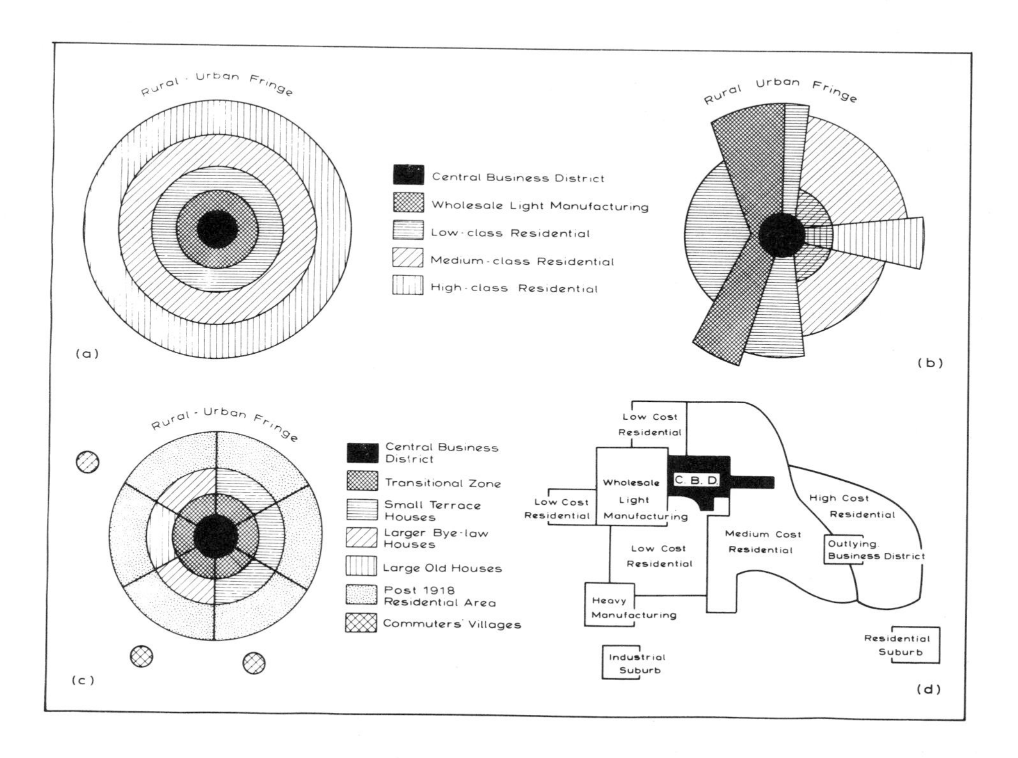

Figure 13–7 Urban land use models: (a) the zonal model (Burgess, 1925); (b) the sectoral model (Hoyt, 1939); (c) the compromise model (Mann, 1965); (d) the multiple-nuclei model (Harris and Ullman, 1945).

Burgess (see Bibliography), which arose from the observation that city land uses tended to sort themselves out into consistent, concentric zones. Its basis is the hypothesis that urban rents and transportation costs are substitutable: since accessibility to the central business district declines with distance outwards, land values must also decline radially. Since the many activities of the city are not all equally susceptible to changes in accessibility, rent curves will vary in slope. Basing his analysis on a study of Chicago, Burgess postulated that within any city, land use was so organized that zones proceeded in the same order from the city outwards. A precondition of the concentric theory is a large number of closely-spaced radial routes. But the further from the city centre, the less likely this condition can be met, since main radial routes diverge. It was one of Burgess's assumptions that all areas of the city at equal distances from the centre will be equally accessible; but it is clear that at any given distance from the centre, a point on a radial route will be more accessible than one which is not—a matter which leads to variations in land value and hence variations in land use at that given distance.

Dissatisfactions like these led to sector models (Figure 13–7 (b)) developed on the premise that the land use of a city is established by the positioning of the radial routeways. Thus Homer Hoyt (1939) argued that similar land uses agglomerate about such routes. Once contrasts arise near the city centre, these are perpetuated along the radial, or axial, routes. Thus a light manufacturing zone, or a high-class residential area, tends to extend by new growth on its outer area. Hoyt backed up his model by an examination of thirty American cities, concentrating attention on how fashionable residential districts had migrated outwards in sectors between the 'tram-car' era (c. 1900) and the arrival of mass private transportation in the form of the automobile (c. 1936).

Both the concentric and sector models were generated in the United States. Latterly, they have been combined (by Peter Mann, 1965; see Bibliography) to illustrate the urban structure of a hypothetical medium-sized British city (Figure 13–7(c)). The main features of this hypothetical city (one large enough to have considerable internal differentiation) are numbered in this diagram. Beyond the city centre (1), is a transitional zone with a variety of urban functions, including commerce and industry (2), followed by a zone of small terrace houses in sectors C and D, larger bye-law houses in sector B and larger old houses in sector A (all in zone 3); inter-war (1918–1939) residential areas with post-Second World War development on its outer periphery (zone 4); and commuting-distance, nearly suburbanized 'villages' (5).

Interestingly, Mann incorporated the notion that the prevailing wind was from the west, and in this circumstance is much of the explanation for sectoral differences. Higher class residential developments, shunning the smoke, grime and unsightliness of coal-burning factories, tend to locate upwind, to the west of the commercial centre. This sectoral tendency is pursued on the outward edge of the city to the west. Sector A is therefore the 'middle class' sector, B is the 'lower middle class' sector and C is the 'working class' sector where, in the outer areas, are found the main municipal housing estates. Most industrial developments are sectored in D where, because of factors already noted, only the lowest working class homes are found among factories, workshops and commercial premises.

As James H. Johnson points out (in *Urban Geography: An Introductory Analysis*, London: Pergamon Press, First Ed., 1967): 'The concentric and sector theories have the advantage of attractive simplicity, but the situation in most cities is possibly too complex to be enfolded in such easily comprehended generalizations'. It is not surprising, therefore, that from time to time geographers have attempted to provide more elaborate theories, especially with regard to large cities. Both the zonal and sectoral theories assume a single centre from which all outward growth takes place. Many large cities however, are multi-centred, with many growth nuclei; an observation which led Harris and Ullman (1945) to advance a multiple-nuclei theory. They recognized that some urban activities had specialized needs which limited their locations and contributed to their clustering at particular sites; others tended to agglomerate to take advantage of shared services or close contacts with suppliers or markets. They also observed that some land-using activities repel others, leading to the formation of distinct homogeneous districts. Some activities could co-exist in high-rent areas, but there were other activities with a low-rent paying ability and therefore more limited locational opportunity.

We may look more closely at the Harris and Ullman multiple-nuclei theory by referring to Figure 13–7(d). For commercial land use, one central nucleus and many outlying, smaller nuclei are noted. Industrial land is broken into three categories: (1) light industry and wholesaling, which occupy more central locations, (2) heavy industry, found in a sector further removed, and (3) separate, isolated nuclei of unspecified industrial use. Residential land use consists of homes that are classified according to value as low, middle, and high class (found, respectively, at increasing distance from the central business district) as well as an outlying nucleus of suburban development. Note that the low-class residential area is near the wholesaling and light industry area, on 'the wrong side of the tracks', whereas the high-class residential area is found on the opposite side of the city. High-class residential land is a particularly opposite position vis-à-vis heavy industry land. The general tendency is for high-class residential areas to move away from the low-class and industrial areas toward places offering more amenities. Ironically, then, it appears that the people least able to cope with the physically most obsolete parts of the city are occupying just those environments.

In summary, several spatial features are evident: (1) centralization or concentration in the most accessible places, (2) general zoning with decreasing intensity and density as one proceeds outward from the point of maximum accessibility, (3) general sectors of land use similarity along various corridors, and (4) a tendency for similar land-use types to form clusters or nucleations. Thus the composite land use pattern of a city is a compromise among a number of opposing forces, some tending to pull functions to outlying areas (centrifugal forces), and others tending to attract certain activities to a more central position (centripetal forces).

Many geographers agree that the multiple-nuclei theory provides the most satisfactory indication of the way large urban areas develop; but it must also be stated that many authors, including Garner (see Bibliography), suggest that the several models are not incompatible with each other and that they may be incorporated to provide better explanation in theory. If we are to rely on the Harris and Ullman multiple-nuclei theory, perhaps we should be aware that the most detailed evidence these authors considered was based upon Salt Lake City, Utah, in the early 1940s. This is important not only if we are to ask the question 'How well does this model represent the major land use types in American cities today?', but also 'How well does it fairly reflect the internal spatial arrangements of cities elsewhere?'

The full answers to these questions cannot emerge quickly, but in brief it seems possible to suggest that although more suburbanization and nucleation have occurred in cities since the early 1940s, in basic form Harris and Ullman's diagram still remains accurate. Indeed, it anticipated many of the land use trends and patterns still current in cities. Most evident in this regard was the recognition of detached clusters of various business, industrial, and residential uses.

Nonetheless, such a model provides a far too simplistic portrayal of general land use patterns in American cities today. First, there is far more nucleation and sectorization of functions. Second, many of the functions, such as high-class residential areas, were bifurcated when growth could not continue in a particular path. Also, the major influx of blacks to the central cities, particularly the migration into the formerly low-class residential areas, was not anticipated. Nor could the impact of the federal housing programmes and freeway development on urban sprawl and suburbanization be accurately foreseen. Examination of a modern pattern, however, reveals enough spatial consistency with the original model to lead to the conclusion that the basic forces have not been much affected. Thus, as Harris and Ullman postulated, the four major forces at work still seem to be: (1) accessibility and need for special facilities, (2) cluster affinities among like functions, (3) disaffinities among unlike activities, and (4) the inability of certain activities to afford the high rents of the most accessible sites.

Figure 13–8 reflects the continuation of these trends in the land use pattern of American cities today. Residential land has spread outward at great distance from the central city. Most of this might be considered middle-class, suburban, single-family homes, but there is also considerable representation to high-class residences and multiple-family dwellings in outlying areas as well. High-class housing has been particularly attracted to site amenities like wooded lots and views. Views are particularly prized in Seattle so that most high-value housing runs in sectors along major waterfront areas. By contrast, the industrial land is primarily concentrated in the opposite end of the city in a broad, flat, river valley, which is well served with rail and highway transportation. Also evident is the linear or sector development of commercial activities, particularly along major arterials. Thus the commercial patterns in a metropolis now consist of at least three major components: a central business district, linear commercial strip development, and major nuclei or outlying development mostly in the form of planned shopping centres.

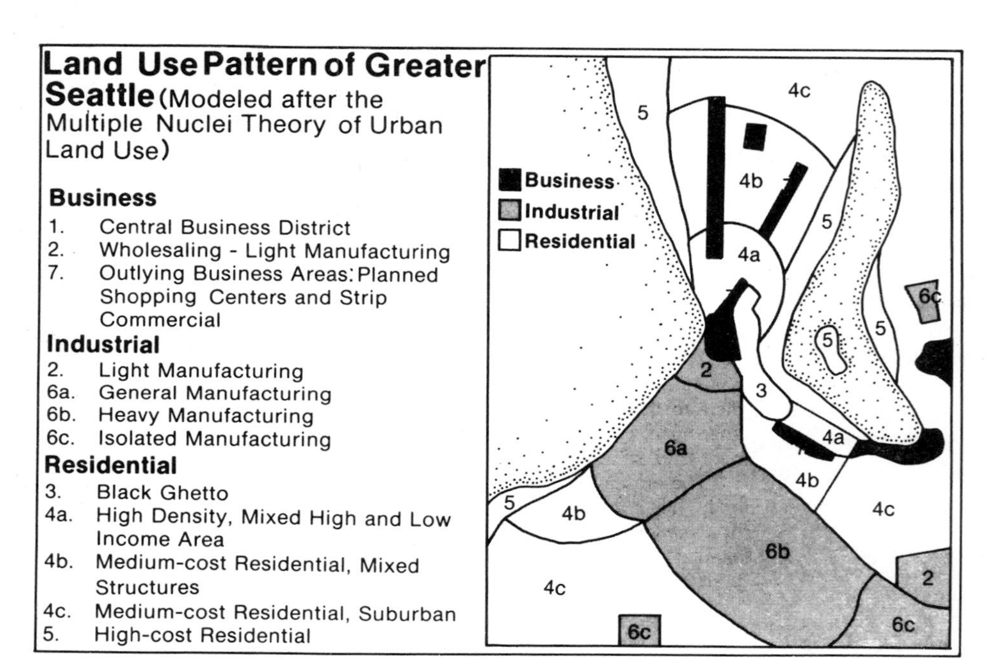

Figure 13–8 Land use pattern of greater Seattle, Washington, as modelled after the multiple-nuclei theory of urban land use.

A limited test of the wider applicability of the multiple-nuclei theory may be to see how far it fits a particular British city. Figure 13–9, modelled after the work of Stedman (1958) and Giles (1976) (see Bibliography), shows the business, industrial and residential districts of Birmingham. Again, we see not only that residential land has spread outward from the central city, but that attenuated zones, or areas, of industrial and business functions are also in evidence. Industries which early in Birmingham's industrial history were developed close to the core of the city spread outwards in directions best served by a dense network of canals and railways. Heavier industrial activity in particular spread out along such lines of communication, especially in the shallow valley bottoms. The inner city retained its maze-like zones of workshops and services to the metal trades which made the city famous, but other industrial nuclei were developed at a distance from these, to transform industrial villages into mixed zones of housing and industry. Tramway-linked, inner nineteenth century residential neighbourhoods clustered around the inner city, and outer nineteenth century neighbourhoods began their spread from suburban railway lines. As in the North American example, it is possible to discern the central business district, linear commercial strip development, and major nuclei or outlying development in the form of district shopping centres, twentieth century municipal housing estates and large private housing developments of varying quality and density. Note the apparently northward extension of industrial developments and the greater proportion of private housing developments across the southern half of the city, also, the nineteenth century, low density and high quality residential area (Edgbaston) close to the inner city on its southwestern side.

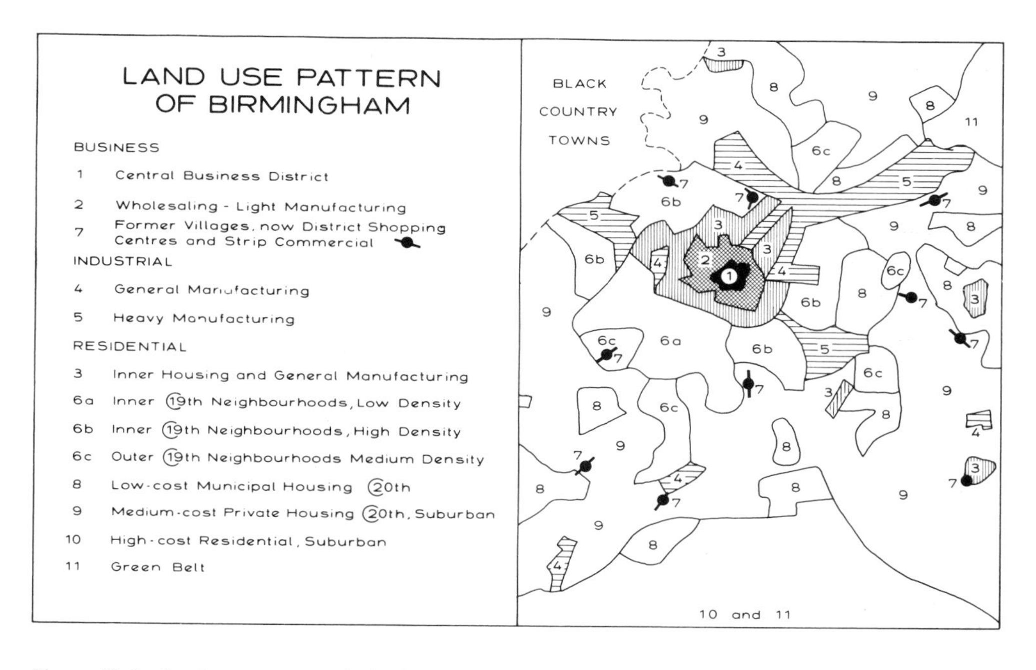

Figure 13–9 Land use pattern of Birmingham, England, as modelled after the multiple-nuclei theory of urban land use (based on Stedman, 1958, and Giles, 1976; see Bibliography).

URBAN LAND CONSUMPTION

It is dangerous to make general and unqualified statements as to the amounts of land occupied by different uses in a typical city. An industrial city will have more industrial land than will a commercial market centre. Moreover, different types of industrial uses require differing amounts of land. Such factors as topography, government control, and the like also may play an important role.

Another difficulty in providing information on the proportion of land occupied by different uses among cities results directly from the nature of the classifications themselves: industrial land, for example, may be tabulated in various ways, according to the system of classification used. In many cases, industrial land includes not only land occupied by manufacturing plants, but land occupied by warehouses, wholesaling, and even railways. Other classifications may also include potentially usable vacant sites as industrial land. Similar discrepancies occur among other categories such as residential land. In this category residential streets, which usually account for almost one-third of the total area, are often included.

The key to understanding such discrepancies lies in paying particular attention to the difference between net and gross space used. Net, of course, refers to the actual amount of land directly used for the purpose. On the other hand, gross land use estimates often include vacant land, nondevelopment land, streets, and in some cases even rural land.

The amount of land used by different functions also varies greatly over time. Thus it is not very meaningful to compare the proportion of land used for a particular function in one city with that in another if the surveys were taken at different times. Generally, the amount of land actually used for any given function has been increasing over time, hence the tremendous spread, or sprawl, of cities. Industry has shifted from a vertical to a horizontal scale and favours large sites that offer parking facilities and other advantages. Likewise, business structures are now one or two storeys high, with ample parking space nearby. But perhaps nowhere has the space change been more prominent than in residential land.

In the United States, the single-family detached dwelling has become the dominant residential structure, but larger and larger lots have been the rule until very recently. Nonetheless, multifamily structures and condominiums are presently the most prominent type of new construction. In the United Kingdom, the dominant residential structure became the semi-detached house, featured in rows or in entire estates, the characteristic development of the inter-war years carried through to the present. In the post-war decades, in the large city, they were joined by civic-built, multi-storey flats, surrounded by two-storey 'semis' and modern-type terrace housing characteristic of civic responsibility for slum clearance, urban renewal and housing expansion generally. Such developments have taken place in both the middle and outer zones of the large city, with the most extensive schemes built on the perimeters on compulsorily-acquired land. As in the United States, most new single-family detached houses today are built privately for the higher-income groups, but in much smaller numbers and for the most part as infill or extension to select residential zones which already characterize this type.

As variants on the model of the 'Western City', further comparisons might be made between the North American and the British city. Briefly, if we can see many similarities in zonal composition and extension, there are considerable differences with regard to urban development controls, restrictive land use zoning practices and the degree of local authority or central government involvement (discussed below). There are considerable differences, too, in townscape (or urban appearance) occasioned by contrasting life-styles, methods of home financing, scale of disposable incomes and attitudes to the home, office, factory, or public buildings.

If our concern, however, is principally with the comparability of land use arrangements, then there is greater contrast to be noted between the cities of the Western World as a group and those of the Third World. The first difference is primarily one of scale because the density of population is higher, and hence the area occupied for any given population is less, in other cities of the world. This is particularly true in cities like Shanghai, Bombay, and Calcutta (over 40 000 per square km). Nevertheless, Japanese, Australian, and even some European cities, such as Moscow, occupy space somewhat comparable to American cities. There is also a tendency in non-US cities to mix residential use with commercial and especially industrial use to a much greater extent. In cities like Calcutta, the old home-shop pattern is still much in evidence. Nonetheless, most large cities of the world, such as Buenos Aires and Mexico City, have definite industrial sectors. On a per capita basis US cities have led the world in intraurban space and consumption. By contrast, European and non-Western cities are far more compact and have far higher densities than do their US counterparts. The automobile has allowed great freedom of locational choice and has made possible great horizontal extent. Sprawl and spread and low-lying structures are becoming the rule not only in the United States, but in much of the world (usually under 4000 per square km).

There is also considerable variation in land uses and their intensities within large and small cities. Large cities have higher densities and taller structures. Therefore, they are far more parsimonious users of land than are small cities. On a per capita basis, more land is used in small cities than in large. Thus, if one's policy were to conserve on land, it would naturally follow that more people should be encouraged to live in larger and larger cities. But the dictum to live in larger cities, as with most urban policies, carries with it negative aspects as well.

Despite warnings about generalizations, some composite estimates which apply to the geographic or built-up area of the city rather than to the central city only are useful. Regardless of the area taken, residential land is by far the largest single land use within a metropolis. More than one-third of all occupied land is taken by this use. By contrast, commercial use accounts for the smallest amount of land occupied by any one function, usually only about 5 per cent. However, inasmuch as commercial land use is often the most intensive in the city, most space is found above street level.

Another large user of city space is transportation routes. Streets alone account for one-quarter or more of all urban land. In fact, the aggregate public and quasi-public land (streets, municipal golf courses, hospitals, churches, government buildings, utilities, cemeteries, airports, parks, and so forth) in a city amounts to well over one-half of its total. Because land thus used is generally nontaxable, the concentration of such facilities within the central city of large metropolitan areas adds greatly to the tax revenue problem.

Of all the land uses, industrial land is the most variable within cities. In fact, some cities (state and national capitals, for example) may have practically no industrial land. On the other hand, cities found in the manufacturing belts of Europe and the USA may have more than one-fifth of their total occupied land in such use. Thus the amount of land taken by manufacturing has a considerable range, but as a crude average may be thought of as being about 10 per cent of the occupied urban land space.

ACCESSIBILITY AND LAND USE ARRANGEMENT

The general land use pattern in any given city appears to be the result of four major forces: (1) centrality or accessibility, (2) competition for sites, (3) interconnections among functions, and (4) public policies. Each of these forces can be best understood within a particular theoretical framework in which other factors are held somewhat constant. It is further understood that such forces operate within a particular economic, social, and political framework and that the examples given below apply primarily to cities in the Western World.

Accessibility is generally considered to be the key determinant of the structure and form of the city. This assertion is based on the simple truth that in any given area, one place has greater potential for interaction with all other places than has any other; this is the prime place from which all other locations can be measured. The importance of centrality of location was explicitly noted by Hurd as early as 1924: 'As first laid down, the theory of agricultural ground rents emphasized fertility as a source of rent. Later, when it was noted that it was not the most fertile lands that were first occupied but rather those nearest new settlements, accessibility or proximity to cities was recognized as an important factor in creating agricultural ground rent. *In cities, economic rent is based on superiority of location only, the sole function of city land being to furnish area on which to erect buildings*.' (Hurd, R. M. *Principles of City Land Values*, New York: The Record and Guide, 1924, p. 1. Emphasis added by present authors.) While this undoubtedly claims too much for the 'standing room only' factor, accessibility is surely a dominating force, giving particular form and shape to cities.

Accessibility is simply a measurement of the degree of potential connection or interaction any given location has with all other parts of the city. Thus 'nearness' and 'proximity' are often near appositives for accessibility. Accessibility is a reflection of the 'friction of distance' among places and is commonly measured in reference to time and/ or cost.

The most accessible place in any given city is the place that provides the greatest potential interaction with all other places at least 'cost'. Such cost is computed by determining the cost of getting from one location to all others, in terms of time-distance. The most accessible point in American cities is characteristically at the focal point of major arterials. This is generally the central business district, or CBD.

Once an area has achieved a position of prime accessibility, such as the central business district, it tends to be maintained for a considerable time. In the United States various programmes have been and are being initiated to maintain the central business district as the most accessible place within the city. Until recently these efforts have been successful. As cities have grown outward, however, the central business district has usually become more off-centre. Moreover, new crosstown roads and outer circumferential expressways have given strategic advantage to outlying areas. Ironically, the heavy congestion of routes and traffic in the inner city has also made it less accessible to other parts of the city. As a consequence, the former, almost monopolistic spatial advantages of the central business districts have been lost, and various business and employment functions have decentralized to outlying locations.

The immediate implication of the accessibility concept is that different locations have different potentials for interaction and hence for development. For any given activity, all locations in the city can be ranked according to their degree of suitability relative to the prime point of accessibility. In more theoretical terms, any given site can be rated in terms of its degree of substitutability with the chief site. The implication is that the more any particular site nearly approaches the prime site in terms of the accessibility measure, the higher is its potential for that particular use (Figure 13–10).

The spatial result of this substitutability feature is a drop-off in the accessibility index with increasing distance from the point of highest accessibility. This point will be occupied by the function paying the highest rent and hence will be the place where the city's peak land value is found. Thus a general curve reflecting decreasing substitution capabilities with the peak land value place can be constructed. For example, a site

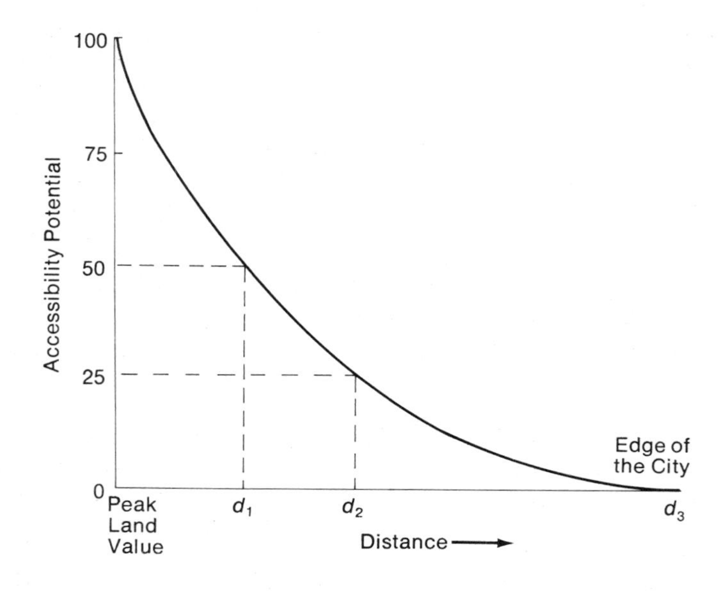

Figure 13–10 Accessibility potential and peak land value. Note that at any given distance some substitution for the prime accessibility site can be achieved until distance d_3.

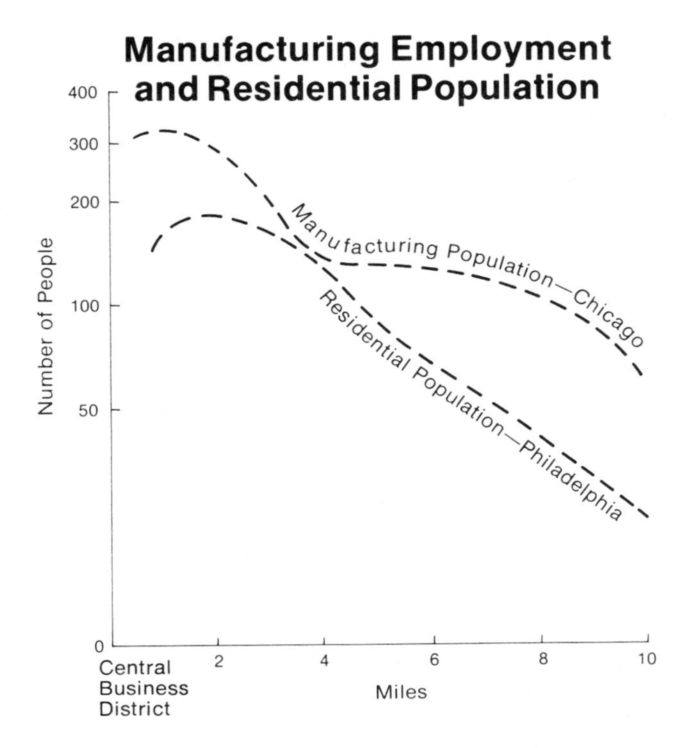

Figure 13–11 Manufacturing employment and residential population. (Adapted from Blumenfeld, H. *Journal of the American Institute of Planners*, Winter 1954, pp. 3–14; and Duncan, O. D. *Cold Spring Harbor Symposia on Quantitative Biology*, 1957, pp. 357–371.)

at distance d_1 has about one-half the accessibility value of the prime site. The site at distance d_2 has only one-quarter of the accessibility of the highest point. Finally at distance d_3 no substitution capabilities exist. Theoretically, d_3 might be assumed to be the edge of the built-up area of a city. Thus it is accessibility that provides the basic explanation for the fact that cities occupy so little area and are so dense and compact. Without the restraint of accessibility, cities could expand outward almost indefinitely.

If accessibility declined outward at an equal rate in all directions from its centre, cities would have a circular shape. However, inasmuch as the transportation route system is rarely radial or identical in all sectors of the city, some parts of the city at any given distance from the centre are more accessible than others. As a result, the shape of cities is rarely circular but elongated outward along the most speedy routes. Suburban tramways and railways were particularly influential in enhancing this stellar pattern in the late nineteenth and early twentieth centuries, and it has been extended in the later twentieth century by the arterial road system. By the same token, growth is restricted in other sectors because of poor transportation. A crude way to confirm this result is to plot a series of isochronic lines at equal time-distance from the centre of a city outward. It will be noted that these lines, connecting points of equal time, generally correspond to similar levels of land-use intensity. In large American cities the outward edge generally conforms with the 60-minute time-distance line, whereas in most European cities, because of their greater compactness, the outward edge may be reached in about half this time-distance.

Further evidence of the importance of accessibility can be empirically derived by noting the intensity of land use and its pattern of decrease with increasing distance from the peak land value place. Thus land values, population density, employment density, and the size of lots all reflect the degree of accessibility at any given location. In American cities the central business district has characteristically contained the highest accessibility site, and all measures of intensity fade off from this centre. Note that both the density of population and the density of industrial employees fall off with distance from the centres of Philadelphia and Chicago, respectively (Figure 13–11). In Philadelphia net residential population drops from almost 450 persons per hectare in the inner part of the city to about 75 persons per hectare in its outlying parts. Likewise, in Chicago, the density of industrial employees per net industrial hectare falls from a high of more than 750 persons per hectare in the inner zone to about 125 in the outer zone. It should also be noted that the density of population and industrial employees is almost nonexistent in the very centre of these cities inasmuch as these functions have been squeezed out by the higher-rent-paying business functions.

This decline in intensity of urban land use with distance from the centre of the city is a feature characteristic of all cities. Presumably because of better transportation within American cities, however, the density and intensity decline is less steep than in non-Western cities. American cities are therefore lower in overall density and more spread out for any given population than are cities in other parts of the world.

RENT-PAYING ABILITY AND LAND USE ARRANGEMENT

The fact that the amount of land acceptable for any given activity is so limited because of the accessibility requirements results in a highly competitive bidding for sites among various urban activities. Theoretically, each urban function is in competition with all others for the most accessible, that is, the best sites. Functions that can pay the highest price (rents) will be able to obtain the choicest sites. The functions that are

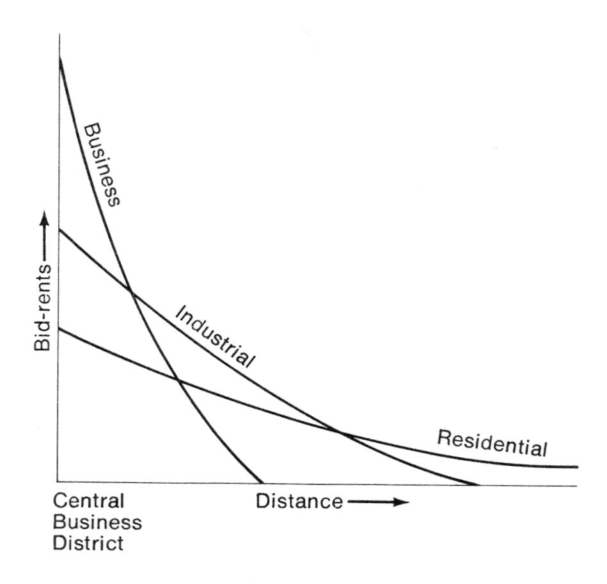

Figure 13–12 Bid-rents among business, industrial, and residential land use with distance from the central business district. Note that the highest bidder for sites prevails. Business use would be found nearest the most accessible place (for example, the central business district), and residential land use would be found at the more remote locations from the city centre.

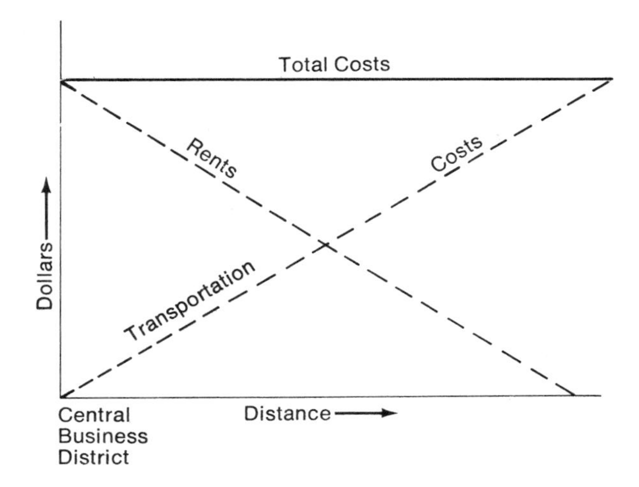

Figure 13–13 Transportation costs to city centre and residential rents. Note that close to the centre transportation costs are low but rents are high, whereas in outlying areas just the reverse situation holds.

outbid will consequently be squeezed outward to less accessible sites. This process is made explicit by the concept of *urban rent theory*. This concept is based on the differential ability of various urban activities to bid for land (pay rent) because of diminishing utility of various sites with increasing distance from the urban centre. The operational process is, in fact, highly analogous to the Von Thünen agricultural rent models presented in Chapter 9.

Among the major categories of urban land use (commercial, industrial, residential, and public) commercial activities are the highest bidders for sites. Thus they usually occupy the most accessible sites within a city. Residential land uses are more prominent in the outer parts of a city but are also represented in the inner and middle parts (Figure 13–12). While some users of residential land, such as millionaires, may be able to outbid commercial functions for land, there seems to be little inclination to do so.

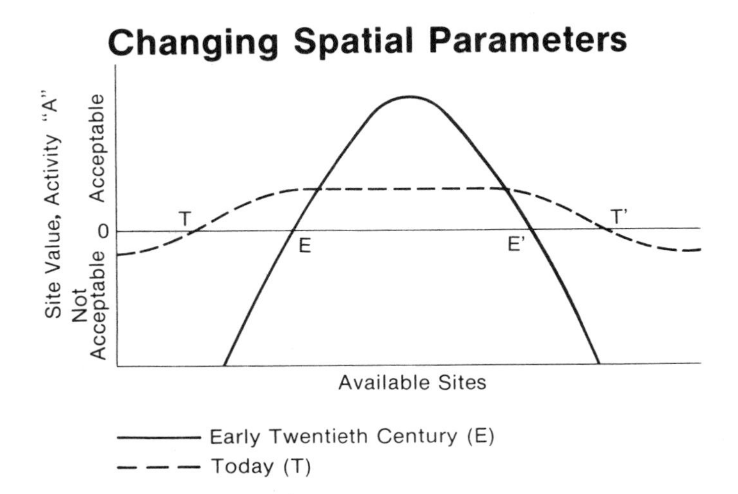

Figure 13–14 Changing spatial parameters. Note that in the early twentieth century there were fewer acceptable sites (E to E[1]) than today where the acceptable sites range from T to T[1]. The greater abundance of site choices, however, tends to reduce the monopolistic advantages of any particular site. Hence there is less variation in price among sites today than formerly.

In fact, the general rule in Western-world cities is for the high-income population to occupy outlying sites and for low-income population to occupy sites fairly close in.

One reason the poor can occupy inner-city sites is that the value and the rent-paying ability of the land can be enhanced through the building of taller structures whereby space is occupied more intensely. For example, a four-storey building may yield several times as much rent as a one-storey building on the same site. In this way some residential sites when used intensely, as for apartments, may actually outbid single-storey commercial uses for that property. Thus the type of structure placed upon a site is a function of its accessibility potential.

For residential land, there is also a play-off between rents and transportation costs. A family may pay less rent in a less accessible place but will have to pay more for transportation, or a family may pay high rents close in, but incur lower transportation costs (Figure 13–13). Theoretically, as based on these two expenses, both families might be paying the same total 'housing' costs.

The variation in rent-paying abilities among functions also appears to be related to the range of acceptable sites for that function. The more limited the range of choice, the higher will be the competition for acceptable sites and the higher will be the rents paid (Figure 13–14). Of all functions, business appears to be the most restricted as to viable or acceptable sites. As a consequence, business must be located fairly close to highly accessible points in order to operate. By contrast, many locations are acceptable for residential use; hence less value is placed on close-in sites for residential use.

AFFINITIES AND DISAFFINITIES AS LOCATION FACTORS

Other important factors in intraurban spatial understanding are the affinities and disaffinities among functions. Some activities are highly interconnected with, or linked to, other establishments and to one another, whereas others have few interaction needs. To understand the location of any one activity, many others must be simultaneously considered.

Such linkage requirements are more profound than is the case in agricultural or industrial location. For example, people may locate near employment, stores, services, and one another. Likewise, businesses may locate relative to customers, to other businesses, and to various suppliers of goods and services. In short, every function appears to be linked somewhat, and hence influenced in its location, by relationships with all others.

This results in considerably more clustering of similar land use types than would be anticipated according to accessibility and rent restraints. For example, whole districts of used car dealers, clothing stores, supermarkets, and other activities are common. Business districts, residential neighbourhoods, and industrial areas are often found rather pure of dissimilar functions and are separate from one another. The reason, to be more specific, is that like functions generally derive more benefits from being near one another than do unlike functions. In contrast to the law of physics whereby unlike particles attract each other, it is the similar types of urban land uses that are attracted to one another. Activities that are unlike commonly repel each other and are found in separate locations.

Residential uses are particularly affected by intrusions of nonresidential use. People in single-family residential neighbourhoods are often much concerned about maintaining their homogeneous residential character. Business and especially industrial uses are commonly zoned out of the higher-income, single-family neighbourhoods. Single-family neighbourhoods are even at variance with other types of residential use such as renter-occupied units and apartments. The age bracket of renters is often quite different from that of homeowners as are income and other occupational and social factors. In short, the life patterns of the various residential status groups rarely mesh.

From the foregoing it is evident that clustering can occur for different reasons within an affinity or disaffinity framework. Most activities locate in clusters because of the advantages such purity affords them. Some locate near other activities because it benefits both. And some functions cluster together because they are ostracized, discriminated against, or otherwise forced to occupy isolated and cast-off positions relative to other functions. A useful way to understand this 'business clustering' is in terms of their kinds of linkage. John Rannells, in *The Core of the City* (New York: Columbia University Press, 1956) recognized four linkage types: (1) *competitive*, (2) *complementary*, (3) *commensal*, and (4) *ancillary*.

Competitive linkages are very numerous among business activities and are easily understood. In such cases like activities cluster near one another in order to compete for a common market. There is usually no direct interaction among establishments; they are simply each trying to get their 'fair share' of the same market. Service stations are a good example of this type of linkage. If one service station locates on one corner of an intersection, others are likely to follow on the other three corners. Each station is competing with others for common customers. The same is true with regard to many other clusters of similar-type stores. In fact, this is probably the most widespread reason for clustering of like functions.

Competitive linkages are beneficial to both customer and function. The duplication of stores in a single area provides the customer with possibilities of comparative shopping, thus enabling him to compare goods and services offered by the various stores. Such competition is so keen that the customer often pays lower prices than he would if there were no competition. Clustering of similar activities also provides benefits to the stores inasmuch as they are able as a group to extend their trade areas and draw customers from a wider territory and in greater amounts than they would individually. The rule is that the whole cluster is greater than the sum of the establishments (its parts).

Complementary linkages result in related functions clustering together. In this case different establishments supply the same market with different but somewhat similar and interlocking kinds of goods or services. A common example is found among speciality clothing shops where one may sell dresses, another shoes, another hats, and still another purses and various accessories that make up a complete outfit. Another common grouping of complementary store types is found at the neighbourhood shopping centre level, where a supermarket is often found near a drug and variety store.

But nowhere is the complementary linkage more apparent than among industrial land uses. Here, various small manufacturing firms supply bits and pieces needed to manufacture a larger product. Thus the small firms locate near the larger one (perhaps an assembly-type manufacturing firm) in order to supply it with components.

Commensal linkages are less common but no less important in spatial effect. In this case two or more establishments may be dependent on a single supplier of materials, or perhaps different types of firms may get their goods delivered from the same warehouse distributor or from the same wholesale area. Thus both firms might be somewhat similarly affected in location but will have little else in common. Similarly, residential areas may be created in a particular place because it affords good opportunities for the residents to get to work.

In another example of commensal linkages, different firms may be found together because of the common use of facilities. Office buildings provide such a common facility to lawyers, real estate firms, physicians, and others. These various office uses are simply found near one another because of similar facility needs.

Ancillary linkages also usually lead to a mixing of unlike establishments in an area. This commonly occurs when a number of different businesses supply a common residential area. Another example is the large office centre or employment complex around which cluster small shops, cafes, cigar stores, and other services geared to the nearby work force. These establishments are subordinate to a larger function. In some respects they are parasitic inasmuch as they live off the customers provided by another firm. This is too strong an indictment however, inasmuch as almost every activity is related in some way to all others. In fact, if such ancillary services were not provided, the parent firm may have to provide them (particularly eating facilities) at its own expense.

PUBLIC POLICY AND LAND USE CONTROL

Another feature that strongly affects the location of activities within cities is public control over land use. This may be in the form of direct measures such as land use regulation and general zoning and planning, or indirect measures such as taxation and financial policies. In the first instance, the effect is straightforward in that various functions are restricted to specific areas of a city. Such functions commonly include those that are deemed to be a public nuisance, to present a threat to public safety, health, morals, or general welfare, or to be otherwise obnoxious or undesirable except when restricted to special areas. Such publicly regulated activities include many types of business and industrial use; in fact, almost every type of activity in a city is regulated in some way.

Business uses that are often carefully regulated include public houses, betting shops, clubs of entertainment and amusement arcades. Many of these may be regulated by rules that no longer make much sense to many in the community: nonetheless, the spatial result is profound. It is commonly the case in North American cities that 'liquor establishments' and various 'honky-tonk' functions, which the natives presume are there to cater to tourists, are also often restricted to a small area of the city. Pawnbrokers' shops (sometimes euphemistically called 'jewellery and loan companies') are tightly restricted in most cities to a small downtown area. Functions regulated for moral reasons—taverns, for example—are also usually restricted to particular sites within the city. Zoning laws usually keep them at a 'respectable' distance from 'good' residential neighbourhoods, schools, and churches.

But regulation of industry is perhaps the most widespread, long-standing, and reasonable of direct governmental controls. Historically, undesirable and highly incompatible industries have been restricted to particular parts of the city. The early zoning test cases in the United States concerned the regulation of Chinese laundries (a disagreeable function, given the technology of the time) in the midst of a shopping area. And slaughterhouses, with their putrid smells and waste problems, were commonly relegated to an area outside the city boundaries of the time. Increasingly today many industries that emit water, land, and air pollutants are being examined critically to determine whether they should be allowed at all in or near some cities. Industries currently being inspected with a jaundiced eye include the iron and steel industries, oil, copper, and other refineries, cement plants, pulp and paper industries, and a large variety of chemical operations. Conversely, some clean industries are now obtaining new-found admittance into residential areas. Many of these campuslike and highly research-oriented industries are even being solicited by many communities in both North America and Western Europe. The electronics industries are among the most highly favoured in this respect.

Public control of land sometimes imposes restrictions even more severe than those regulating the type of function that will be permitted in a particular area. For example, residential land is commonly distinguished through zoning on the basis of ownership, lot size, and other measures that directly relate to density. Thus, residential areas are of many different types and vary greatly in residential density. It might initially be assumed that municipalities would encourage high densities in order to receive high tax revenues. Conversely, many small communities protect and maintain a low density of single-family structures; business and industrial uses are often kept out of such municipalities altogether because of their disaffinities with residential use.

INDIRECT GOVERNMENT CONTROLS: THE AMERICAN EXPERIENCE

Surprisingly, the many direct land use controls discussed are less important in shaping overall land use patterns within US cities than are the indirect and more subtle measures undertaken as a part of general governmental policy. These indirect controls, which are generally not specifically designed to affect land use within cities, nevertheless trigger tremendous spatial responses. Unlike the direct control measures such as zoning, which can be set aside by variances, the indirect controls are largely unaffected by local governmental manipulation.

Indirect governmental influences upon intraurban land-use patterns occur frequently and are many-faceted. The two most powerful tools for exercising control in recent years have been (1) governmental financial policies, and (2) general highway and transportation policies. Until recently, the financial policies have favoured home ownership and the highway programmes have provided the means whereby land for such structures could be reasonably assembled. They are important to understand because they go far to explain the characteristic sprawl of the American city, for they favoured growth of the outlying and peripheral areas rather than the inner parts of the city. More recently, new financial policies and direct government aid have somewhat reversed this imbalance.

After World War II the federal government, through the Veteran's Administration (VA) made no-down-payment, low-interest, and long-term (25 and 30 years) residential loans available to veterans. The Federal Housing Administration (FHA), in operation since 1933, then responded with a similar policy for the general public. Both favoured the construction of new single-family housing units. A significant change in housing ownership patterns resulted. Many Americans who otherwise would have occupied rental housing became homeowners for the first time. The spatial result was that the cities began to expand rapidly in area. Suburbanization, or sprawl, was the result. This in turn led to decentralization of business and industry. Thus the present spread pattern was largely triggered by federal financial policies because these made construction and purchase loans easy to obtain for new single-family structures.

This spatial spread was augmented even further by the 1956 Interstate Highway and National Defense Act whereby toll-free expressways were constructed in most large cities. Consequently, agricultural land around cities, now near such expressways, was made accessible and valuable for urban use. With massive residential suburbanization, the outer-circumferential highway, designed initially as an urban bypass, became heavily used as a cross-town feeder and as a major magnet for business and industry

The combined results of these governmental financial and highway policies was a less intensive utilization of land within the built-up area of cities. Given the growing abundance of raw land on the edge of cities now made accessible through improved mobility, and given the VA and FHA financial policies that favoured single-family home ownership, the one-storey rambler and a subdivided land development system developed. The new policies required lots of seventy or so front feet, almost double the minimum size of city residential lots before World War II. Naturally enough, net residential density dropped drastically.

Business and industry likewise decentralized and used the land less intensively than formerly. Business developed along lines more attuned to the escalator than the elevator and provided several times more area for parking than for ground floor space of buildings. Large shopping centres, a new feature developed during the 1950s, often took more than 40 hectares each, an area about as large as the central business district of the city in which the shopping centre was located. Industry became more dependent on a horizontal, or continuous line, production system rather than on the more intensive vertical system. Thus 40-hectare sites for some industries also became common.

Moreover, the automobile and its related needs resulted in a considerable addition of urban land needed on a per capita basis. First, direct space was needed for parking both at the residence and at work, business, and other sites. This latter requirement alone doubled the land needs of many business and industrial uses compared to the 1930s. Even more important, in terms of lowering overall density of land within cities, was the creation of new and augmented automobile-oriented uses such as service stations, automobile parts stores, garages, and other automotive service establishments. Many other traditional establishments became modified to suit the automobile customer. Drive-in theatres, supermarkets, banks, and so forth sprang up. These functions likewise generally required more space to accommodate the automobile than they did formerly.

It should be remembered, however, that the automobile would not have become so widespread had it not been for the necessity of owning an automobile as the result of suburbanization and decentralization. Residences in outlying locations could no longer be served efficiently by mass transit. First, most suburban developments were located beyond the central city boundaries, that is, beyond the limit of most city bus systems. Second, because of low residential densities and dispersed employment centres, the mass transit was physically unsuitable for assembly and distribution of workers and shoppers in outlying areas. Finally, increased incomes during this period allowed people to spend more on transportation, which meant living farther from work and owning one or even two automobiles. By the 1970s the automobile had become a necessity for the majority of people who lived in cities. Those living in suburban areas and beyond were completely dependent on the automobile for all manner of movement.

However, the new, automobile-oriented life style found in outlying areas could not be emulated in the older and inner portions of cities. In the inner city, freeways required so much scarce space and were so disruptive to nearby residences that in many cities they were discouraged. Thus a dichotomy was created whereby each city began to consist of a new up-to-date city and an older, perhaps obsolete city: a dual city and perhaps a dual society had been fostered in large part by indirect governmental policies.

CITY EXPANSION AND GOVERNMENT INVOLVEMENT: ASPECTS OF THE BRITISH EXPERIENCE

The rapid growth so characteristic of the American city has, of course, also affected British cities, although government involvement in steering this growth, and in checking urban sprawl, is of longer duration. Rapid growth was experienced in the 1930s on the fringes of most towns and cities, but it was particularly marked in the prosperous home counties. Slough's population increase of 150 per cent between 1921 and 1939 was 'unplanned', as was much of the suburban sprawl around many urban communities. To accommodate an increase of population of one-third, from 6.5 million to 8.5 millions, the built-up area of London expanded threefold. Much of the suburban expansion at this stage was achieved by the erection of owner-occupied houses by speculative builders, estate developers and local authorities.

The housing supply was conditioned by a whole series of social and economic factors, in particular in the private sector the availability of credit for house purchase through the building societies and in some cases through the local authorities. Sometimes of mean construction and in inefficient street layouts, the rash of semi-detached houses nevertheless provided a higher standard of comfort than most of their purchasers had previously experienced. Such developments still survive relatively unchanged to provide a visible indication of the revolution in urban living that accompanied the provision of lower residential densities for a substantial proportion of the urban population. Unfortunately, much of this development was unwisely allowed to follow the existing roads, clutter-up town by-passes, or leap-frog other suburban clusters to spill into the open countryside.

The Restriction of Ribbon Development Act in 1935 was an attempt to check this tide of linear sprawl. There followed the first Green Belt legislation in 1938, conceived both as a method of preserving agricultural land from urban encroachment and as a means of providing a recreational girdle for London. Later, as other major conurbations and cities were similarly girdled, the green belt was to become a most powerful agent in the 'containment' of urban England. But while green belt zoning attempted to prevent the further blurring of the distinction between town and country, population growth and rising aspirations still created a huge demand for rehousing in better conditions, especially after the ravages and neglect of the war years, 1939–1945.

Successive Town and Country Planning Acts from 1947 led to local government land-allocating development plans intended to steer both residential and industrial expansion. The rebuilding of the commercial sectors of the central city and the rehabilitation of industry and communications were strong priorities. But the highest priority was accorded to the rapid erection of low-cost, low-density, civic-built housing for single or multiple occupation. Many thousands of hectares were compulsorily appropriated for this purpose on green sites on the outer edges of the major cities. Many of these, in such large cities as Glasgow, Manchester, Birmingham, and Bristol, were developed at a considerable distance from the workplace of the majority. Business and industry could not so freely decentralize in this period as in the United States, although many suburban or edge-of-town trading estates (some of which owed their origin to national or local government involvement pre-war) were expanded by the adhesion of a growing range of light manufacturing, warehousing, distributive trading and service activities.

From the 1950s, bold steps were taken to locate urban overspill populations in New and Expanded Towns, where the opportunity could be taken to provide 'model' urban facilities. Under New Town Commissions with their own special planning powers and massive government financial support, it was easier in the early stages to build houses and shopping centres than to attract 'footloose', labour-intensive industries. Yet there is no doubt today that most of the New Towns have been very successful in their aim to provide higher housing standards, better environmental conditions, and a more desirable spatial segregation of urban functions than the alternative or parallel solutions. For these were either to graft still more people on to the perimeters of the older urban communities or 'package' them up in the high density, high rise flats, which were liberally sprinkled about the otherwise worn-out inner ring areas of the city, most often on pockets of land made available by slum clearance or war damage.

In spite of New Towns, so many of the land use problems of the big British city went unrelieved, and urban development strategies still needed to be better related with those of the rural areas which they bordered and encroached upon. Suburbanization and decentralization did not proceed altogether on the unrestrained American model, yet the marked demands, industrial pressures and people's aspirations which encouraged it were the same. One result was that as much as 60 per cent of daily commuting began to take place between one suburban zone and another. Rising car ownership encouraged still further dispersal. Increasingly, the cross-currents of personal movement and lengthening distance separation between homes and jobs were unsupportable by public transportation. This, and the increasing dependence of industry on road freight distribution, put pressure on the inadequate urban and interurban roads.

By the 1960s, many cities reacted to their growing road traffic problems of congestion and poor access by embarking upon bold new urban radial and ring road systems which reflected the provision already made in the North American city. By the early 1970s, however, the realization came that rising urban traffic volumes could not be accommodated by even the most ambitious new road provision. Most British cities have turned, therefore, to a mixture of public and private transportation encouragements and restraints. Unlike the American city, the British city has arrested the decline of its bus transportation. Outlying areas of the city, however, and especially its 'rurban fringe', are as difficult to serve by mass transit systems as its North American counterpart.

Overall, it is the saving grace for the British city that it has remained more compact than cities in North America, Australia, or many other 'Western' cities in South America or Southern Africa. And it is in further contrast with the large American city in that it has not yet discernibly turned its back on the supremacy of its CBD (see below). Indeed, we may suggest that the planning involvement of government at the various levels—local, regional, and national—has enabled the British city to avoid the worst excesses of function-dispersion evident in the modern American city. But it is also fair to say that this is a challengeable conclusion, especially when we take into consideration the many alternative and conflicting land uses which compete for land in the relatively small British Isles.

The Second Land Utilization Survey, which was carried out in the 1960s, has shown that the 'rurban fringe' now occupies an area twice as great as the townscape itself. It is estimated to take up to three times as much farmland as it actually converts to urban uses, but it fragments the remainder and may subject it to intolerable urban pressures to the point of abandonment by the farmer (see Chapter 9). But further to this, it is claimed that if the rurban-fringe were to be extended at the overall rate of advance actually achieved during the last forty years, then Britain's farmland would totally disappear within 200 years! While no-one believes that this will come to pass, it is salutary to quote the Director of the Second Land Utilization Survey on the dilemmas that still face Great Britain in the late 1970s after forty years of government intervention in town and country planning, much of which is devoted to the zoning of land for specific purposes or to protect it from undesirable use:

> *'. . . we are now faced with an unprecedented crisis of urban decay and rural erosion. The situation can be summed-up as a great chain-reaction of land use pressures spreading outwards like ripples on a pond, but with far less rapid and transitory effect. For 'ripples' read 'shock waves', which have a profoundly dislocating effect upon one zone of land use after another.*
>
> *Great holes of dead space have been caused by the mismanagement of the inner city. These have converted formerly functional townscape into 'rurban' inliers of dereliction and disuse. The displaced people, sometimes as many as three-quarters of the 1901 population, have often moved to greenfield sites in former farmscape.'* Coleman, A. 'Last bid for land-use sanity', *The Geographical Magazine*, September 1978, p. 820.)

A view such as this seems dire, the severest indictment of our public and private failure to restrain the land-hunger of the city. Primarily it is a view looking in at the city from the outside, and it may be interpreted as a plea for conservation of rural land. As such, it may be considered entirely justifiable in a country where land is a much scarcer commodity than it is in the United States. Whether there should be more or less direct government involvement in the land available was highlighted with the passing of the Community Land Act 1976, which enabled local authorities to buy land in order to make it available for private housing and industrial development. This new instrument enables planning authorities to act more positively than in the past, for although the immediate impact may be small, it can provide for idle urban land being brought forward for development in a sensible and orderly fashion.

PROBLEMS AND DEVELOPMENTS IN THE INNER CITY

In both the United States and Great Britain, current governmental polices of recent date would appear to have the effect of favouring the central parts of the city over the suburban areas. In the United States, the 1970s saw house-purchase interest rates reach such heights that home ownership, especially for low-income families, became difficult. A dramatic shift occurred in favour of constructing apartment and rental housing rather than single-family owned housing as was the case earlier. Thus a reversal in land intensity had begun. Ironically, however, the major change was in the suburban and outlying areas rather than in the older, inner portions of the city, where structures often had to be demolished before new construction could occur. Because of the widespread use of the automobile, even by persons in the inner city, apartments could easily be rented in outlying locations. The 'tight money' policy thus led to further suburbanization and a further drop in central city population.

Recently, the federal government has begun to subsidize rapid mass transit for the central city. Theoretically, this should encourage higher densities and limit outward expansion. In this way the inner city might be expected to regain some of its lost population, business, and industry. Whether such transit will be able to reverse the suburban tide, however, is highly doubtful. Certainly those now beyond the central city, for the reasons given above, cannot be served well with a mass transit system. Nonetheless, such a system will surely tend to make the central city less anachronistic than is currently the case.

The problems which face the innermost parts of the contemporary, large American cities have recently been reported by a team of North American geographers. Volume I of this survey (Adams, J. S., and Abler, R., *Contemporary Metropolitan America: Twenty Geographical Vignettes*. Cambridge, Massachusetts: Ballinger Publishing Company, 1976) considers cities all over the nation as reported on by a variety of authors. Here, it is instructive to quote a most perceptive reviewer of their findings as they relate to land use and economic functions of the centre of such cities:

> *'They do not forget about employment or the need for urban centrality in the modern metropolis, though the nature of that centrality may be very different from Walter Christaller's. Several monographs worry about a rising skyline or the lack of it. In fact, the social concerns focus mainly on the inner cities and on the uncomfortable situations arising there from the proximity of central business and monumental districts to a spreading welfare camp of newcomers who are seeking refuge from less generously endowed rural areas and smaller towns.*
>
> *The destiny of the central city in economic terms remains a key factor for the future because it joins together the various systems of the nation. The medium-sized metropolis is still building up its skyline and other components of the center. Even in some of the larger central cities, such as Chicago, Los Angeles, or San Francisco, the business districts are expanding in physical terms, though the vicinity may be losing residents and even jobs. The very nature of urban centrality in large units is undergoing fundamental change. . . .*

> *. . . The new types of buildings and land uses provide significant leads: manufacturers, warehouses, and establishments trading in goods are replaced in 'central districts', old or new, by offices, hotels, hospitals, convention halls, universities, research institutions, museums, sport stadia, and recreational establishments. Instead of manipulating goods, the central district administers or entertains abstract transactions and collective rituals. This trend expresses a basic change, gradual but irreversible (short of a complete breakdown of modern civilization) in the ways people work and spend their time. This change replaces various forms of hard physical labour with other kinds of work, mainly processing information.'* (Gottman, J. 'The mutation of the American city: a review of the comparative metropolitan analysis project', *The Geographical Review*, April 1978, pp. 206–207.)

Broadly, in the United Kingdom, central city areas remain more traditional in their functions, which also remain in better heart than in the United States. The CBD reveals fewer signs of losing its commercial and business supremacy to outer city zones, and the tendencies noted above by Gottman are either absent or incipient. As he further reports, 'the same technology is usually available to several nations that use it differently. Labour and tax legislation vary geographically' (and) 'A society controls its way of life largely through the conditions of employment, the rewards assigned to various occupations, and the constraints imposed upon movement.' Where America leads, it might be concluded, it does not necessarily mean that other countries will follow.

BRITAIN'S INNER CITY HOUSING AND EMPLOYMENT PROBLEMS

Beyond the CBD, the inner city in Britain has undesirable problems, both of a social and economic kind. For twenty years or more, many have urged that the regeneration of inner urban areas should be made a top priority. In older industrial cities, the outworn housing stock, unemployment problems (through the loss of local trades and industries that have fled to the suburbs or else decayed through falling demand for their products and services) and various social dislocations occasioned by patchy, ill-sorted developments are at the seat of their problems.

Most large cities now have sizeable immigrant coloured populations which cluster themselves partly by their overseas origins in the decaying Victorian villas and terraces of the inner wards. Both longer and shorter term residents of the inner middle rings of the city are better housed in a variety of single- or multiple-occupation houses which vary greatly in age, condition, and access to public facilities such as nearby shops, schools, surgeries and recreation space. And the inlying, high-rise flats already alluded to, mostly built between 1950 and 1970, have not proved to be as satisfactory an answer to public rehousing as they were once thought to be.

High-rise residential living for 'blue-collar' working populations seem to owe their origin to the 'Radiant City' visions of the Swiss architect Corbusier, who conceived the idea of the high-rise residential unit with its own shops and services. The school of architects and planners who would have pursued his ideas also stated their faith in the creation of spacious expanses of lawns, trees, and parks. It was very clear to this international school that multistorey housing for such city dwellers would require maximum privacy with a stimulating outlook upon the public domain outside, as well as collective facilities put to maximum use by their siting and quality. Open spaces would be needed for all age groups, both indoors and outdoors to compensate for elevated living; covered walks would be needed to reach adequate garaging, and caretaker skills required to service such facilities.

All these things, however, were incompatible with the scale of post-war housing demand, scarce materials, limited national and civic capital, and—above all—low rents. As cheap public constructions, therefore, high-rise developments fell far short of the ideal, and with the social problems they have caused, they are no longer in vogue. In Glasgow, Liverpool, and London especially, many tower blocks have all too soon replicated the slum conditions they replaced. Oscar Newman, in his book *Defensible Space*, reports tests of how far these and other types of housing design and layout are vandal- or crime-prone habitats in the USA, and his findings are confirmed by a Home Office study of fifty-two housing estates in London which document the inability of high-rise to generate any sense of community.

Inner city palliatives in Britain have a long history, and in the last ten years there have been government programmes attending to its problems in piecemeal fashion: the Urban Aid Programme (1968) was followed by the Education Priority Areas Programme, the Community Development Projects scheme, General Improvement Areas programme, Housing Action Areas and Inner Area Studies. Now, however, the problems of the inner city are being tackled comprehensively with a view to win back people, stimulate employment, improve living conditions and provide better services and amenities of all kinds. It is too early to say what this new spirit of urban regeneration may achieve. But it has been well-stated that what happens to the decaying hearts of many of our towns and cities could decide the health of the country as a whole.

REGIONAL PROBLEMS IN OTHER CITIES OF THE WORLD

Of such patterns and problems is the city thus made, and its economic importance to mankind the world over is undeniable. In many respects large cities throughout the world reflect some of the same problems of ageing and peripheral growth that plague American cities. They also reflect, in some cases even more severely, the disaffinities with regard to certain land uses. An exception to the fringe problem occurs in many European cities, where as in the United Kingdom land use around cities is very tightly zoned and controlled for rural purposes. However, this creates severe crowding in many of these cities and relegates most of the population to apartment living. Moreover, European cities in particular are even more obsolete with regard to their ability to handle the automobile than are American cities. Thus parking and traffic problems are possibly even more severe in them.

Low-income populations have difficulties in all cities of the world. In American cities poor people are forced to occupy the more dilapidated neighbourhoods and structures in the metropolis. Nevertheless, they do occupy fairly centrally located property. In contrast, much of the low-income population in other world cities, outside of Europe, live on the edge of the metropolis in shantytowns. Moreover, these squatter settlements are almost totally lacking in any of the utilities of the main city, such as water, sewerage, fire protection, and in many cases even streets. Such settlements are built almost overnight on land that belongs to very large landowners or to the government.

Many of the people in these shantytowns, like people centrally located in American cities, are newcomers to the city. They come primarily from the agricultural areas of these underdeveloped countries. Yet jobs are scarce in large cities. Moreover the jobs are located mostly in the central portion of cities in underdeveloped countries rather than in outlying areas. Thus these people must face a long commuting journey every day in order to work. This is double jeopardy for such populations. Because of the relatively poor transportation, however, the more accessible residential sites are occupied by the higher-income population. There is an analogy with American cities, however, in that here too jobs are moving away from the low-income population. Over the long run, undoubtedly the low-income populations, if they are to improve their accessibility to employment, must move closer to the centre of cities in underdeveloped countries and closer to the edge of American cities.

Finally, in this chapter where we have considered both the locations of cities in the context of the economic landscape as a whole (interurban) and in the context of the ways in which economic and other forces shape the city itself (intraurban), we can see that the search for order is fully justified. We have demonstrated above all, perhaps, that land use patterns within cities are worthy of study, are spatially orderly, and can be understood within a theoretical framework much like other activities studied. A skeleton framework by which an appreciation of the spatial order within cities can be perceived is achieved by examination of the principles of accessibility, rent-paying abilities among functions, and various affinities and disaffinities among urban activities. Finally, order is seen to be reflected in various direct and indirect governmental policies.

Nonetheless, these principles leave one unsatisfied. This is because they do not jell into a neat and comprehensive model of urban land use arrangement and form. Instead, each of the principles is a piecemeal explanation. Moreover, because of the ceteris paribus manner in which these principles must be treated, they still leave unanswered many questions about urban location. Therefore each of these principles only illustrates a particular shaping force; other approaches and models are needed before a satisfactory replication of an existing urban land use pattern can be achieved.

Particularly needed is a more dynamic approach to urban land use. The placement of things within a city is in constant flux so that the spatial pattern is one of continuous change in location. A satisfactory location for an activity at one time may be an obsolete or anachronistic location at another time.

BIBLIOGRAPHY

(This reading list is in two parts, corresponding to the two-part division of the chapter.)

(a) The city in general, including an interurban emphasis:

Abrams, C. *The City is the Frontier*. New York: Harper & Row, 1965.

Bell, T. L., Liber, S. R., and Rushton, G. 'Clustering of services in central places'. *Annals of the Association of American Geographers*, June 1974, pp. 214–225.

Berry, B. J. L., and Garrison, W. L. 'Functional bases of the central place hierarchy'. *Economic Geography*, 34, 1958, pp. 145–154.

Berry, B. J. L. *Geography of Market Centres and Retail Distribution*. Englewood Cliffs, N.J.: Prentice-Hall, 1967.

Berry, B. J. L., and Meltzer, J., eds. *Goals for Urban America*. Englewood Cliffs, N.J.: Prentice-Hall, 1967.

Berry, B. J. L., Barnum, H. G. and Tennant, R. J. 'Retail location and consumer behaviour'. In *Readings in Economic Geography* (Ed.) Smith, R. H. T. et al. Chicago: Rand McNally, 1967, pp. 362–384.

Borchert, J. R. 'America's changing metropolitan regions'. *Annals of the Association of American Geographers*, June 1972, pp. 352–373.

Christaller, W. *Central Places in Southern Germany*. Trans. Baskin, C. W. Englewood Cliffs, N.J.; Prentice-Hall,1966.

Coleman, A. 'Last bid for land use sanity'. *The Geographical Magazine*, September 1978, pp. 820–824.

Cullingworth, J. B. *Town and Country Planning in Britain*, London: Allen & Unwin, 1976.

Davis, K., ed. *Cities: Their Origin, Growth, and Human Impact*. San Francisco: W. H. Freeman, 1973.

Dickinson, R. E. *City, Region and Regionalism*, London: Routledge & Kegan Paul, 1947.

Doxiadis, C. A. *Ekistics*. New York: Oxford University Press, 1968.

Futtermann, R. *The Future of Our Cities*. Garden City, N.Y.: Doubleday, 1961.

Getis, A., and Getis, J. 'Christaller's central place theory'. *Journal of Geography*, May 1966, pp. 220–226.

Golany, G., ed. *International Urban Growth Policies: New Town Contributions*. Chichester: Wiley, 1978.
Gottman, J. *Megalopolis: The Urbanized North-eastern Seaboard of the United States*. New York: The Twentieth Century Fund, 1961.
Hall, P. *The World Cities*. London: Weidenfeld & Nicolson, 1966.
Hall, P., Thomas, R., Gracey, H., and Drewett, R. *The Containment of Urban England*, 2 vols. London: Allen & Unwin, 1973.
Harris, C. D. 'A functional classification of cities in the United States'. *Geographical Review*, 33, 1943, pp. 86–99.
Harris, C. D., and Ullman, E. L. 'The nature of cities'. In *Readings in Urban Geography*. Ed. Mayer, H. M., and Kohn, C. F. Chicago: University of Chicago Press, 1959, pp. 277–286.
Hart, J. F. 'The changing American countryside'. In *Problems and Trends in Geography*. Ed. Cohen, S. B. New York: Basic Books, 1967, pp. 64–74.
Harvey, D. *Social Justice and the City*. London: Arnold, 1973.
Hicks, U. K. *The Large City: A World Problem*. London: Macmillan, 1975.
Hoyt, H. 'The utility of the economic base method in calculating urban growth'. *Land Economics*, February 1961, pp. 51–58.
Johnson, J. H. *Urban Geography, An Introductory Analysis*, London: Pergamon Press, 2nd edition, 1974.
Jones, E. *Towns and Cities*. London: Oxford University Press, 1966.
Lomas, G. M. 'The conurbations and the countryside'. *Journal Town Planning Institute*, 54, 1968, pp. 275–280.
Lösch, A. *The Economics of Location*. New York: Wiley 1967.
Mayer, H. H. 'Urban nodality and the urban economic base'. *Journal of the American Institute of Planners*, Summer 1954, pp. 117–121.
Morrill, R. L. 'The development and spatial distribution of towns in Sweden: an historical-predictive approach'. *Annals of the Association of American Geographers*, 53, 1963, pp. 1–14.
Murphy, R. E. *The American City*. New York: McGraw-Hill, 1966.
Owen, W. *Cities in the Motor Age*. New York: Viking Press, 1959.
Pahl, R. E. *Urbs in Rure: The Metropolitan Fringe in Hertfordshire*. London: London School of Economics, 1965.
Pahl, R. E. *Whose City? And Further Essays on Urban Society*. Harmondsworth: Penguin 1975.
Pred, A. R. *City-Systems in Advanced Economics*. London: Hutchinson 1977.
Putnam, R. G., Taylor, F. J., and Kettle, P. G., eds. *A Geography of Urban Places*. Toronto: Methuen Publications, 1970.
Ratcliffe, J. *An Introduction to Town and Country Planning*. London: Hutchinson 1976.
Ratcliff, R. U. 'The economics of urbanization'. In *Urban Land Economics*. New York: McGraw-Hill, Chapter 2, 1949.
Shaw, D. J. B. 'Urbanism and economic development in a pre-industrial context: the case of Southern Russia'. *Journal of Historical Geography*, Vol. 3, No. 2, 1977, pp. 107–122.
Sjoberg, G. *The Pre-Industrial City: Past and Present*. Glencoe, Illinois: Free Press, 1960.
Smith, R. H. T. 'Method and purpose in functional town classification'. *Annals of the Association of American Geographers*, 55, 1965, pp. 539–548.
Stewart, M., ed. *The City: Problems of Planning, Selected Readings*. Harmondsworth: Penguin Books, 1972.
Ullman, E. 'A theory of location for cities'. *American Journal of Sociology*, 46, 1941, pp. 853–864.
US National Resources Committee. *Our Cities: Their Role in the National Economy*. Washington, D.C.: US Government Printing Office, 1937.
Vance, J. E. *The Merchant's World: The Geography of Wholesaling*. London: Prentice-Hall, 1970.

(b) The city mostly considered in intraurban aspects

Adams, J. S., and Abler, R. *Contemporary Metropolitan America: Twenty Geographical Vignettes*. Cambridge, Massachusetts: Ballinger Publishing Company, 1976.
Alonso, W. *Location and Land Use*. Cambridge, Massachusetts: Harvard University Press, 1964.
Applebaum, W., and Cohen, S. B. 'The dynamics of store trading areas and market equilibrium'. *Annals of the Association of American Geographers*, March 1961, pp. 73–101.
Berry, B. J. L. *Commercial Structure and Commercial Blight*. Chicago: University of Chicago Press, 1963.
Berry, B. J. L., Simmons, J. W., and Tennant, R. J. 'Urban population densities: structure and change'. *Geographical Review*, July 1963, pp. 389–405.
Berry, B. J. L., and Horton, F. E. *Geographic Perspective on Urban Systems with Integrated Readings*. Englewood Cliffs, N.J.: Prentice-Hall, 1970.
Bourne, L. S., ed. *Internal Structure of the City: Readings on Space and Environment*. New York: Oxford University Press, 1971.

Buchanan, C. *Traffic in Towns (The Buchanan Report)*. London: HMSO, 1963.
Carruthers, W. I. 'Service centres in Greater London'. *Town Planning Review*, 33, 1962, pp. 5–31.
Carsey, R. *The Debate on Urban Policy: Decentralisation versus Improvement*. Cambridge: Retail Planning Associates, 1977.
Carter, H. *The Study of Urban Geography*. London: Arnold, 1972.
Clark, C. 'Urban population densities'. *Journal of the Royal Statistical Association*, 1951, pp. 490–496.
Clark, C. 'Transport—maker and breaker of cities', *Town Planning Review*, 28, 1957, pp. 237–250.
Cresswell, R. (Ed.) *Passenger Transport and the Environment: the Integration of Public Passenger Transport with the Urban Environment*. London. Leonard Hill, 1977.
Dantzig, G. B., and Saaty, T. L. *Compact City*. San Francisco: W. H. Freeman, 1973.
Diamond, D. R. 'The central business district of Glasgow'. In *Proceedings of the I.G.U. Symposium on Urban Geography, Lund 1960*, Ed. Norberg, K. 1962, pp. 525–534.
Dorau, H. B., and Hinman, A. G. *Urban Land Economics*. New York: Macmillan, 1928.
Edge, G. 'The suburbanization of industry'. In *The City as an Economic System*. Milton Keynes: The Open University, 1973.
Everitt, J. C. 'Community and propinquity in a city'. *Annals of the Association of American Geographers*, March 1976, pp. 104–116.
Fellmann, J. D. 'Land use density patterns of the metropolitan areas'. *The Journal of Geography*, May 1969, pp. 262–266.
Garner, B. J. 'Models of urban geography and settlement'. In *Models in Geography*, Ed Chorley, R. J., and Haggett, P. London: Methuen, 1967.
Giles, B. D. 'High status neighbourhoods in Birmingham'. *West Midlands Studies*, Vol. 9, 1976, pp. 9–33.
Gittus, E. 'The structure of urban areas: a new approach'. *Town Planning Review*, 35, 1964, pp. 5–20.
Gottman, J. 'The mutation of the American city: a review of the Comparative Metropolitan Analysis Project'. *Geographical Review*, April 1978, pp. 202–209.
Hansen, W. G. 'How accessibility shapes land use'. *Journal of the American Institute of Planners*, May 1959, pp. 73–76.
Harris, C. D. 'Suburbs'. *American Journal of Sociology*, 49, 1943, pp. 1–13.
Herbert, D., and Johnston, R. J. *Social Areas in Cities: Processes, Patterns and Problems*. Chichester: Wiley, 1978.
Horwood, E. M., and Boyce, R. R. *Studies of the Central Business District and Urban Freeway Development*. Seattle: University of Washington Press, 1959.
Hoyt, H. *The Structure and Growth of Residential Neighbourhoods in American Cities*. Washington, D.C., 1939.
Hurd, R. M. *City Land Values*. New York: The Record and Guide, 1924.
Johnson, J. H. 'The suburban expansion of housing in London, 1918–1939'. In *Greater London*, Ed. Coppock, J. T., and Prince, H. C. London: Faber & Faber, 1964, pp. 142–166.
Johnson, J. H., ed. *Suburban Growth*. London: Wiley, 1974.
Johnston, R. J. *Urban Residential Patterns*. London: Bell, 1974.
Keeble, D. E. 'Industrial decentralization and the metropolis; the North-West London case'. *Transactions of the Institute of British Geographers*, 44, 1968, pp. 10–54.
Korcelli, P. 'A wave-like model of metropolitan spatial growth'. *Papers, Regional Science Association*, 24, 1970, pp. 127–138.
Levin, C. L., Little, J. T., Nourse, H. O., and Read, R. B. *Neighbourhood Change: Lessons in the Dynamics of Urban Decay*. New York: Praeger, 1976.
Mann, P. *An Approach to Urban Sociology*. London: RKP, 1965. (Chapter 4, Urban Society.)
McGee, T. G. *The Southeast Asian City*. New York: Praeger, 1967.
Morrill, R. L. 'The Negro ghetto: problems and alternatives'. *Geographical Review*, July 1965, pp. 339–361.
Murphy, R. E., and Vance, J. E., Jr. 'Delimiting the CBD'. *Economic Geography*, 30, 1954, pp. 189–222.
Murphy, R. E. *The American City*. New York: McGraw-Hill, 1966.
Newling, B. E. 'Urban population densities and intra-urban growth'. *Geographical Review*, July 1964, pp. 440–442.
Park, R. E., Burgess, E. W., and McKenzie, R. D. *The City*. Chicago: University of Chicago, 1925 (Reprinted 1967).
Putnam, R. G., Taylor, F. J., and Kettle, P. G., eds. *A Geography of Urban Places*. Toronto: Methuen, 1970.
Quinn, J. A. 'The Burgess zonal hypothesis and its critics'. *American Sociological Review*, 5, 1940, pp. 210–218.
Rannells, J. *The Core of the City*. New York: Columbia University Press, 1956.
Robson, B. T. *Urban Growth: An Approach*. London: Methuen, 1974.
Rose, H. M. *Black Suburbanization: Access to Improved Quality of Life or Maintenance of the Status Quo?* Cambridge, Masschusetts: Ballinger Publishing Company, 1970.
Shaw, D. J. B. 'Planning Leningrad'. *Geographical Review*, April 1978, pp. 183–200.
Smith, B. M. D. 'Economic problems in the core of the old Birmingham industrial area'. In *Metro-*

politan Development and Change: The West Midlands, A Policy Review, ed Joyce, F. Chapter 9, pp. 148–163. Westmead: Saxon House, 1977.

Stafford, H. A. 'The dispersed city'. *Professional Geographer*, 14, 1962, pp. 4–6.

Stedman, M. B. 'The townscape of Birmingham'. *Transactions of the Institute of British Geographers*, No. 25, 1958, pp. 225–238.

Thomas, D. 'London's greenbelt: the evolution of an idea'. *Geographical Journal*, 129, 1963, pp. 14–24.

Thomas, D. 'Urban land evaluation'. In (Ed.) Dawson, J. A. and Doornkamp, J. C., *Evaluating the Human Environment: Essays in Applied Geography*, Chapter 4, pp. 88–108. London: Arnold, 1973.

Thomas, R. *London's New Towns: A Study of Self-Contained and Balanced Communities*. London: Political and Economic Planning Broadsheet 510, 1969.

Timms, D. W. G. *The Urban Mosaic: Towards a Theory of Residential Differentiation*. Cambridge: Cambridge University Press, 1971.

Tulpule, A. H. 'Dispersion of industrial employment in the Greater London area'. *Regional Studies*, 3, 1969, pp. 25–40.

Ullman, E. L. 'The nature of cities: reconsidered'. *Proceedings of the Regional Science Association*, 1962.

Vernon, R. *The Changing Economic Function of the Central City*. New York: Committee for Economic Development, 1959.

Vernon, R. *The Myth and Reality of Our Urban Problems*. Cambridge, Massachusetts: Joint Center for Urban Studies of Harvard and M.I.T., 1962.

Voorhees, A. M. 'Factors influencing growth in American cities'. *Highway Research Record*, 24, 1965, pp. 83–95.

Willatts, E. C. 'Planning and geography in the last three decades'. *Geographical Journal*, 137, 1971, pp. 311–338.

Wingo, L., Jr. *Transportation and Urban Land*. Washington, D.C.: Resources for the Future, 1961.

Wingo, L., ed. *Cities and Space*. Baltimore: Johns Hopkins Press, 1963.

14

Intraurban Patterns

The industrial, residential, and commercial land use patterns within cities will now be examined separately. Unlike the preceding chapter, in which the principles of land use location were discussed in general terms, this chapter will examine particular land-use types and their spatial arrangements. Thus the concepts previously presented will here be applied to fit specific activities. Instead of approaching the question of intraurban land use arrangement from the standpoint of concepts, particular land use patterns are viewed from the standpoint of general spatial principles. Such treatment should offer important insights into the variety of land use patterns within cities and provide a further demonstration of the necessity for matching theory and reality carefully to gain satisfactory spatial understanding.

A fundamental purpose in much of geography is to understand the basic differences in spatial patterns among functions. This same goal is applicable to activities within the city. The major uses of the land within the city, therefore, are the building blocks by which such patterns are composed. In this respect seven basic needs within cities are noted: (1) living (residential), (2) selling (commercial), (3) making (manufacturing), (4) movement (transportation), (5) governmental, health, and safety, (6) land for the future, and (7) others, such as recreation. Nonetheless, these logical breakdowns may obscure or mix spatial patterns because the subcategories of each major type may have different locational forces and restraints. For example, commercial land use does not have one pattern but several. Second, the value of the classification depends on the purpose at hand—in our case, pattern dissection and analysis. Finally, the purpose here is not to explain all activities but to recognize a variety of spatial arrangements within cities, especially focusing upon those which relate to economic forces. Although much time could be spent on land use classification, for our purposes it seems wisest simply to follow three generally accepted land use categories: industrial, residential, and commercial. This trichotomy is consistent with the general theories in Chapter 13 and provides an adequate grouping for discussion.

INDUSTRIAL LAND USE PATTERNS

In many cities industry is the triggering agent for urban form and growth. Industry provides the fundamental employment base by which money is brought into the community and by which other people not engaged directly in industry are supported. This concept is made explicit by *economic base theory*. According to this theory, there are only two types of employment activities in a city: (1) *service* activities or industries, that is, those that locate in a city primarily to serve the existing urban population (labelled *ubiquitous* industries in Chapter 11); and (2) *basic* activities or industries, that in those that locate at strategic places within cities in order to better serve areas and persons outside the city in question (labelled *sporadic* industries in Chapter 11). The ubiquitous type of manufacturing is generally market-oriented; it occurs in a particular city only after a certain threshold level of demand has been reached. Thus this type of industry is also often referred to as a *city-serving* industry. In contrast, the sporadic types are *city-forming* activities and are largely responsible for the location and growth of cities themselves. The city-forming type of industry has an effect similar to those retail and service functions found within central place cities (see Chapter 13). For example, central place cities interact with people in rural territory. Likewise, city-forming manufacturing plants sell their products to markets at great distances from the city in which the plant is located. Thus the size of any urban place and its importance is implicitly assumed to be related to some activity that brings money to the city in question.

According to economic base theory, each city must have one or more city-forming functions. These functions provide the triggering mechanism for urban growth. From these initial functions other employment, activities, population, and of course land uses develop. A simple example of how growth occurs is demonstrated by a manufacturing plant that sells its products outside its urban area. This plant is thereby a city-forming or *basic activity*, and its workers are called *basic workers*. Let us assume that this manufacturing plant employs 100 workers, who are hired from outside the city. These workers immediately add to the city's population because they are added to the city's employment force. Their wives and children are a further addition to the population of the city. Given an average family size of 3.5 per worker, then some 250 persons are added to the community through the initial employment of 100 workers. Thus some 350 persons (workers and their families) are new to the community (Figure 14–1).

The major growth impact of such basic workers, however, is through their families' demand for various goods and services within the community. Such basic workers and their families need homes, schools, stores, health and government services, and many other things. Inasmuch as these things must be provided by service workers like building contractors, teachers, merchants, doctors, and civil servants, such service workers and their families are largely supported by the basic workers. If the service workers have families, even more persons will be added to the community through an initial basic worker input of only 100 persons, and the cycle of growth continues. The service workers and their families, in turn, require services similar to those of the basic workers and their families. Thus a second-round effect is set in motion. Of course, with each round the number of additional service workers is reduced. Nonetheless, the process continues through several rounds. Eventually, the catalytic effects are dissipated as fewer services are provided locally.

From the above, another relationship is also evident: the ratio of basic workers to service workers. The higher the basic-service ratio, the higher the growth potential from any given basic worker input. This magnifying feature of basic employment is often called the *multiplier* effect. In small cities the ratio is often less than 1, but in large metropolises it may be almost 3—that is, 3 service workers for each basic worker.

The nature of the multiplier effect is clearly related to the nature of the basic workers. Obviously, the family coefficient is critical. If the basic workers are mostly single, as is commonly the case in mining, lumbering, and military operations, then there will be little demand for housing, schools, and the like. Instead, as in the mythical Wild West town, most of the basic workers' money is spent in saloons, hotels, and in recrea-

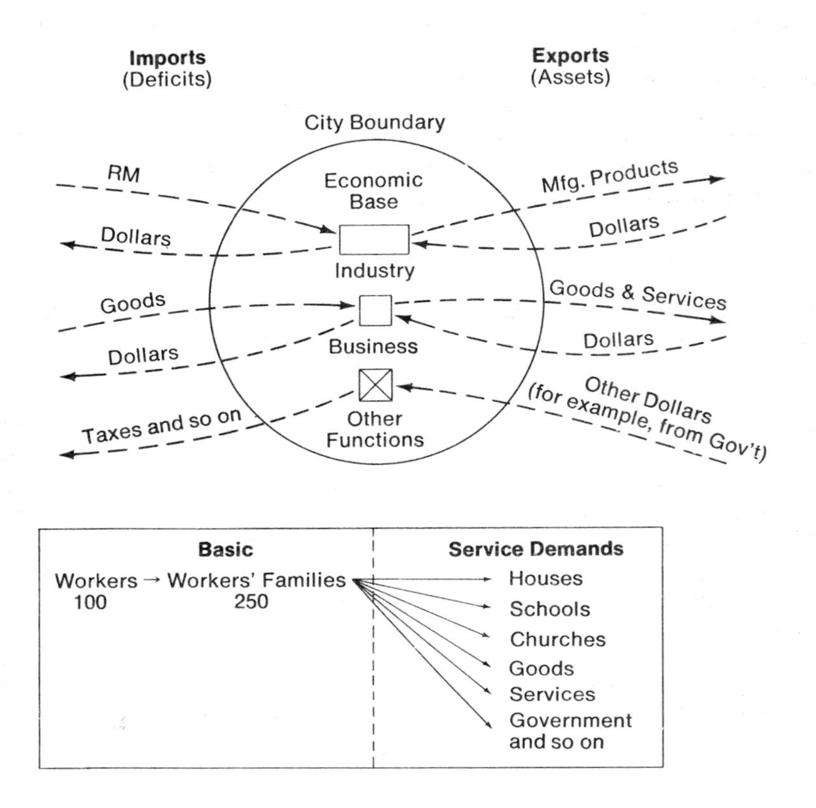

Figure 14–1 Urban economic base theory. Within this framework all imports to the urban community are viewed as deficits because money flows out of the community. Conversely, any activity that brings money into the community from outside is viewed as an asset to the urban economic base. The connection between basic workers and service demands is shown schematically.

tional pursuits. And if, in turn, the saloon and hotel keepers have no wives or children, or send their profits out of town on the Wells-Fargo Express, the multiplier effect will be small indeed.

The economic base concept can also be used to provide estimates as to the amount and type of new land that might be added in a city, given a basic activity input. Just as basic employees act as a service employee multiplier, they also act as catalysts for new urban land uses. Given assumptions similar to those in the manufacturing case above, the amount of land generated by such an initial input can be roughly determined if population and employee densities are known. Such information is best based on the densities required for new functions, rather than on average existing densities of residential land or of the number of workers per unit area for commercial land.

Data on marginal land density—the density at which new additional workers and their families will occupy the land—is needed. Information like this is always hypothetical, but estimates can be made based on the most recent land use changes. These figures are commonly called *land absorption coefficients*.

Assume that the land absorption coefficients are 0.034 workers per acre for industrial land, 0.047 persons per acre for commercial land, and 0.059 for residential land. (Coefficients are based on the calculations of John H. Niedercom and Edward F. R. Hearle in 'Recent land use trends in forty-eight large American cities', *Land Economics*, February 1956, 105–109. They increased during much of the 1960s but diminished during the 1970s. Consequently, the coefficients still appear to be correct in 1978.) Given the fact that our new plant adds 100 workers, some 3.4 acres of industrial land will be added to the community through the building of the industrial plant on a 3.4-acre site. Given a residential density of 16.9 persons per acre and a family size coefficient of 3.5, then some 20.7 acres will be added in residential land. If the number of required commercial service workers is 80, then some 3.8 acres of new commercial land will be added. Thus with the addition of 3.4 acres of industrial land, more than 24.5 acres of urban land will be added in the first round (Figure 14–2).

Many geographers feel that far too much importance has been attached to growth. True, a new basic activity may make a city grow in population, but it may not necessarily make the city a better place in which to live. In fact, some urban-affairs specialists think that many cities are already too large. Certainly the indiscriminate solicitation of new industry is rarely beneficial. It may cause further pollution and traffic congestion, require the addition of municipal services, and make many functions, such as schools, inadequate, all of which would require the initiation of new spending programmes.

A balance-of-payments approach to the urban economic base also reveals that service activities and imports are as critical as export activities. In fact, it is argued that the service activities are composed of things that make a city distinctive, a good or bad place in which to live (see Hans Blumenfeld, 'The economic base of the metropolis', *Journal of the American Institute of Planners*, 1955, pp. 114–132). It is reasoned that if a city has good housing, schools, public services, and other amenities that reflect a high-quality urban environment, there is no need to worry about the basic activities. The latter will take care of themselves without any chamber of commerce ballyhoo and land-giveaway programmes. Moreover, it is argued that the basic activities may come and go as a city changes its character and responds to new opportunities, but the service activities are the truly permanent and distinctive features.

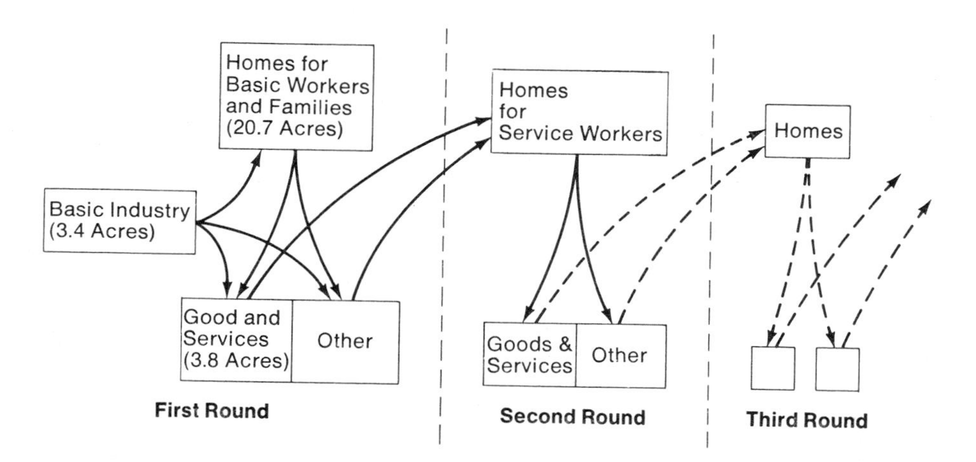

Figure 14–2 Basic service activities. Assuming a basic industry occupying 3.4 acres and employing 100 workers, and by using standard land absorption coefficients and making certain multiplier assumptions, the amount of land for various uses is shown for three 'rounds' of development.

THE GENERAL PATTERN OF INDUSTRY

Economic base theory sheds much light on the general relationships among land uses, with particular reference to industry. There is still too little understanding of the locational pattern of industry, but in broad form, the pattern is fairly clear. In most American cities one part of the city is generally dominated by manufacturing and the other by residential land use. The primary locational factor in almost every city is proximity to major transportation routes. Historically, this has meant proximity to water at ports and proximity to railroads. Such a pattern is strikingly evident in St Louis (Figure 14–3).

A more careful examination of the pattern, however, indicates considerable variation. In simple terms, three types of patterns are noted: (1) firms occupying small sites fairly close to the centre of the city; (2) firms heavily geared to railroad and port facilities and (3) certain large-scale sites found in rather isolated outlying positions.

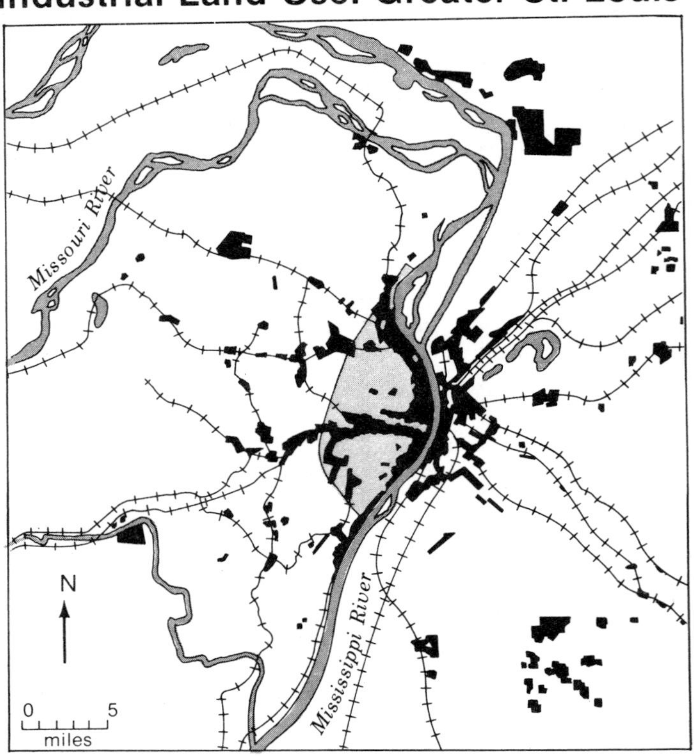

Figure 14–3 Industrial land use pattern in greater St Louis. Note the affinity of industrial land to railroad routes and to the Mississippi River. (From *St Louis Metropolitan Area Transportation Study.* St Louis: W. C. Gilman & Co., 1959. Reprinted with permission.)

The reasons for the general location of different industrial types within the city can be better appreciated from the vantage point of urban rent theory. In this regard, we may distinguish between *light industry* and *heavy industry*. This broad differentiation is related to such factors as intensity of site used, size of site, size of finished product, and the density of employment per unit area. The light industries generally occupy the more accessible positions within the metropolis as compared with heavy industry, because of labour input requirements, needed associations with other similar industries, and other factors. They also, fortunately, can pay much higher per-square-foot rents than can heavy industries like chemicals and steel, which have massive space requirements and a low labour input.

Even so, for light industries there is a play-off between horizontal accessibility and vertical proximity. Traditionally many light industries occupied upper floors of wholesale and other commercial establishments. Consequently, these have been referred to as loft industries. Today with the development of planned industrial parks in outlying areas, many kinds of light industry are decentralizing.

From the standpoint of transportation. it is noted likewise that two scales of industrial site selection are involved. On a small scale it appears that industry requires good access to routes within the metropolis. Aside from this, site features such as flat land are often critical. Insofar as highways and railroads generally seek out the flat land, a complementary relationship in this regard is easily achieved. On a large scale the location of industry appears to be related to transportation routes that offer particular advantages for intercity movement. (For an excellent treatment of industrial location patterns and industrial classifications, see Allan R. Pred, 'The intrametropolitan location of American manufacturing', *Annals of the Association of American Geographers*, vol. 54, June 1964, pp. 165–180.)

The placement of industries with reference to strategic position relative to intercity or intracity transportation requires some further clarification. In the first instance, the location of the firm is determined by exogenous (outside the city) locational forces, whereas in the second instance the location is endogenously determined. It would appear that industries most dependent on rail and port facilities are of the exogenous types. They also commonly require large amounts of space and are usually found in outlying locations or near port facilities. Undoubtedly the location of exogenously oriented industries is much influenced by transportation cost advantages. This is particularly the case with industries that have large or bulky finished products or that require extensive inputs of raw materials. As might be expected, such industries gravitate toward the side of the metropolis facing their largest market, if suitable land is available.

Those industries that primarily sell their wares within a single metropolis are endogenously oriented and are subject to some of the same locational principles as retail business—they attempt to minimize distribution costs to markets. For manufacturers of many types of small and highly perishable products, the choice location has traditionally been near the central business district. From this point they have the greatest distributional advantage for the entire metropolitan area. Firms that manufacture bread and other bakery products are prime examples. These firms also commonly have wholesale distribution outlets and thus favour locations in or near the traditional wholesaling area of the city. Nonetheless, this locational pattern is being broken by suburbanization and decentralization. For example, a wholesaling or manufacturing firm that distributes food products primarily to supermarkets often finds a location near a circumferential expressway highly advantageous. Other industries that serve primarily a local area might be fixed in their location because of their input source of raw materials. This is particularly the case for firms that receive raw materials by ship. When raw materials like marble, sand and gravel, petroleum, and timber come in by water, the manufacturer often finds it more profitable to locate near these import sources than at other places in the metropolis.

Finally, some industries are closely linked to others in the metropolis. These firms may be using the finished products of another firm as their primary raw material. Thus, industries that use steel products may congregate around steel mills. Chemical industries may cluster about refineries. And various canning and meat-packing plants may seek out locations near slaughter yards.

RESIDENTIAL LAND USE PATTERNS

Based on urban rent theory, it might appear that residential spatial patterns are relatively easy to understand. Residential land use simply is squeezed outward to the least accessible sites. Thus it might be expected to be found on the edge of cities and at other less accessible places within the city. That there is general confirmation of this relationship is evidenced by examining the land use map of any city.

Nonetheless, residential land is also found in some prime positions such as in the inner portions of the city. At first this seems to present some contradiction to the urban rent theory, but in fact it can be partially explained by the varying intensity of space use among various residential structures, ranging from single-family dwellings through high-rise apartments. By building taller structures; that is, increasing the ground floor rents through the vertical compounding of units, some residential uses are able to outbid many other use types. Forty families on 1 hectare paying £50 a month rent are using the land more intensely than four families paying £100 a month rent, and hence are able to outbid them. The amount of space being used per unit is substitutable for rent per unit area. Given the high priority placed on land space for residential use,

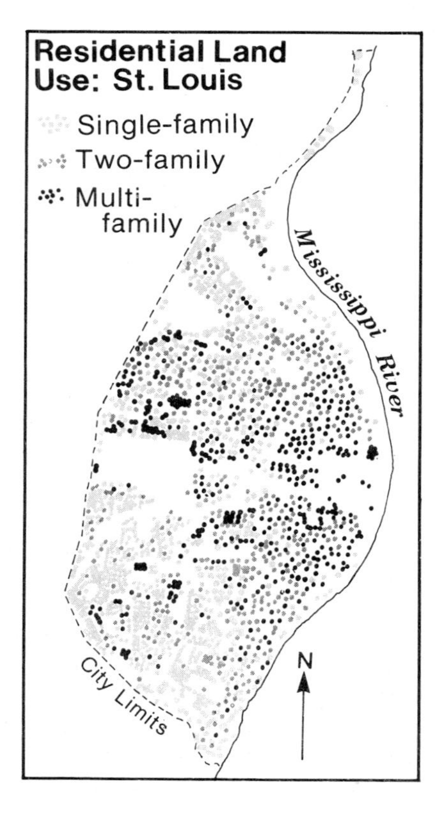

Figure 14–4 Residential land use in St Louis. Note the general tendency of multifamily housing areas to be located closer to the centre than two-family and single-family housing areas. This 'squeezing out' of less intensive residential uses is largely the result of the operation of urban rent theory. (Adapted from Harland Bartholomew, *Land Uses in American Cities*. Cambridge, Massachusetts: Harvard University Press, 1955.)

the more well-to-do prefer to have larger lots, but not perhaps at the high cost of land in close-in areas.

As expected, multiple-family units dominate the inner zone, double-family units then appear and single-family residences predominate in the outer zones (Figure 14–4). This is clearly seen in St Louis even though the map shows residential land within the municipal limits only—a territory only one-quarter that of the built-up area. Today the suburban area, although predominantly single-family, low-density residential in nature, is intermixed with a good many multifamily structures. Thus urban rent theory, which has heretofore been rather effective in giving meaning to population density at any given distance, breaks down as an adequate explanation when residential land is viewed more closely.

Nonetheless, in aggregate terms a clear density gradient within cities does exist (Figure 14–5). One interesting facet of this is the manner in which this gradient changes

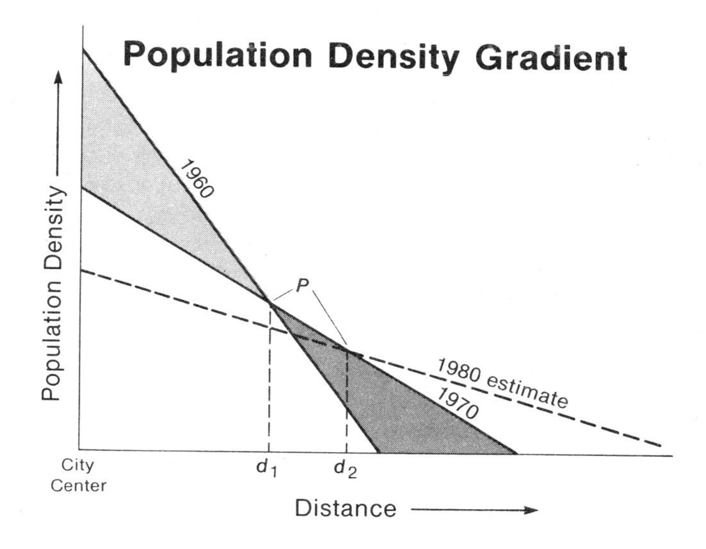

Figure 14–5 The population density gradient within cities over time. Population density falls off with distance from the centre of the city in all cities. In the United States over time, the inner portions of cities experience a zone of decline, whereas the outer portions experience a zone of growth. At the fulcrum point (*P*) little density change is experienced between any two time periods. Note also that the edge of the city moves outward over time.

over time. The trend in all so-called 'Western' cities is for the gradient to flatten out over time so that a fulcrum point (*P*) is created. From this fulcrum point toward the centre of the city, the population generally declines, but outward from this point the population density increases over time. In the past this decline occurred because the inner zone was invaded by nonresidential uses, which replaced residences. Recently, however, the drop in population here has not been offset by commercial functions. Instead, an actual drop in the intensity of land use in the inner zone of cities has occurred. The rise in population density from the fulcrum point outward occurs primarily as a result of a residential filling-in process. Of course, as raw land is converted from rural to urban uses, densities will rise in the outermost areas and the general built-up area will be extended.

An important implication of the fulcrum principle is that the fulcrum moves outward over time (this movement is well documented by Bruce E. Newling in 'The spatial variation of urban population densities', *Geographical Review*, April 1969, pp. 242–252). Moreover, the area near this fulcrum during any given time period remains relatively unchanged. In Figure 14–5 note that in 1970 the fulcrum point (P) was at distance d_1, but by 1980 it will have shifted outward to distance d_2. As will be demonstrated in the following discussion, this principle is related to the general population change for any given residential neighbourhood (see Figure 14–7, stage II, which appears later in the chapter.

From the foregoing it is fairly evident that residential land use has a fairly wide range of locational choice. In fact, residential land appears to have more spatial choice than any other type of land use because it is the most widely distributed within the city. For example, commercial land appears to be much affected by strategic points and is limited to highly accessible places in conjunction with proximity to residences. Industrial land is much affected by access to major transportation routes so necessary in the assembly and distribution of materials. Historically, residential land use was highly restrained by proximity to work. However, this proximity is no longer necessary or perhaps even desirable, as long as the worker lives within about one hour's time from his place of employment. Thus most of the metropolis has become one large, generalized housing market. The workplace can no longer be used as an effective way to explain the location of residential land.

The complexity of the residential classification problem increases as one examines the various elements found in residential areas. Residential land may be differentiated as follows, according to such features as structure, occupants, and neighbourhoods:

By structure
1. Type (brick, frame, one storey and so forth)
2. Family accommodation (single, double, multiple)
3. Condition (sound, dilapidated, and so forth)
4. Proportion of site covered
5. Technological up-to-dateness

By occupants
1. Family size
2. Income
3. Age of family members
4. Marital status
5. Occupation
6. Education

By neighbourhood
1. Low-class (value) residences
2. Middle-class (value) residences
3. High-class (value) residences
4. Density of population
5. Mixed or pure housing types and values
6. Quality (parks, schools, and so forth)

Fortunately, there is considerable contamination among all such measures. For example, a high income is highly correlated with a high-priced house and a high-quality neighbourhood. This is, in turn, correlated with still other features like family-size, type of structure, occupation, and education. Since there is considerable redundancy among the various features, it is necessary to examine only a few in order to reveal the various patterns. The primary basis for residential distinction, as common sense would suggest, stems from the nature of the residential occupants. Two of the most critical variables in this regard are income and ethnic characteristics.

VARIATIONS IN INCOME PATTERNS

Incomes are more alike within residential neighbourhoods than is any other variable. Perhaps one reason for this results from the simple fact that houses within any given neighbourhood are all built at about the same price level. To purchase housing in any given neighbourhood, the buyers must all have about the same payment potential. Of course, there will be some variation inasmuch as some persons are willing to pay more for housing than others, but with financing as standard as it is, income provides a reasonable explanation. (Earnings of the wife are increasingly being taken into consideration by lending institutions.)

On the other hand, there is some question about which is cause and which is effect: homogeneous housing types or a voluntary tendency of people with similar incomes to cluster together. Certainly, the similarity of housing value in any given area is a prime consideration. Nevertheless, there are other good reasons that people with similar incomes might prefer to live in the same area. People with similar incomes are also often similar in age, occupation, and other areas.

The income pattern for Seattle, Washington, is shown in Figure 14–6. Note that again, the high-income areas are generally found farthest out from the centre of the city. Nonetheless, some qualification should be made. Incomes are highly clustered within areas so that there are pockets of high-income population in certain inlying areas and

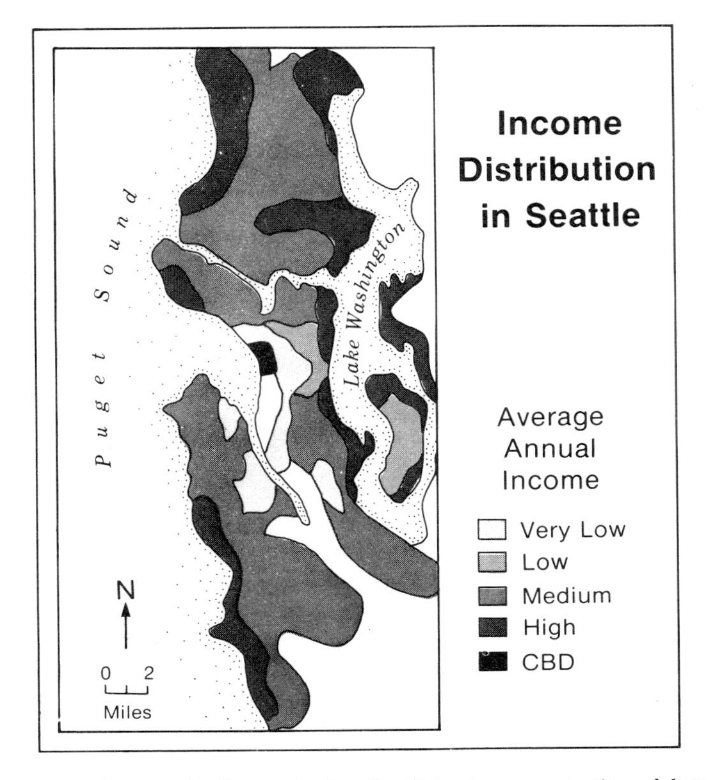

Figure 14–6 Average income distribution in Seattle. Note the concentration of low-income population near the centre of the city and the tendency of high-income population to be sectorized along water-based amenity areas.

pockets of extremely low-income population in outlying areas. One spatial pattern is very evident, however. The high-income populations appear to spread out in sectors, particularly along those accessible corridors that offer views of the water. Likewise, hilltops generally have higher-income population than do the valley areas around them. Of course, the latter areas may be adversely affected by their proximity to arterials, which in Seattle generally run along the valleys. Here they cause disaffinities (noise, lack of safety, and congestion) with residential land. Other high-income areas appear to be created in places offering wooded or lake settings. The lot size is another variable, not shown on the map, which varies considerably among income classes; the high-income populations generally have larger lots and hence occupy the land at lower densities for any given location as compared with low-income areas.

It is evident that wealthy persons could, if they desired, occupy close-in sites within a metropolis. Yet in American cities it is the lower-income populations that occupy the most accessible residential sites. The important paradox to many is how the poorest people in American cities can occupy some of the prime land. Basically, this can occur for two reasons. First, housing stock wears out, and secondhand structures are occupied in a kind of hand-me-down process in which the poor occupy the cast-off housing of the next higher income level. Second, the land is occupied intensively through crowding and high-rise rental structures. In American cities the wealthy are found primarily in selected outlying locations. By contrast, the wealthy more often occupy the inner zones of non-Western cities, where many of the poor are found on the periphery in squatter shacks. A common explanation for this, aside from historical precedent, is the high friction of distance in non-Western cities. Transportation is too time-consuming for the wealthy in such cities. Consequently, the poor are faced with the burden of daily commuting.

Although much can be said about the crowded, dilapidated, and slum conditions of the poor in American cities, until recently they have at least been closer to their jobs than have the more affluent. This is changing rapidly, however, and America's poor increasingly find that their jobs have moved to outlying areas of the city. Public transportation is of little value for getting them to outlying areas.

VARIATIONS IN SOCIAL AND ETHNIC PATTERNS

Another very noticeable pattern of differentiation among the residential population is according to ethnic background. In Seattle, as in most cities in the United States, the black population, other minority groups, and the very poor occupy areas close to the central business district. These areas were once occupied by some of the more affluent city population. Ethnic concentrations are evident among many groups. The Ballard district of Seattle is predominantly Scandinavian; there are Polish districts in Chicago, German districts in Milwaukee, and Jewish and Italian areas in many cities. There would thus appear to be a tendency for people of the same ethnic background to cluster together. How much of this is strictly voluntary and how much a result of general custom, deliberate prior policies of discrimination, or voluntary ethnic grouping varies from place to place.

Clusters of minority religious groups are also common in many cities. Thus Mormons will often locate in the same area of a large American city. Here they have mutual social relations and good access to their churches. Occupational clustering is rare but is perhaps more pronounced among university professors than among other professions. Captains of industry and high-echelon business executives often settle in particular districts. These high-income areas, perhaps almost by definition, would be expected to contain similar types of occupations. Finally, there is a tendency of young people, married and single, to congregate in the same vicinity of a city. This stage in the family cycle (no interest in nearby elementary schools) and the need for rental housing (to maintain mobility) are most apt to be the motivating factors. Inasmuch as such housing is often localized in particular areas, this may be much of the cause.

THE TRICKLE-DOWN PROCESS AND LOW-INCOME GROUPS

Explanation for income and ethnic variation within cities is also based on historical processes. Traditionally, each higher-income group casts off its housing to the next lower-order group. Thus a trickle-down process is set in motion whereby definite movement paths occur throughout the city. Under this system the lowest-status income group consistently occupies the oldest, most dilapidated, and least up-to-date housing stock in the city. Generally, this is the housing in the more central locations. Given such a hand-me-down phenomenon, the general locality of the various income and status groups within a city is fairly well predetermined.

Until recently in American cities, the most obsolete housing was occupied by the newest immigrant groups. These groups landed in the central portion of cities, became 'Americanized', and then began a long series of residential changes in their long, upward climb to the newest housing stock. Thus the occupancy of such inner-zone land by the black population, newly arrived from rural areas, was consistent with past patterns. Several factors, however, caused the trickle-down process as a route to upward mobility to become ineffective.

The first major change occurred immediately after World War II, when various housing programmes made available to young families new, moderately priced housing on the edges of cities. Thus these people broke out of the long cycle of rentals and the general secondhand housing process as a method of improving their residential lot. Such families still moved to higher-status neighbourhoods where many occupied used homes, but the basic cycle of inner-city living had been broken. Even with more potential used housing available, the process did not work for blacks. Because of housing and employment discrimination, blacks were unable to achieve upward housing mobility within the city and therefore seemed relegated to occupying slums permanently.

In an attempt to alleviate such a housing stalemate, in the late 1950s urban renewal and redevelopment programmes were initiated. Unfortunately, these did not operate to bring black people into the mainstream of housing changes and to effectuate residential mobility within the metropolis. First, a situation primarily resulted in which the black population, which was occupying highly accessible land in the inner city, was merely pushed to nearby neighbourhoods, which also became overcrowded. This resulted also, in part, because of the political desire of city administrations to build middle- and high-value apartments on the old slum sites, a move calculated to entice the affluent back to the central city. Although such attempts largely proved futile, they did circumvent any adequate housing programme for blacks. In fairness it should also be pointed out that many high-density public apartments for low-income populations were built. (Low-density housing was ruled out because of the presumed high value of the land.) These, however, were generally unwanted by the low-income population, and many became slums of a new type. In fact, most of the low-income populations preferred living in inadequate private housing and apartments rather than in the massive, new public housing projects.

During the past few years, housing discrimination has been legally eliminated. Nevertheless, there are few new houses priced for low-income population. Many reasons for this have been given, including high construction costs, high land costs, and, of course, high interest rates. The truth is that very little profit could be made in low-cost housing by private contractors, in comparison with housing built for the middle- and upper-income groups.

Government effort has been placed on improving the areas now occupied by the low-income population. Local enrichment programmes nonetheless primarily amounted to 'situs quo' status for the low-income population. Little attention was given to the question of accessibility to rapidly decentralizing jobs. Instead, emphasis was placed on finding employment in the inner city area, and in some cases even on trying to bring industry to the residential areas, rather than vice versa. Not surprisingly, this approach has not been too successful.

Today, the old trickle-down process has once again been set in motion. The difficulty is that hundreds of acres of formerly middle-income housing has become technologically obsolete in almost every city. Thus the 'middle zone' of the city is faced with a glut of such housing. Should current government programmes to provide new single-family housing on the edge of cities for the low-income population succeed, much of the older housing stock will become highly marginal. Because of nonoperation of the hand-me-down housing system for several decades, vast areas of the central city now find themselves with little opportunity for residential survival. However, the current energy crisis and housing shortage now appear to have given new hope to many middle zone housing areas.

RESIDENTIAL LAND USE CYCLES

Any residential area one might choose to examine is probably in a stage of change. In order to assess the nature of the change, close examination of the residential population is required. The two keys to stage of development are the age in which the development first occurred and the location of the settlement within the city. Generally,

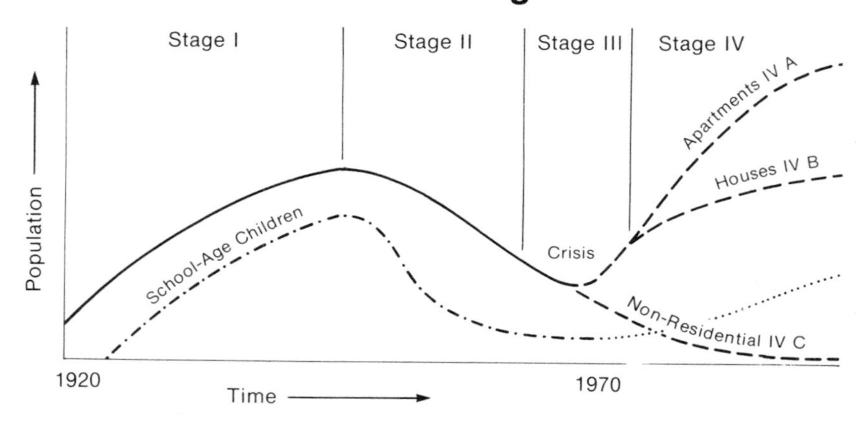

Figure 14–7 A typical population cycle of a residential neighbourhood. In any given neighbourhood over time, the population goes through four stages as shown. At stage IV a critical crisis point is reached inasmuch as structures are physically and technologically obsolescent. In most cities this cycle lasts about 50 years.

the closer to the centre of the city, the older the development and the more advanced it is in the residential cycle.

In the United States the cycle for residential change appears to be about 50 years, the general economic life of residential structures. Thus a typical pattern for residential areas built in the 1920s and early 1930s might evidence three stages (Figure 14–7). Stage I is usually characterized by a physical filling-in process of single-family-owned houses. Such houses built during the 1920s were usually of mixed quality and size within any given neighbourhood but were generally built to fit the technological requirements of that time. Consequently, only the better homes had garages, and these were single-car garages. The lots were small, averaging about 30-foot fronts, and most houses were of one and a half or two storeys. The occupants during the initial settlement stage were young marrieds with school-age children. Consequently, schools were built within walking distance of home and small neighbourhood businesses were built close by. Thus stage I is characterized by a process of general growth. New functions are added, population grows, and the general future of the neighbourhood is one of solidarity and growth. By the end of the first phase, the population for single-family houses has reached its peak. The area is mature.

After an indefinite period of little change, the second stage begins. Density drops as children grow up and leave home (stage II in Figure 14–7). The families get older and widows occupy some single-family homes. Houses are put into rental status, and dilapidation begins to occur among some of them. It also becomes apparent toward the end of stage II that the neighbourhood is becoming technologically obsolete. Some houses have to be replaced, generally with apartment houses. The rental property further increases, thus emphasizing the growing obsolescence of the neighbourhood. The parking problem becomes severe as the public streets become filled with automobiles in the evenings, because suitable parking accommodations are not available at residential sites. Moreover, most of the houses need remodelling and updating, particularly kitchens and bathrooms. By the end of stage II most of the original families have left the area. Some have died, others have moved to new residential areas, and still others have moved into apartments in other areas. A new group of residents have arrived, and a large percentage of the houses are occupied by renters rather than owners.

Depending on the location and nature of the original neighbourhood, change could then go in several different directions. During stage III it might grow in population as single-family homes are torn down or cut up for apartments (stage IVA). In this case there would still generally be an oversupply of schools in the area because the new

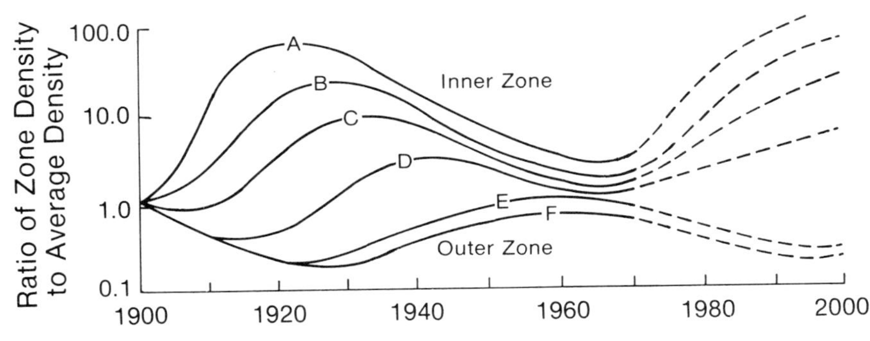

Figure 14–8 Relative density change over time for various zones within the city. Each zone contains a period of relative high density followed by a decline. The inner zones of American cities have peaked first, followed by successive zones. Note that there is less density variation within cities now than at any time since the turn of the century. (Adapted from Hans Blumenfeld, *Journal of the American Institute of Planners*, Winter 1954, pp. 3–14.)

residents have fewer children. On the other hand, the area could continue on a path toward more serious dilapidation and obsolescence. This would be the case if the area proved unattractive for other land uses. In prime areas the neighbourhood might be razed and invaded by commercial uses (stage IVC). At any rate, it is clear that all residential areas go through some cycle effects. They have a period of growth, a peak of maximum development, a period of general decline, and the beginning of a new cycle of development.

American cities today have vast areas of housing, between suburbia and the inner zone, that are well into stage II. The housing is such that it is not ready for complete clearance, but it is not able to make a viable adjustment to technological change. Thus one of the major problems in cities results from the inherently changing and cyclic nature of their residential land use.

Documentation of the cyclic nature of residential areas has been noted by several specialists in urban affairs (see Hans Blumenfeld, 'The tidal wave of metropolitan expansion', *Journal of the American Institute of Planners*, Winter 1954, pp. 3–14). It has been shown that each zone had a period of growth and a period of decline in population density (Figures 14–8 and 14–9). The inner zone (A) had growth first, followed by the zones next in sequence. However, also note that the more distant zones experience a lower density peak (possibly because of less accessibility potential) and a less steep fall-off in density over time than do inner zones.

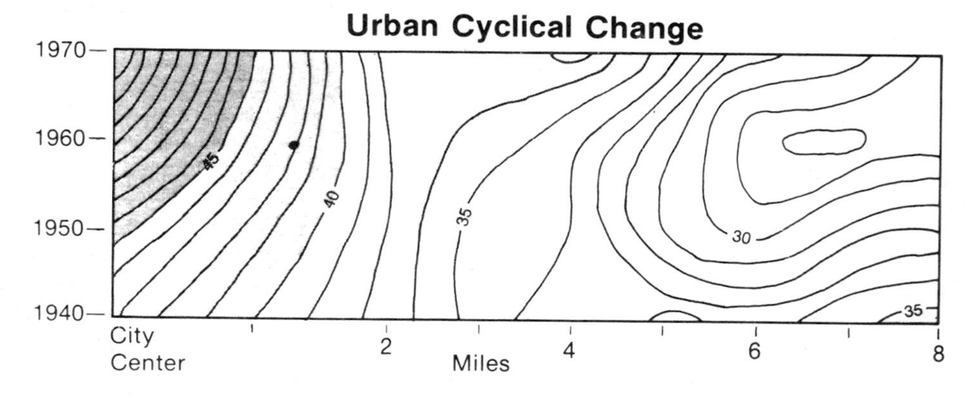

Figure 14–9 Mean age isolines, 1940–1970. The pattern of mean age within an urban area is cyclical. In the oldest zone (1) the mean age is increasing. In the newer zones (6, 7) the mean age decreases until urbanization is reached and then begins to increase. The zone of youngest mean age continually moves outward over time. (From David C. Johnson, *The Population Age Structure of an Urban Area*. PhD. diss., University of Washington, 1977. By permission.)

RESIDENTIAL CHANGE ON THE FRINGE OF CITIES

Another dynamic feature of residential land use occurs on the edge of cities. Here, land is transformed from rural to urban use. However, the transition does not occur in one fell swoop. The profile of change is fairly complicated. First, there is a series of waves of penetration during which changes in parcel size, ownership, and routes are laid in veneerlike fashion over the original rural pattern. This might be termed a 'precession' wave in that it precedes actual development (Figure 14–10). This wave of preparatory development is generally found ahead of, by 15 to 30 minutes in driving time, the actual edge of suburban development. As this suburban edge, or tidal wave, moves over the area, land is assembled, plotted, and otherwise prepared on paper for eventual urban use. Although much of the action is visible only at the county courthouses, title offices, and real estate firms, it is nonetheless very real indeed. In fact, the nature of the paper plans (called *plats*) and land assembly patterns often have great effect on the future density and settlement patterns of the fringe area.

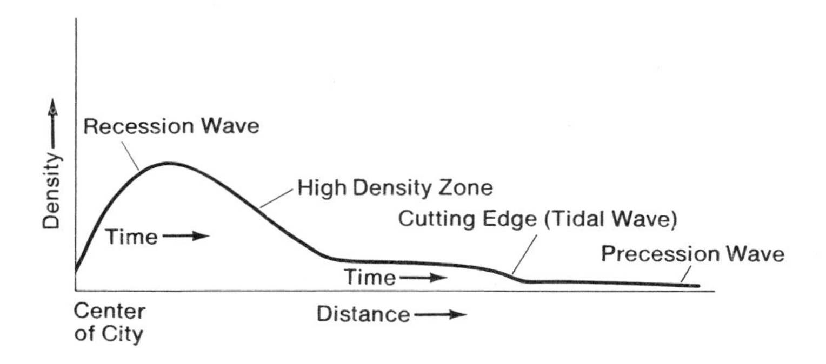

Figure 14–10 The density profile within American cities. Density is highest in the inner portions of the city, except for the central business area. Over time, central density declines as it moves outward—the recession wave. The edge of suburban invasion of rural areas is shown by the cutting edge. The recession wave is characterized by changes in land ownership, land parcel change, and extensive land speculation.

COMMERCIAL LAND USE PATTERNS

A city's pattern of commercial land use is confusing to the casual observer because it is actually composed of a number of business location arrangements. Some commercial uses are aligned along arterials, some are in clusters of varying sizes, and still others are found widely scattered throughout the metropolis (see Figure 14–11). We might thus categorize the first type as highway-oriented, a type in which commercial strip development apparently benefits from automotive traffic. We might categorize the second type as accessibility-oriented, a type in which business is clustered at a hierarchy of strategic sites. The third pattern might be assumed to reflect low threshold establishments that can be accommodated by neighbourhood markets. Therefore, when the general pattern of commercial land use is dissected into various subpatterns, a meaningful distribution can be discerned.

The type of businesses that are found in clusters or nucleations might be considered *intensive* businesses. These are high-rent-paying functions by comparison and are characteristically pedestrian-oriented. Common examples include drug, variety, jewellery, and clothing stores. According to urban rent theory, the intensive establishments can easily outbid the arterial-oriented, or *extensive*, establishments for sites. Hence the intensive functions occupy the most accessible sites. They also often benefit greatly by cluster economies achieved as the result of multiple purpose shopping.

In contrast, the businesses located along arterials use a large amount of space but pay low rent. Characteristic functions include many drive-in establishments, service stations, automotive sales and repair establishments, supermarkets, and furniture stores. These are usually single-stop functions, that is; shoppers go to only one store and have a definite purchase in mind.

The density of commercial development varies greatly within any given city. As might be expected, it is much more prominent near major employment sources, such as the

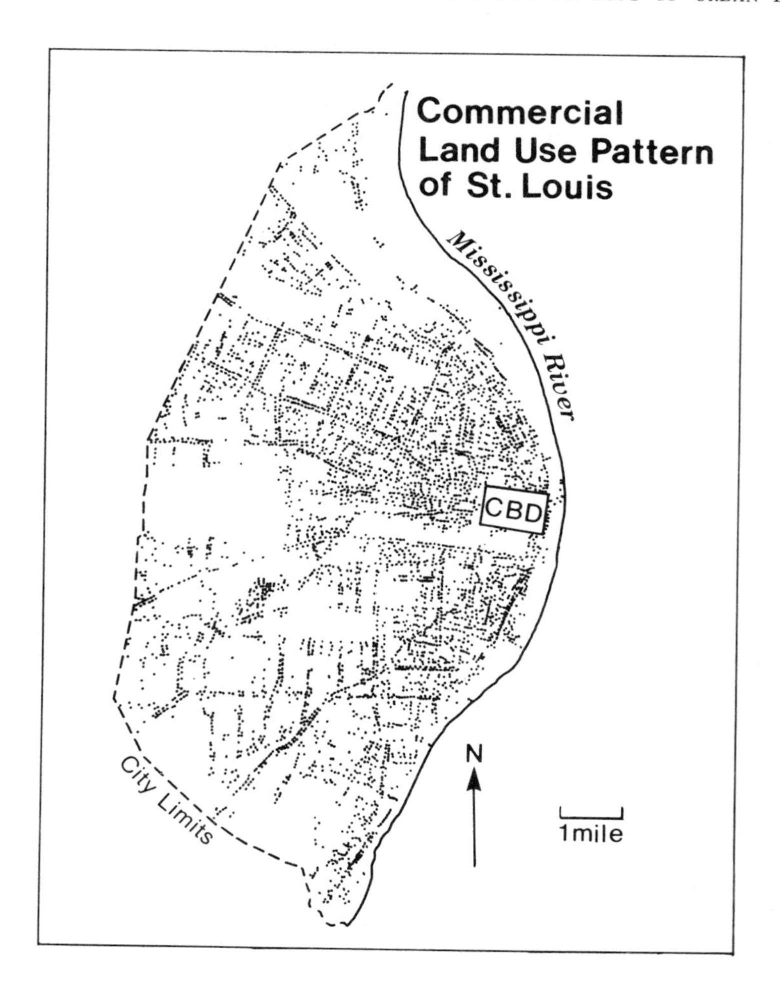

Figure 14–11 Commercial land use pattern in St Louis. The linear strip development of commercial activities along major streets is most evident. With the development of planned shopping centres, a much more nucleated pattern of business is evident in those areas beyond the central city. (Adapted from Harland Bartholomew, *Land Uses in American Cities*. Cambridge: Harvard University Press, 1955.)

central business district, than in predominantly residential areas. It is also severely limited in the heavy industrial sector of cities (see Figure 14–11). In the residential sector of the city commercial development is found in a fairly regular hierarchical pattern, much as might be postulated according to *central place theory*. Especially prominent is the rise of major regional, planned shopping centres near strategic interchanges of the freeway system. The strip commercial developments are based on automotive traffic and are perhaps related to the commuting work trip.

CENTRAL PLACE THEORY AND THE PATTERN OF BUSINESS

Much explanation of the distribution of commercial activities can be achieved by applying the principles of central place theory, discussed in Chapter 13, to tertiary activities within cities. Each business requires that a certain threshold or market be met before it can be viable. Thus business activities that require small numbers of customers are the most widespread within the metropolis. These lower-order functions (grocery stores, service stations, drugstores, and neighbourhood-level services like barber shops) are found in centres of almost any size and in many different locations. In fact, the smaller centres will be limited almost exclusively to such low-threshold functions. The next order of functions, those that require a larger number of persons to draw upon, will be found only in centres at the community and regional level. At the highest-level centres, functions that require a large trade area are found. Such activities characteristically include department stores and various clothing establishments.

Remember that the total number of customers of some high-order functions is not necessarily larger than that of a low-order function. The difference relates primarily to the number of persons within any given population who will purchase an item. For example, every family requires groceries, and a high proportion require gasoline, and they require these items frequently. The demand, moreover, is relatively inelastic. Higher-order establishments, on the other hand, commonly find that only a small percentage of the total potential customer market will purchase goods at any one time. Thus it takes many more people in a trade area to supply a viable number of customers for such an establishment than it does for the low-order type, where almost all customers frequently require such goods and services.

COMPETITION AMONG BUSINESS CENTRES IN THE USA

In the United States, many of the smaller and older unplanned business (shopping) centres are in trouble. Many businesses in such centres have either moved to planned centres or have become marginal. The villain appears to be the rise of the mighty, planned shopping centre. These large centres have affected small business districts, planned and unplanned alike. There appear to be two reasons for this. One is that such centres are simply more attractive and more attuned to the automobile than are the older centres. The second reason concerns the change in mobility within cities, which has had an adverse effect on smaller centres in much the same way that it has placed small towns in competition with metropolises.

The large centres have benefited more from improved transportation than have the small centres. The manner in which this operates is demonstrated below with regard to two business centres, one small and one fairly large and new. Each is in competition with the other for customers in the common territory separating them. Let us assume that the smaller centre has 5 functions, the larger has 20 functions, and that they are 30 blocks apart. Now by using the standard procedure for computing the breaking point (D) between centres, their respective tributary areas can be estimated (Figure 14–12). If the general accessibility, or friction of distance exponent, is distance squared, then the breaking point between the centres will be 10 blocks from the smaller centre and 20 blocks from the larger centre. (Recall that according to the calculation formula or Reilly's Law, distance from the smaller centre (D) equals

$$\frac{d}{1 + \sqrt{\frac{P_2}{P_1}}}$$

where d is the distance between centres, P_2 is the size of the larger centre, P_1 is the size of the smaller centre.)

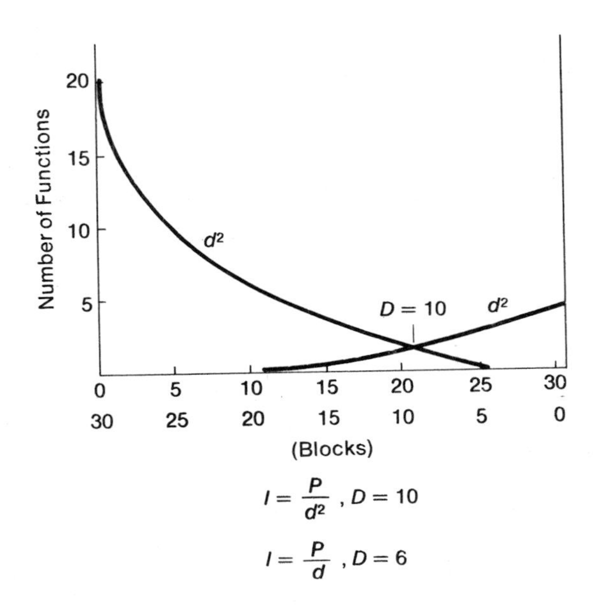

Figure 14–12 Trade areas between business districts. Given one business district with 20 functions and another with 5 functions, the breaking point between the two districts is 10 blocks from the smaller centre. With an improvement in transportation such that the friction of distance coefficient is no longer d_2 but d_1, the breaking point between districts would shift in favour of the larger centre by four blocks.

Now let us assume that accessibility has generally improved between the two centres to the extent that the friction of distance exponent is no longer distance squared, but distance with an exponent of 1, which is a reasonable assumption. In this case the breaking point will have shifted four blocks toward the smaller centre so that the trade area boundary (*D*) is now only 6 blocks from the smaller centre. The result is a drop in customers formerly frequenting the smaller centre. In fact, such a drop may result in loss of a function in the smaller centre because of inadequate market to support it. If so, then another adjustment would occur, and an even further shift toward the smaller centre would occur so that the breaking point would now be only 4.8 blocks from the smaller centre. Thus an entire series of adjustments results from one change in accessibility. First, the customer-trade-area breaking point (*D*) is shifted toward the smaller centre. Second, the size of the smaller centre is made even smaller. Third, other shifts in trade area follow, until some equilibrium is reached. However, quite an opposite effect has occurred with regard to the central business district of cities. In this case a movement of customers away from this business district has more than cancelled out any improvements in transportation.

Thus it can be seen that the pattern of business nucleation is constantly in flux. The general effect of improvement in mobility has been an increase in the competitive position of the larger centres, the primary exception being the central business district. In addition, such new innovations as the planned shopping centre, with its ability to accommodate the automobile, have rendered the central business district and many of the older, nonplanned centres technologically obsolete.

COMPETITION WITHIN BUSINESS CENTRES

An inverse relationship exists between amount of land used and intensity of urban land use. This relationship is logical if it is assumed that the intensity of use results directly from the bidding process whereby prime land can be occupied only in an intensive manner. In fact, the intensity of retail use decreases with increasing distance from the peak land-value intersection or '100 per cent corner' as it is commonly called. Typically, the larger users of space per establishment such as grocery, furniture, drive-in, and automotive establishments are found near the outer edge of the business district. By contrast, functions using relatively little space (barber shops, tobacco shops, jewellery stores, drugstores, and the like) commonly occupy central positions.

There is also variation in rent-paying ability for the same function. The higher the point of accessibility, the higher will be the rent-paying potential. For example, a drugstore located in a large shopping area and thus drawing customers from a wide area may be achieving a more intensive use of its site than a drugstore located in a small neighbourhood centre and dependent on a more limited number of customers. Therefore the drugstore in the larger retail centre (greater accessibility to a larger market) will be able to pay higher rents than the drugstore in the smaller centre. The former will also have to pay higher rents than the latter in order to achieve the same relative position within the centre in competition with other functions.

Thus a first clue to the location of any particular business within any given centre is a function of the accessibility potential of that centre. (This does not apply to planned shopping centres, which will be discussed shortly.) In small neighbourhood centres (N), the highest-level function will often be a supermarket (Figure 14–13.) This

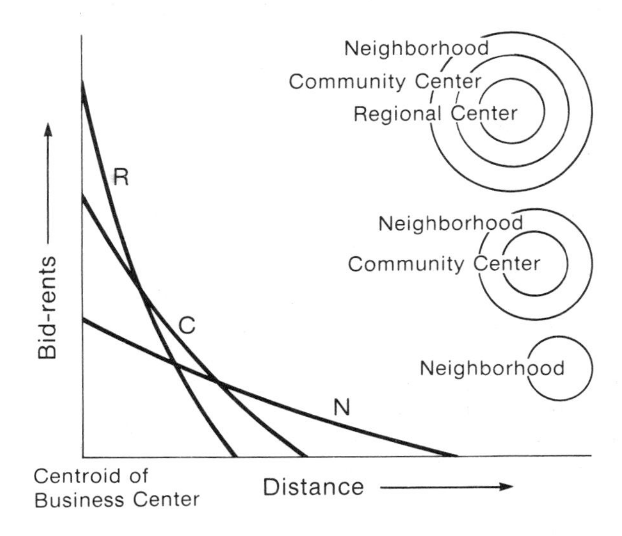

Figure 14–13 Neighbourhood, community, and regional level functions within various centre types in the USA. Regional level functions can bid higher for sites than community and neighbourhood level functions and thereby take prime sites in regional level centres. Wherever a higher level function is absent, as in lower centres, that function which is highest in the bid-rent hierarchy will occupy the prime sites.

function will naturally occupy the prime or 100 per cent site. Within community centres (C) functions at a higher level than supermarkets will be found, and these will outbid the supermarkets for prime sites. This will occur even though the supermarket can and will pay more rent in the community centre than it would in a neighbourhood centre. In the regional business districts (R) only the very highest order functions will be found in the core positions. In summary, the general rent-paying ability of a firm is based on the amount of return that such function can reap at one site versus another, and its placement within a particular centre is based on its rent-bidding power relative to other functions within that centre.

If it is assumed that a prime reason for differentiation in the location of business types within 'unplanned' business centres is rent-paying ability, then it is proper to investigate the reasons for such rent differential in far greater depths. The factors enumerated below affect any retail establishment's ability to pay rent and hence to bid successfully for the most accessible sites: (1) the value of merchandise per unit area; (2) the turnover and markup of merchandise, and (3) the various operational cost peculiarities, such as labour costs, among firms. First, we will see how these operate to affect rents and produce various spatial patterns. Second, we will attempt to determine why there should be variations among retailing establishments with regard to such features as markup, clearly a key item directly affecting profits and ability to pay rents and bid for sites.

The value of merchandise per unit area is a very crude but often effective way of classifying firms on the basis of their potential rent-paying abilities. The method consists simply in estimating the value of the merchandise on site and dividing that value by the square foot of floor area. Establishments with high square-foot merchandise values will generally be found nearer the prime corner than will those with lower values. Nevertheless, there are many exceptions. For example, automobile sales establishments have fairly high values per square foot, whereas variety stores have fairly low values per square foot. Yet, automobile dealers are characteristically found on the periphery of centres and variety stores are found near the 100 per cent corner.

Turnover (or volume) and markup also provide meaningful measures of rent-paying ability. Thus firms with the highest turnovers and markups presumably can pay higher rents, other things being equal, and occupy closer-in locations in contrast to activities with opposite features. Unfortunately, since the same amount of money can be gleaned through substitution of turnover and markup, confusion often occurs. Usually a firm has a high turnover and a low markup or a low turnover and a high markup. For example, dealers in new cars characteristically have low turnover and rather substantial markups, presumably because of salesmen's commissions and the like. Likewise, jewellery stores have fantastically high markups (often 200 or 300 per cent over wholesale cost) and low turnover. Grocery stores have low markups but high turnover. Thus inasmuch as mixing occurs, the locational case is not made clear. Two kinds of firms whose locations do not conform to these measures further demonstrate the problem. Bookshops have moderate merchandise value per square foot, high markups, and low turnover, but rarely occupy prime locations. Music stores likewise have high merchandise value, high markups, and low turnovers but are invariably in much more obscure and peripheral locations than are bookstores. But why, we might ask, do not firms vary their markups to compensate for turnover and merchandise value per square foot, inasmuch as markup is the only factor not fixed for any location? If this were done, a situation would result in which all stores could compete for sites equally.

The typical response to this question is that the markup over wholesale costs relates to opportunities various sites offer for earnings. In that way, functions fortunate enough to occupy prime locations have somewhat of a spatial monopoly and hence are able to enjoy higher markups without fear of devastating competitive responses from non-central locations. Thus we might argue that markups reflect the extent of competition as measured by the number of usable sites for that function. (Note that this argument is analogous to that used with regard to rent-paying abilities of different agricultural crops.) Functions with the least choice of suitable sites can charge extremely high markups when they occupy central sites. Nevertheless, there are glaring exceptions to such arguments. Certainly, tradition also might provide a strong explanatory reason because some markup rates appear to result primarily from historical precedent. Another explanation for differential markup relates directly to the nature of the item being sold and the advantages derived by proximity to other businesses.

The relative location of functions within planned shopping centres is quite different from that in unplanned centres. Given the fact that planned shopping centres are fast becoming the common business medium in most of the newer portions of cities, perhaps it is the 'unplanned' districts that are peculiar. Certainly, it is the older districts that are the most anachronistic and most in trouble. Nonetheless, in many respects, the forces that were already discussed with regard to the older centres and that are most easily seen in such centres are still somewhat operable in planned shopping centres. It is primarily the nature of shopping centre operation that results in a pattern peculiar to this type of business cluster.

On the one hand, the organization of activities within planned shopping centres offers the customer a more advantageous arrangement of goods and services than that found in unplanned centres. First, these centres are commonly free of such nonshopper space use as churches, mortuaries, insurance companies, and similar businesses found within unplanned centres. (While some of these business services are of considerable aid to other businesses in the centre, they are of no direct use to the shoppers themselves. Instead, the customer is forced to walk past such establishments to assemble the desired items.) Because the location of functions within planned shopping centres is not based on independent bid rents, many functions are situated more advantageously for the shopper than they commonly are in unplanned centres. As an example, grocery stores are often found in planned centres even though they do not qualify in unplanned centres of equal size.

On the other hand, some aspects of business arrangement in the planned shopping centre are less advantageous to the shopper. First, a number of functions that are considered beneficial to shoppers are commonly excluded from planned shopping centres. Such excluded functions commonly are those catering to single-purpose trips or those that have low sales/per unit area. Often a large department store can dictate terms to the shopping centre management and thus force exclusion of certain otherwise highly competitive stores. In addition, large numbers of extensive businesses normally associated with an unplanned business district are also excluded. Such exclusions commonly include hardware, furniture, lumber and paint stores, some book and most music stores, and all manner of automobile-oriented establishments. One justifiable basis of exclusion relates to the fact that stores in the centre carry some of these items. The usual basis for excluding furniture stores, however, is the simple fact that they do not contribute many sales to the overall centre inasmuch as most of the items they sell are of a single purchase nature rather than part of a multiple-purpose shopping tour.

Another problem for the shopper in many planned shopping centres is the tendency to place a large department store, usually complementary rather than truly competitive, at either end of the shopping mall. (It will be recalled that in unplanned centres, department stores are usually found adjacent to or very near each other.) The result is that people are forced to make the long trek between these stores in order to do comparison shopping. This is a deliberate strategy on the part of shopping centre management in order to create as much impulse buying as possible in intermediate stores, a number of which are placed between these major magnets. (In very large hypermarket centres with ample parking facilities, the customers have circumvented this somewhat by shopping in only one end of the centre or by driving, rather than walking, to the other end.)

THE VERTICAL LAND USE PATTERN

Commercial land use is found vertically as well as horizontally within cities. In fact, in the central business districts of American and most other 'Western' cities far more square feet of space are found above than at ground level. This has long been so, but as the skylines of central London, Paris, Brussels, and many other European capitals and regional cities testify, it is an important circumstance and in particular it is very characteristic of office space. As might be expected, the vertical placement of various business uses does not occur haphazardly but has a distinctive pattern and operates by much the same spatial principles as does economical land use in the horizontal dimension.

Commercial activities place such a high premium on accessibility that rent gradients occur vertically as well as horizontally. Although only the ground floor space has been discussed thus far, it is nonetheless apparent that some measure of vertical height may be substituted for some measure of horizontal distance. Thus a furniture establishment may be found either on the edge of a business district or on the upper floors of a downtown building, such as a department store. In this regard, some businesses may have a measure of substitution between horizontal and vertical proximity to the 100 per cent location. Thus at distance X rents may be the same as at height A, and at distance X_2, comparable to height B. Of course, not all businesses have this substitution capability, since the latter depends on the nature of the business.

The functions most capable of such substitution, and most prone to favour vertical to horizontal proximity, are office uses. Typically, these functions, which include lawyers, physicians, real estate and insurance brokers, occupy the upper floors of buildings in which the street level is taken by retail uses. It should also be noted that some of these functions are suitable for both ground level and upper level space; ground floor occupancy, however, occurs only at the periphery of a business district.

Nevertheless, unusually high accessibility indexes are maintained vertically, largely as the result of the efficiency of the elevator versus comparable horizontal movement. In fact, the elevator is far superior to movement made horizontally. This can be demonstrated by imagining a high-speed elevator operating on the horizontal scale (Figure 14–14). Assume that the vertical elevator and the horizontal elevator are equal in speed. If so, the vertical elevator would provide access to almost ten times as much space in a given time as the horizontal elevator. (Stop and start time would, of course, diminish this advantage somewhat.) In order to have access to circles of space with 100-foot radii, the horizontal elevator would have to travel 200 feet between stops. But to provide the same access to space vertically, the elevator would have to travel only about 20 feet between stops, that is, the distance between floors.

For any given ground-floor accessibility, there is a fade-off of accessibility upward. The degree of fade-off depends on the accessibility value at ground floor level, the height of the building, and the potential functions that may find such a location suitable. The decrease in rents with height is shown theoretically for three different positions of horizontal accessibility, A, B, and C (Figure 14–15). The ground floor rent is determined by its position in the horizonal accessibility fade-off from the prime site. The rents decrease as one proceeds upward. In fact, the second and third floors show a much steeper fall-off than does any comparable horizontal space extension. Beyond these first few lower floors, however, the drop becomes less.

A special peculiarity of vertical space is the tendency for rents at the very top floors to increase with increasing height (Figure 14–16). This 'penthouse' effect evidently occurs for at least two reasons, one psychological and the other based on the creation

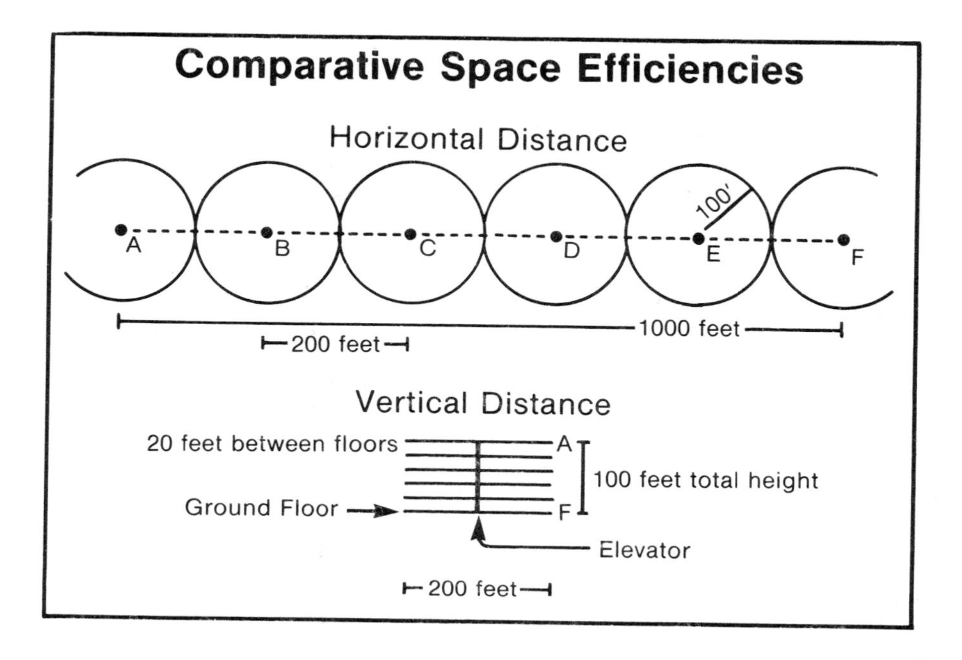

Figure 14–14 Comparative space efficiencies between horizontal and vertical transportation systems. Because of the stacking advantages afforded vertical space, the elevator is considerably more efficient than a one-floor horizontal system for reaching small areas. However, the slower speed of elevators and the high cost of vertical construction largely mitigate this advantage.

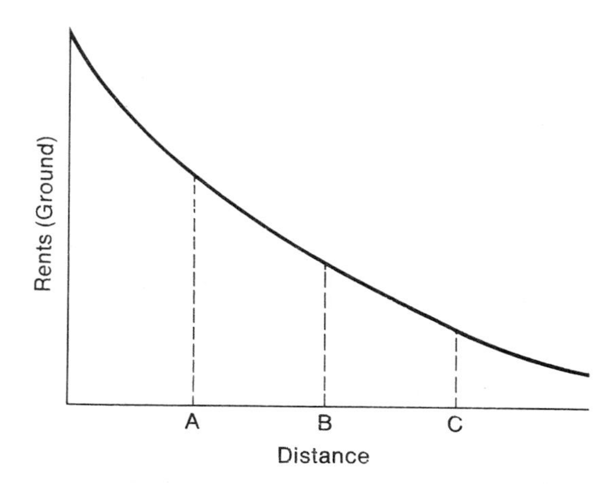

Figure 14–15 The play-off between ground floor rent and distance from the prime rent site. Sites at distance C bring less rent than those at distance A closer to the prime rent site. This is because of the inferior interaction potential of less accessible sites.

of 'site' amenities. In the first instance, there appears to be psychic, and hence prestige value, in occupying the top floor or floors of a very high building. Such height also provides certain advantages of view, lack of noise, and, in some areas, diminution of smog.

With regard to retail goods, the effect of height is severe. Most retailing outlets not already operating on the ground floor find anything above the ground floor unacceptable. However, in certain large establishments like downtown department and furniture stores, much use is made of upper floors. Here again there is a fade-off in intensity of space use analogous to the horizontal scale. The top floor of department stores invariably contains articles that are characteristically single-purchase, or certainly definite-purchase, items, such as furniture, carpeting, and so forth. People will make the trip upward in order to see a particular thing. By contrast, goods usually purchased on the basis of convenience and impulse are placed on the ground floor.

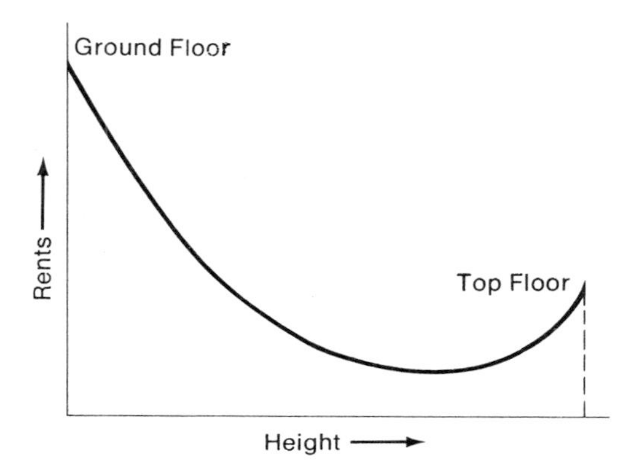

Figure 14–16 The play-off between ground floor rent and vertical distance. The fade-off in rents with height is somewhat similar to that for horizontal distance. However, with increasing height a 'penthouse effect' is often achieved whereby the top floors bring higher rents than those in the centre of the building.

Rent paying ability also differentiates office space use within tall buildings. For example, legal offices are most prominent in the upper stories of office buildings, yet lawyers require considerable interaction with other functions and with clients. By contrast, real estate and insurance offices are characteristically found on the lower floors, even though their need for access to the street may not be greater than that of lawyers. Undoubtedly, rent-paying ability operates here with a double-edged sword. First, some functions, such as lawyers, are high-rent payers yet could more profitably occupy the lower floors. However, perhaps for prestige reasons, they prefer offices higher in the building. The very top floors (penthouse floors) are devoted primarily to legal firms because of their ability to outbid other functions. Real estate offices and insurance offices in downtown office buildings often pay higher rents than they might otherwise have to because, paradoxically, many are squeezed down to lower levels by bidders for space above. Perhaps one result of this is reflected in the increasing tendency for insurance firms to move to ground floor space elsewhere or to separate buildings in outlying locations.

CONCLUSIONS

Despite this rather lengthy discussion, only about one-half of the land uses in a typical North American city was presented. Many other land use types, particularly the public and semipublic uses, were omitted. An interesting discussion could have been focused on particular activities like hospitals, churches, parks, and cemeteries. Nonetheless, such extended treatment of intraurban patterns would yield little additional knowledge about the nature of urban patterns and the nature of urban location problems. In this regard, keep in mind that our purpose is not to discuss all possible activities, but only those that provide additional incremental insights and enlightenment about the nature of various locational patterns and problems. The treatment of industrial, residential, and commercial land use patterns has largely filled this purpose.

Although this discussion has been largely based on the North American city, it is possible to generalize much of what has been written above in terms of the 'Western-city' as a group, including British and European examples. Thus we can conclude that the positioning of activities within 'Western' cities is largely affected by competition for accessible sites over time. All kinds of land use compete with each other, but those with the greatest need and greatest rent-paying abilities, however, usually end up with the prime sites. In this regard, commercial functions are the most competitive bidders for space within cities and consequently are found in most of the strategic locations. The possibilities for vertical-scale benefits are often involved in achieving high rents. Farthest removed from the centre of the city are commonly single-family residential structures. Thus there is a general fade-off of both land value and population density as one proceeds from the centre outward.

Residential land use has become highly independent of accessibility requirements. People can now travel considerable distances to work, thus making available for residential occupancy a vast area well beyond present-day suburbia. Moreover, business functions, schools, and other needed services tend to follow the path of residential development, so that residential areas appear to be setting the stage for the future pattern.

Nonetheless, industrial land is often the triggering element for urban growth. This is particularly the case for those industries that export their products beyond the city. These city-forming industries employ basic workers who indirectly support many other persons and activities within the city. Often these industries select their site within the city based on good access not to areas within the city itself, but to areas outside the city from which they may obtain raw material and ship finished products. Often port locations and railroad proximity, plus flat land, are site requirements that override centrality within the metropolis. In fact, most large land-using primary industries are found on the edge of cities.

There is a strong tendency for like functions to cluster within parts of a typical city. Industry is heavily attracted to other industry inasmuch as one firm's finished products are often another's raw materials. Business also is characteristically found in tight clusters because of the benefits one business offers another as the result of comparative shopping by customers. Even residential areas remain quite pure despite the obvious daily interaction with all types of diverse uses.

The fact that various functions within a city are not mixed but largely pure is responsible for the daily ebb and flow of traffic in the city. Foremost among these is the work trip, but shopping, business, social, health, and other trips are also involved.

Finally, there is a pattern of land use in a vertical as well as a horizontal direction within cities. In fact, there is often a direct substitution between height and distance from the prime centre of accessibility for business functions. Thus a stratification of retailing activities occurs as one proceeds upward as well as outward. For example, furniture is often sold on the top floor of a department store or on the edge of the business district in the downtown areas. Likewise, a vertical stratification, based on the ability to pay rent, occurs within office buildings. Lawyers commonly pay high rents and, perhaps for prestige reasons, usually occupy the topmost floors of office buildings. Nonetheless, there are many kinds of lawyers and many different locational choices. Finally, residential land can also effectively compete for central city uses through high-rise structures.

Increasingly, however, most activities are emphasizing the one-storey or horizontal dimension rather than the vertical dimension. This is true for industrial, residential, and commercial uses. The horizontal production system (continuous flow process) has become standard through the adoption of assembly line techniques in much of industry. Business uses also increasingly emphasize the one-level approach surrounded by ample parking. Residential land use is perhaps the most land-hungry of all. In the United States especially, the rambler concept in home construction is responsible for much of the necessity for larger lots. But the major factor necessitating more residential space is the automobile and the attached garage.

Perhaps the most important and continually true statement that can be made about the intraurban activity pattern is that it is in rapid and constant flux. Consequently, formerly well-placed functions now find themselves in the backwash locations. Shopping centres and businesses within them are in severe competition. In the United States, older, unplanned enterprises are often becoming marginal as new, large, planned shopping centres, strategically situated near major freeway interchanges, become more dominant. Such developments are not so evident in Great Britain (although they are present) primarily perhaps because of the kinds of governmental intervention with land and market forces as noted in Chapter 13. Neither has it happened, as in the United States, that the formerly supreme central business district of so many cities has become technologically obsolete with respect to shopping. Yet, in greater or less degree, in all cities North American or European, we can say that residential neighbourhoods are rarely stable. They are either increasing or decreasing in population. As a general rule, most residential areas in the central city are tending to lose population, primarily because of ageing, whereas those in outlying and suburban areas are still tending to grow.

BIBLIOGRAPHY

Berry, B. J. L., Parsons, S. J., and Platt, R. H. *The Impact of Urban Renewal on Small Business*. Chicago: University of Chicago Press, 1968.

Berry, B. J. L., and Dahmann, D. C. *Population Redistribution in the United States in the 1970s*. Washington, D.C.: National Academy of Sciences, 1977.

Blumenfeld, H. 'The economic base of the metropolis'. *Journal of the American Institute of Planners*, November 1955, pp. 114–132.

Blumenfeld, H. 'The tidal wave of metropolitan expansion'. *Journal of the American Institute of Planners*, Winter 1954, pp. 3–14.

Boyce, R. R. 'The edge of the metropolis. The wave theory analog approach'. *Internal Structure of the City: Readings on Space and Environment*. Ed. Bourne, L. S. Toronto: Oxford University Press, 1971, pp. 104–111.

Boyce, R. R. 'Myth versus reality in urban planning'. *Land Economics*, August 1963, pp. 240–251.

Brodsky, H. 'Land development and the expanding city'. *Annals of the Association of American Geographers*, June 1973, pp. 159–166.

Chinitz, B., ed. *City and Suburb: The Economics of Metropolitan Growth*. Englewood Cliffs, N.J.: Prentice-Hall, 1964.

Colby, C. C. 'Centrifugal and centripetal forces in urban geography'. *Readings in Urban Geography*. Ed. Mayer, H. M., and Kohn, C. F. Chicago: University of Chicago Press, 1959, pp. 287–298.

Davies, R. L. *Marketing Geography: With Special Reference to Retailing*. Cambridge: Retailing and Planning Associates, 1976.

Donaldson, S. *The Suburban Myth*. New York: Columbia University Press, 1969.

Eaton, L. K. 'The American suburb: dream and nightmare'. *Landscape*, Winter 1963–64, pp. 12–16.

Goheen, P. G. 'Interpreting the American city'. *Geographical Review*, July 1974, pp. 362–384.

Gottmann, J. 'The dynamics of large cities'. *The Geographical Journal*, June 1974, pp. 254–261.

Haig, R. M. 'Toward an understanding of the metropolis: the assignment of activities to areas in urban regions'. *Readings in Economic Geography*. Ed. Smith, R. H. T. et al. Chicago: Rand McNally, 1967, pp. 44–57.

Hartshorn, T. A. 'Inner city residential structure and decline'. *Annals of the Association of American Geographers*, March 1971, pp. 72–96.

Lampe, F. A., and Schaefer, O. C., Jr. 'Land use patterns of the city'. *The Journal of Geography*, May 1969, pp. 301–306.

Lipton, S. G. 'Evidence of central city revival'. *Journal of the American Institute of Planners*, April 1977, pp. 136–147.

Mulvihill, D. *Geography, Marketing and Urban Growth*. Princeton: Von Nostrand, 1970.

Nelson, H. J. 'The form and structure of cities: urban growth patterns'. *The Journal of Geography*, April 1969, pp. 198–207.

Nelson, R. L. *The Selection of Retail Location*. New York: Dodge, 1958.

Pred, A. R. 'The intrametropolitan location of American manufacturing'. *Annals of the Association of American Geographers*, June 1964, pp. 164–180.

Reinemann, M. W. 'The pattern and distribution of manufacturing in the Chicago area'. *Economic Geography*, April 1960, pp. 139–144.

Scott, P. *Geography and Retailing*. London: Hutchinson, 1970.

Simmons, J. *The Changing Pattern of Retail Location*. Department of Geography Research Paper No. 92. Chicago: University of Chicago Press, 1963.

Smith, P. J. 'Calgary: a study in urban pattern'. *Economic Geography*, October 1962, pp. 315–329.

Thornes, D. C. *Suburbia*. London: MacGibbon & Kee, 1972.

Vance, J. E., Jr. 'Housing the worker: the employment linkage as a force in urban structure'. *Economic Geography*, October 1966, pp. 294–325.

Whitehand, J. W. R. 'The changing nature of the urban fringe: a time perspective'. In *Suburban Growth*, Ed. Johnson, J. H. Chapter 3, pp. 31–52. London: Wiley, 1974.

Part Five

GEOGRAPHICAL PROSPECTUS

'The closed earth of the future requires economic principles which are somewhat different from those of the green earth of the past. For the sake of picturesqueness, I am tempted to call the open economy the 'cowboy economy', the cowboy being symbolic of the illimitable plains and also associated with reckless, exploitative, romantic, and violent behaviour, which is characteristic of open societies. The closed economy of the future might similarly be called the "spaceman" economy, in which the earth has become a single spaceship, without unlimited reservoirs of anything, either for extraction or for pollution, and in which, therefore, man must find his place in a cyclical ecological system which is capable of continuous reproduction of material form even though it cannot escape having inputs of energy.'

Kenneth E. Boulding, 'The Economics of the Coming Spaceship Earth', in Henry Jarrett, ed., *Environmental Quality in a Growing Economy*, Washington: Resources for the Future, Inc., Johns Hopkins Press, 1966, p. 9.

15

A World of Crises?

This final chapter attempts to provide some insights into the kinds of problems that occur in response to the operation of the spatial principles discussed previously. 'Problems', as now used, will no longer refer to problems of spatial understanding but to problems that relate to various spatial responses and patterns. In a sense we return to some of the initial questions raised in Chapter 1.

Some problems of this type would result regardless of the spatial system devised. Thus the comments that follow are not meant as an indictment of the present system but simply as an important consequence of it. The primary problem is one of spatial inequality.

That the earth is variable in its display of natural resources is obvious. Minerals appear to be concentrated in some parts of the world and absent in others; good soils are prolific in some zones and poorly represented in others; some parts of the earth are composed of vast plains and others with extensive displays of rugged and mountainous terrain; and there is widespread variation in climatic characteristics from place to place. With such rather obvious physical differentiation, it would seem reasonable that the display of people and wealth would likewise vary.

Such spatial inequality would occur even on a homogeneous earth. As one locational economist aptly stated: 'Even in the absence of any initial differentiation at all, i.e., if natural resources were distributed uniformly over the globe, patterns of specialization and concentration of activities would inevitably appear in response to economic, social, and political principles' (Edgar M. Hoover. *The Location of Economic Activity,* New York: McGraw-Hill, 1948, p. 3.)

Second, in contrast to the kinds of problems discussed earlier, those posed below are not subject to easy solution or perhaps to solution at all. These problems are best treated as crises. A crisis might be defined as a problem that does not have a ready solution. Moreover, because the kinds of problems that will be treated here are on a world scale, they are of such an imponderable magnitude that they cannot be grasped completely. Such problems having strong geographic bases and which fall within the domain of economic geography include the following: (1) the matter of feeding the future population of the earth, in light of food capabilities and population growth and demand patterns; (2) the nature of energy consumption and its character, given the dichotomy between the producing and consuming nations; and (3) certain critical resource issues that bode ill for people. In addition, we might add the fragile political situations, in the Middle East and elsewhere, which are of such an incendiary nature as to ignite the world in war.

All these crises have geographical bases and are geographical in content and character. Even so, the apparent solution that calls for geographical restructuring of population, production, and trade is, in most cases, untenable; the problems are of such a spatial dimension as to defy such simple spatial solutions.

POPULATION, ECONOMIC GROWTH, AND URBANIZATION

For most of the time that man has existed his numbers have been small and his powers limited. However, as shown in Figure 15–1, the world's population increased by a half of a billion in the 100 years between 1750 and 1850, by 1.2 billion from 1850 to 1950, and by more than a billion in the twenty years 1950 to 1970. It may double again in the years to 2000, reaching between 6 and 7 billions. Estimates vary, but a further 4 to 6 billion may be added in the first fifty years of the 21st century.

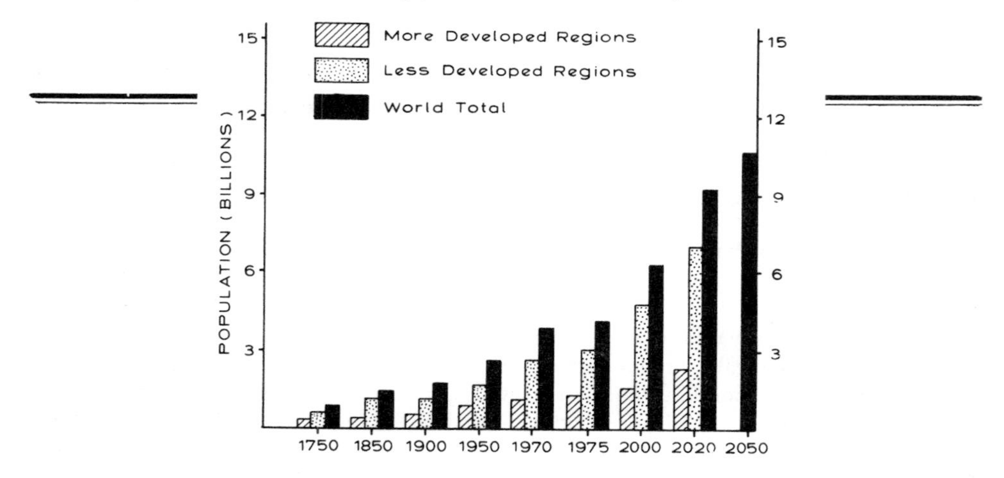

Figure 15–1 World population estimates and conjectures, 1750–2050, based on medium assumptions. (Redrawn from J. W. McNeill, *Environmental Management*, Ottawa: Government of Canada, 1971, p. 28.)

Accompanying this 'demographic revolution' is the national and international pursuit of economic growth. As measured by such indices as gross national product, disposable income, and leisure time, this pursuit has been remarkably successful in the more developed nations of the world, and despite setbacks in the international economy experienced in the late nineteen-seventies, many shorter-term studies suggest that both the goal of economic growth and success in achieving it will continue. The gross world product may not be as high as once believed, but even if the assumed growth rate were only three per cent, this measure would still yield an increase of between two and three times by the year 2000.

Unfortunately, there are few signs that economic growth at even this level (which suggests a slowing down of the world economy) can be shared by the great majority of the underdeveloped nations of the world. It is also clear that a major component of increased economic growth must stem from the expansion of industries that at present are pollution-intensive, and from other activities previously noted in this book which at present ravage the environment. Failure to curb these and other negative effects of growth can mean either a much slower country-by-country growth than forecast, or conceivably, attainment of the forecast growth until the environment is destroyed.

Writing in the early 1970s, one reviewer of the world's population and economic growth wrote:

> *'Looking back over these global projections, one cannot help but wonder about the future of man and his environment. The provision of food, water, minerals, energy, outdoor recreation and other necessities for such vast numbers of people commanding a relatively high purchasing power will place enormous pressures on every corner of the biosphere. There will be no slackening of the growing appetite of the more developed nations, and colossal demands are certain to occur in the less developed regions. A century ago, for example, the production of crude petroleum was negligible. By 1966 it amounted to 1,641 million metric tons per year, having increased sixfold over the preceding 30 years. In the same three decades, world production of motor vehicles grew from 5 million to 19 million per year. Between 1956 and 1966, the total value of all industrial production doubled. Looking to the future, it is estimated that the world will consume more energy in the next 20 years than in the last 100. World food production must increase by 50 per cent over the next 20 years to keep pace with growing populations. Moreover, it is estimated that the world will consume more metals in the next 35 years than it has in the last 2000.'* (J. W. MacNeill, *Environmental Management*, Ottawa: Government of Canada, 1971, p. 34.)

To this problem, caused by rising populations and their rising demands for the world's raw materials and industrial goods, we must also add the problem of rapid world-wide urbanization. This is also a looming crisis of gigantic import. By the year 2000, for the first time in human history, more people will be urban than rural: there will be some three and a half billion of them in settlements of over 20 000 inhabitants. Urban settlements are growing twice as fast as population, and the big cities of the world of over half a million are growing twice as fast again.

In surveying the problem of urbanization on this scale, Barbara Ward (*Human Settlements: Crisis and Opportunity*, Ottawa: Government of Canada, 1974) takes Japan as one example of the dynamics of quantitative change. In 1900, the population of Japan was 40 million, 10 per cent of it urban. By 1945, the population had risen to 80 million, 40 per cent of it urban. By 1970, the rate of increase had slowed down; there were then 100 million Japanese, 65 per cent of whom were urban. By the year 2000, there is expected to be 120 million Japanese, 90 per cent of whom will be urban. In Japan, as in many other developed-world countries, it is a further problem that the great burden of the urban increases will be shared out among very few cities. In Japan, a high proportion of the increase will be living in a single vast megalopolis around Tokyo Bay, much as in Canada the lion's share of the increase is expected to dwell in the 'Golden Horseshoe' from Toronto around the western end of Lake Ontario.

Forecasts made by the United Nations suggest that in developing countries, where population will double before the year 2000, the number living in urban settlements will grow threefold, from 464 million in 1970 to 1930 million in the year 2000 (Figure 15–2). There being no check to the growth of the largest cities, urban 'megaregions' of ten million or more are likely around such cities as Buenos Aires, Mexico City, Bombay, Calcutta, and Djakarta.

It might be said that great cities in Europe and in North America have coped with similar increases during the last hundred years, and that in such cities there has been no breakdown in provision for their inhabitants. But such a view must be tempered a great deal when we learn, as recently, of the municipal bankruptcy of such commercial and industrial cities as New York and Cleveland, while if we remind ourselves of the thousands of poor and destitute people in Dickensian London and the conditions in which they lived, we may have cause to think again. As Ward suggests, the very words

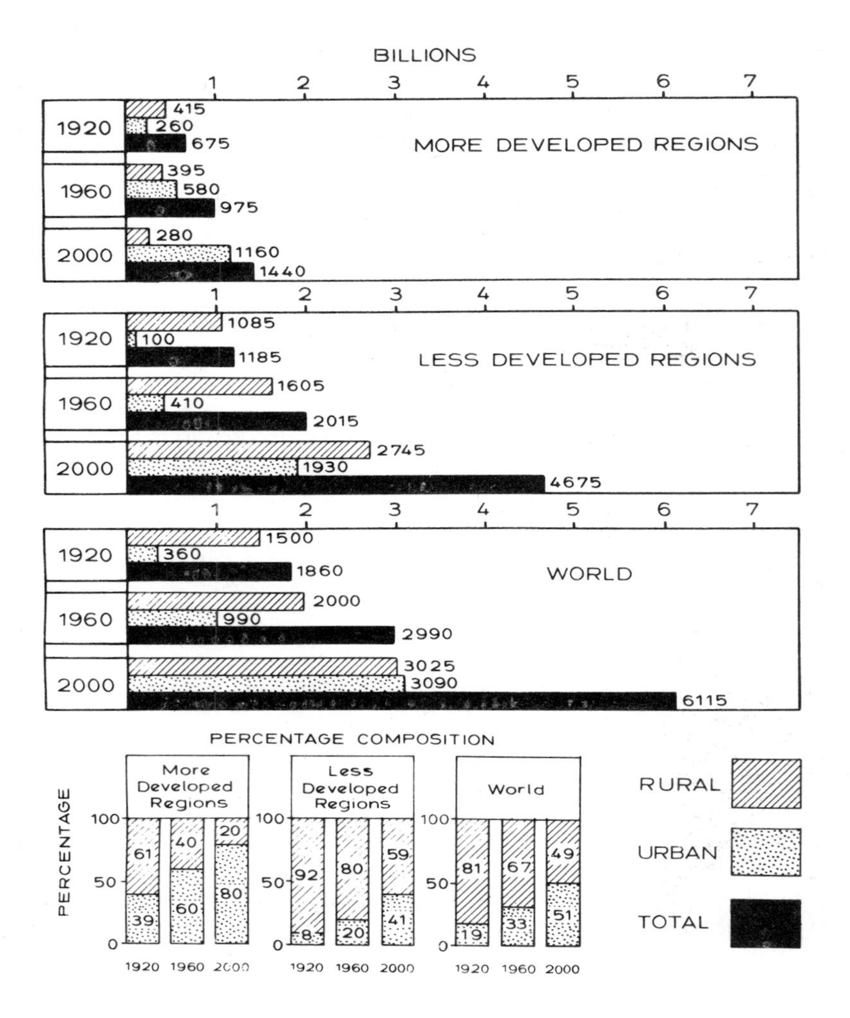

Figure 15–2 World total, rural, and urban population estimates and projections, 1920–2000. (Redrawn from J. W. MacNeill; after UNESCO Population Commission, *World Population Situation*, 1969).

'urbanity' and 'civilization' come in the belief that man-made order, beauty, learning, dignity, and status flourish best in the urban environment, yet '. . . the early industrial cities proved so congested, polluted, violent, noisy, and disease-ridden that the outward movement of wealthier people to suburbia began to define the shape of an "urban region" virtually by the middle of the nineteenth century'. Nations like Britain and the Netherlands became 90 per cent urban in a time span not available to the Third World countries of today, yet even so they failed to evolve a really acceptable, attractive and convenient pattern of settlements. The proof of this to Barbara Ward is that latter-day tourists do not go to the world's industrial conurbations for the pleasure of it: '. . . Manchester, Pittsburgh, Yokohama do not draw them in. They are off to Royal London or the Paris of the Louvre and Baron Haussman or, if they can, to the no-longer forbidden city of Peking. We can thus say that just as man becomes predominantly urban, he invents cities he does not much like'.

Perhaps there could be no better way to illustrate the problems related to the quality of the city environment and to the costs involved in their improvement. Yet the world over a majority of the world's people see in urban settlements the goal of their

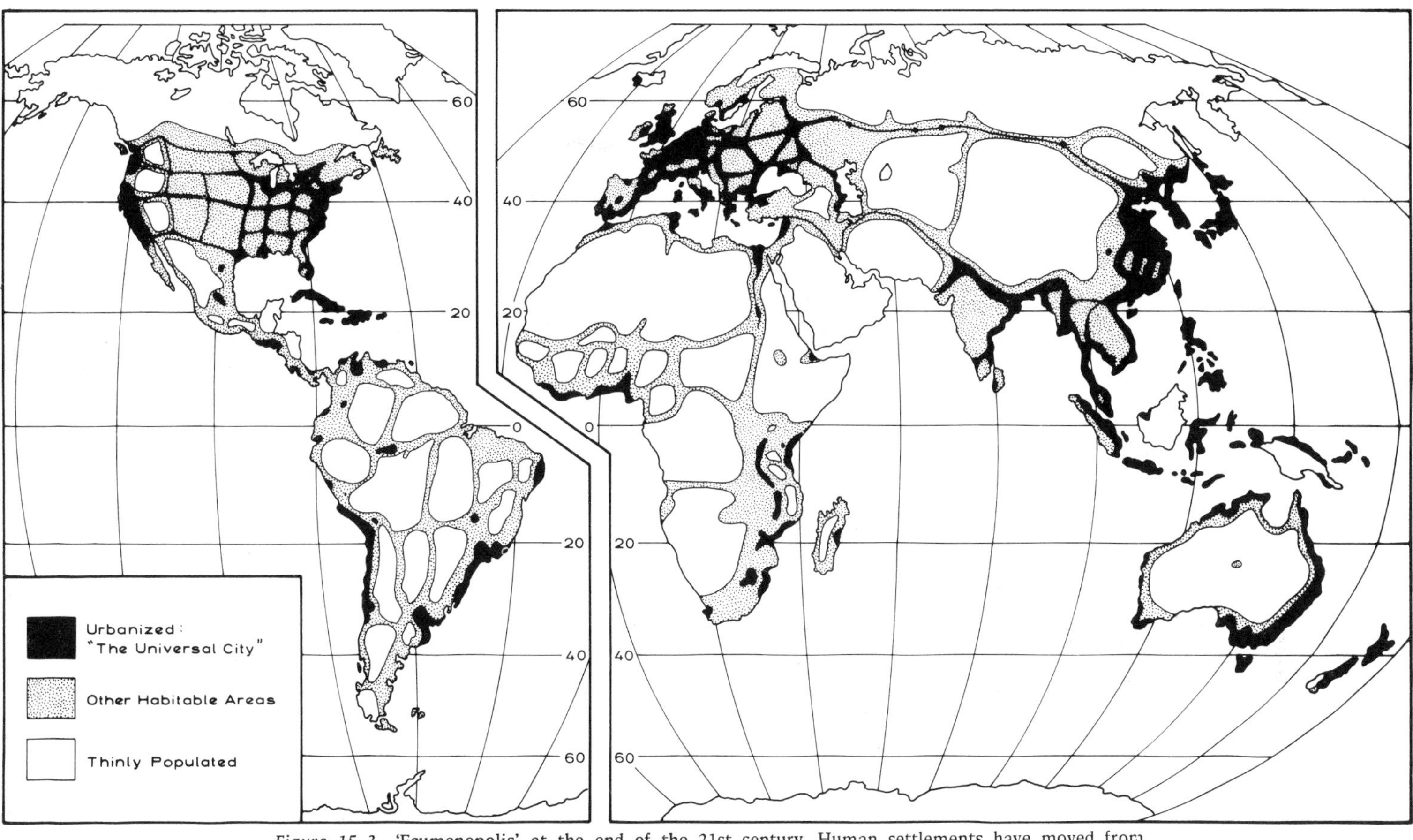

Figure 15–3 'Ecumenopolis' at the end of the 21st century. Human settlements have moved from camp to village, to city, to metropolis, to megalopolis, and as C. A. Doxiadis sees it, we are now moving towards the universal city of Ecumenopolis whose anticipated shape can already be discerned (redrawn from C. A. Doxiadis; see Bibliography.)

ambitions to self-advancement. There is no sign that the bursting cities of the underdeveloped world can avoid the urban problems of the developed world. Even in a developed country like Japan, it is estimated that the coming urbanization of the next twenty-five years will absorb twenty times more capital than has been invested in its cities during the last twenty-five years. In the underdeveloped countries, there are not the resources in capital, skills, and materials to cope. We may again quote Ward:

> *'in developing lands, where, conceivably, different, more decentralized patterns could be evolved—since options are not yet foreclosed in steel and concrete—the forces which drive the people away from the land and into the exploding cities are producing even more skewed and undesirable urban environments. Through sheer poverty and lack of usable resources, these settlements can combine the 19th century pollutions of slums, inadequate sanitation and industrial smog with the 20th century distortions of sprawl, long-distance commuting and automobile emissions. And since they grow at four times the 19th century speed and in an epoch when technology is evolving in a capital-intensive direction, many of the poor world's megalopolises have perhaps a quarter of their population out of work. In fact, as this century draws to a close, the aspirations that draw millions upon millions away from the farming sector to the spreading urban regions are daily contradicted by what may well be the worst environment ever endured by man.'* (Barbara Ward, *Human Settlements: Crisis and Opportunity*. Ottawa: Government of Canada, 1974.)

Thus the urban settlements of the world will expand, many of them in parts of the world where there will be crippling difficulties because of the shortage of resources. 'Megaregions' will grow unplanned, with but little opportunity to counter increasing problems. If we look further ahead towards the end of the next century, it is possible to envisage the city clusters of the world coalescing (Figure 15–3)—a state of urbanization that C. A. Doxiadis calls 'Ecumenopolis'. Ecumenopolis, however technically achieved and socially and politically arranged, would evidently consume vast tracts of the world's agricultural land. This, in addition to the degradation and waste which accompanies our present-day urban-industrial society multiplied many times, suggests a world which must have lower (or perhaps different) expectations than the world's peoples have today.

THE WORLD ENERGY CRISIS IN SPATIAL PERSPECTIVE

To the crisis in human numbers, deepened by uneven distribution and growing concentration in urban areas unable to cope with rapid increase, we must add the problems caused by the diminution of the world's nonrenewable materials. This is brought home to us most clearly in the energy sector. If we expect continued economic growth in the world's richest countries then we may also expect a strong increase in the use of energy: hence the need to produce more of it from mines and wells, or to harness it from the force of falling water, from the atom, and even from the sun. All energy derives from the sun, and perhaps we shall at some future time be able to obtain all our needs by the direct use of solar energy, but such a time seems far off.

It is instructive to note the change in primary energy forms over time. Historically, we have relied on such primitive energy forms as people and animals, wind and water, wood and charcoal, and peat. These are still available, and some like wind and water are reusable and of continuing significance, but by and large our primary source of energy in the world is coal and petroleum. Although there is much talk about harnessing the tides, the winds, and solar energy, such possibilities are extremely limited spatially and totally inadequate for the kinds of power requirements for the future. One of the problems appears to be that high capital outlays prevent the implementation of known technologies. Tidal energy is an example. The harnessing of rise and fall of the tides is still only a reality at La Rance (France, 1967), even though other, larger sites with a greater tidal range might be employed (as the Bay of Fundy, Canada, and the Severn Estuary, England). Electricity can also be produced by harnessing the internal heat of the earth, but geothermal energy can only make an important contribution where geological conditions are suitable: the major producers at present are New Zealand (Wairakei) and Italy (nr. Florence), and it is also produced on a large scale in Mexico (Imperial Valley), USA (north of San Francisco), and Iceland.

The fuel provided from fossils stored up in the earth from prehistoric periods are our basic energy hope. Sources of petroleum for the foreseeable future are finite in nature; they were accumulated over great periods of geological time. They are nonrenewable resources. Coal, petroleum, natural gas, and uranium are in the category of nonrenewable energy sources, and our dependence on them is very considerable indeed. The industrial countries of Europe, including Great Britain, rose on coal, but they now have a much greater consumption of oil. Since mid-century, for so many countries, oil has been a cheaper source of energy than either coal or, where it is possible, hydroelectric generation. Indeed, even in those countries which we familiarly associate with great hydroelectric dependence, the low price of imported oil has precluded further hydroelectric schemes, with their high capital outlays, from being built: Norway, Sweden and New Zealand, for example, have become more dependent on fossil fuels than on hydropower. Within the EEC countries, the fuel mix proportions changed most dramatically between about 1960 and 1970. In the former year the proportions were approximately 53 per cent coal, 6 per cent lignite, 28 per cent petroleum, 3 per cent natural gas, and 9 per cent hydroelectric and nuclear power. In 1970 they were approximately 30 per cent coal, 3 per cent lignite, 52 per cent petroleum, 7 per cent natural gas, and 7 per cent hydroelectric and nuclear power.

Geographically, our continued reliance on the fossil fuels has resulted in three things: (1) a frantic search for new and larger sources, usually in places remote from markets; (2) the development of larger transportation modes, such as the super oil tankers, and the development of world-wide and international routes; and (3) the recent but now perennial problem of international fiscal and economic imbalance as the result of the Middle East oil superiority. Some 60 per cent of all the known petroleum reserves in the world are situated in the Middle East. There are few more severe geographical imbalances in the world, and it is one to which we have already drawn attention (Figure 5–2, p. 61).

Seven nations now account for 75 per cent of the world's petroleum production. Of this amount, $\frac{1}{4}$ comes from the United States and over $\frac{1}{5}$ from the USSR. However, over $\frac{1}{3}$ per year comes from the Arab nations in Southwest Asia and Northwest Africa. These countries, in cartel with Venezuela, which produces nearly $\frac{1}{10}$, make the spatial nature of the problem evident.

With the production and utilization of the world's energy rising at the rate of about 5 per cent for more than two decades, it is not surprising that both the control of oil and its depletion have become causes for world concern. In the long term, it is the reserves of oil and of other fossil energy types which will determine the relative strengths of countries in the world energy market. Table 15–1 and Table 15–2 show the reserves of oil, natural gas, coal, and lignite estimated to be held by the leading countries. It may be noted that estimates of uranium reserves are more difficult to establish. There is, too, a considerable difference between the 'reasonable assured' uranium reserves which can be mined economically and are technologically usable, and the much greater 'additional' reserves surmised to occur close to assured reserves. In the

Table 15–1 Oil and natural gas, approx. reserves, major countries, 1974[a]
m. metric tonnes coal equivalent

Oil			Nat-Gas		
Rank	Country	Reserves	Rank	Country	Reserves
1	Saudi Arabia	20 200	1	Iran	10 200
2	Kuwait	9 800	2	USA	9 300
3	Iran	9 200	3	Algeria	4 000
4	USA	5 300	4	Netherlands	3 500
5	Iraq	4 800	5	Saudi Arabia	1 900
6	Libya	4 000	6	Canada	1 900
7	Abu Dhabi	3 300	7	UK	1 900
8	Nigeria	3 100	8	Venezuela	1 600
9	Neutral Zone	2 700	9	Nigeria	1 500
10	Venezuela	2 100	10	Australia	1 400

[a] To the 'top ten' lists, we must add the USSR (oil: estimated at about 12 200 mmtce and natural gas; probably in excess of 26 600 mmtce) and China (oil: estimated 3100 mmtce). The 1974 oil estimate for the UK was 1500 mmtce (*The Oil and Gas Journal*, Dec. 1973).

Table 15–2 Coal and lignite–brown coal, total measured and inferred reserves, major countries, 1959–1967[a]

million metric tonnes

Coal			Lignite and brown coal		
Rank	Country	Reserves	Rank	Country	Reserves
1	USSR	4 121 600	1	USSR	1 406 400
2	USA	1 100 000	2	USA	406 000
3	India	106 200	3	Australia	95 600
4	S. Africa	72 500	4	W. Germany	62 000
5	W. Germany	70 000	5	E. Germany	30 000
6	Canada	61 000	6	Yugoslavia	26 600
7	Poland	45 700	7	Canada	24 100
8	Japan	19 200	8	Poland	14 900
9	Australia	16 000	9	Czechoslovakia	9 900
10	UK	15 500	10	Hungary	5 600

[a] China has extensive reserves of coal but no recent estimates are available (UN Statistical Yearbook, 1972).

former category, the USA, Canada, and South Africa possess the lion's share, followed at a distance by Australia, Niger, France, and Gabon, with the USSR and China unknown.

The existence of reserves of energy, however, is an incomplete measure of the energy-potential. The ability of any country to utilize its store, big or small, is of great significance, and this largely depends on economic factors. It also depends, too, at least in the countries that can afford the luxury of delay or alternative, on environmental circumstances. This is particularly true of open-cast mining or areas where spoil heaps might be created, while increasingly vocal environmental lobbies resist the adoption and spread of nuclear power plants in Britain, France, Japan, and the USA. What matters most of all in the national sense is the price that has to be paid for energy, while in the global setting this depends on how much there is to share around.

During the last decade, a short-term crisis has been caused by the OPEC countries' decision to more than double the price of their oil, as well as to place quantity restrictions. Essentially, this was made possible by the industrial countries' dependence on oil that was previously supplied cheaply. All these industrial countries were taken unawares (even though the giant oil companies had given ample warning of its inevit-

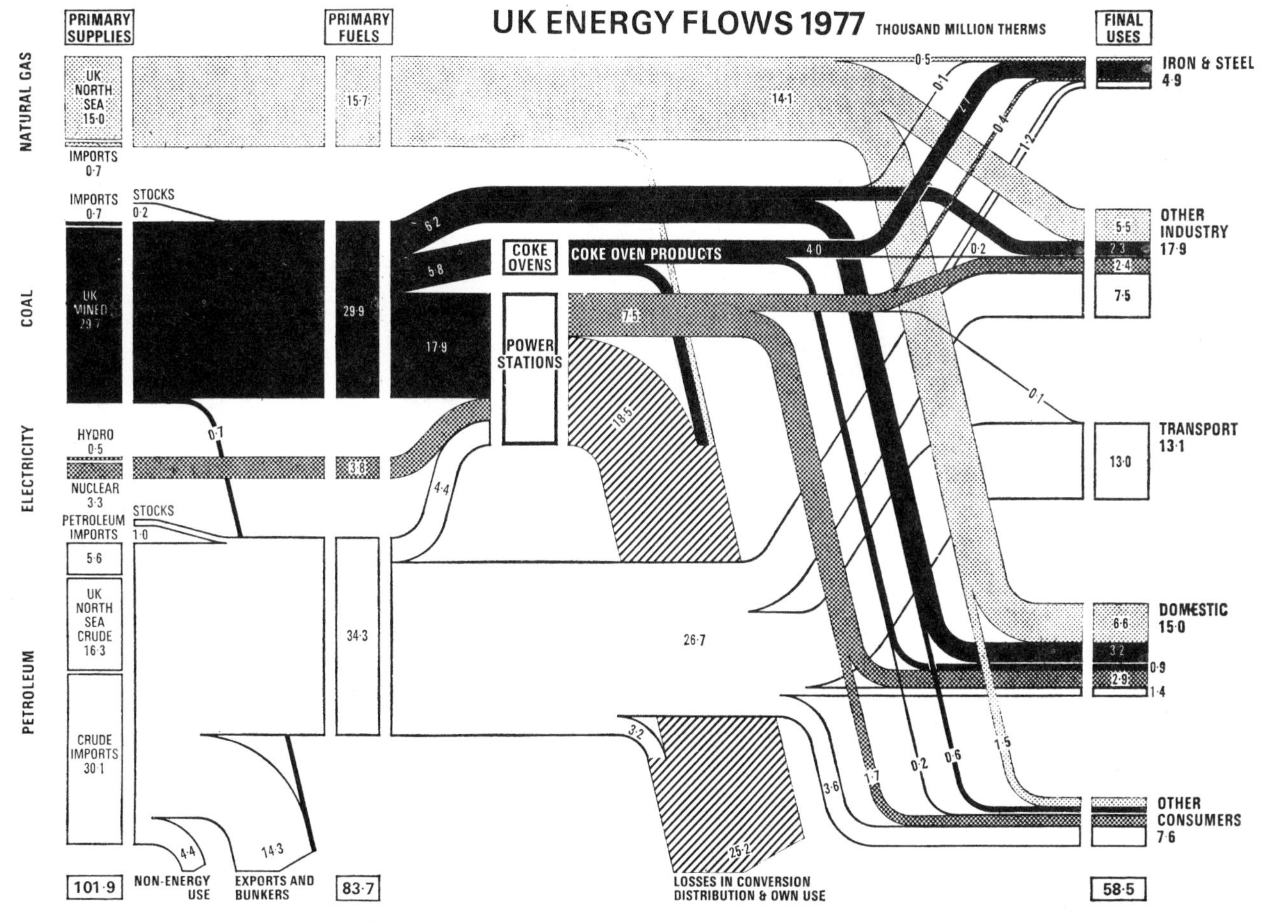

Figure 15–4 A classification of energy consumption in the United Kingdom (*UK Digest of Energy Statistics*, September 1976, reproduced from *The Times Energy Supplement*, October 24, 1978, p. 26).

ability), and most have set about the attainment of a new energy mix together with strong conservation measures. Among the developed nations only the USA and the USSR were unaffected directly mainly because of their existing policies to develop indigenous production. But in the United States, the new situation was more serious than it first appeared. In 1925, the United States produced 71 per cent of the world's petroleum. By 1969 it produced some 22 per cent. In 1970 the United States was self-sufficient in petroleum. Today, it imports more petroleum than it produces, and the imports are growing. In Europe, particularly Great Britain and northern Europe, the situation is improving as the North Sea oil and gas come into full production; but the general strategy again must be one of promoting energy conservation. However, the uses of energy are so significant and widespread, as Figure 15–4 demonstrates, in the case of the United Kingdom, that it is difficult to determine just how much energy can be conserved. Factories consume a significant portion of power, as do homes, for both air conditioning and heating. Offices and businesses provide another major user, and exterior lighting, especially for advertising, is also significant. But one of the heaviest users is transportation, not only for private passenger movement but for the shipment of freight.

The problem with energy curtailment is much like that for transportation reduction. The more transportation that is used, the more freedom of location choice is evidenced. For example, the private car gives average homeowners great choice in where they can live, work, and play. Thus there are real advantages to the continued consumption of transportation.

Likewise, energy use is considered indicative of a higher standard of living. To curtail energy use for such things as air conditioning, heating of buildings, and shipment of goods is to reduce the quality of modern life. Of course, there are secondary gains from energy curtailment as far as the environment is concerned, for example, decreased air pollution and noise and increased safety, but the gains might be offset by tremendous losses.

In summary, energy and the power that goes with it are not likely to be curbed in our modern industrialized society. In fact, increased consumption of energy might be considered a forecast of economic advancement. It is the consequence of affluence, and wasteful. Indeed, the whole of our industrial society has been viewed as a complex machine for degrading high-quality energy into waste heat, in the course of which we extract the energy needed for the creation of an enormous catalogue of goods and services.

In the developed industrial countries, slowing down the rate of increase in energy consumption can be achieved in the short-term, although it seems that in the process there must be difficulties and quite probably a perceptible levelling or lowering of living standards to go with it. But for the developing nations of the Third World there are immediate difficulties with quite drastic effects. Some have indigenous fossil-fuel resources they can turn more attention to—coal in India, for example, or oil shales in Brazil. But for most, the main constraint is likely to be the shortage of financial resources for much-needed investment in the building up of their own agriculture, industry, education, medical and urban programmes, the more especially when their foreign exchange budgets are already stretched paying for the high-cost international oil.

As the world's oil declines, however, in the longer-term, all nations will experience difficulties, which brings back the question: how long will the world's fossil energy reserves last in an increasingly energy-intensive world? Scientists, engineers, geographers have all tried to tackle this question. The reserves being finite, the problem resolves itself in theory around the relationships between consumption, remaining known reserves, and reserves yet to be discovered. The alternative scenarios presented in Figure 15–5 (a) and (b) suggest these relationships. Figure 15–5(a) shows a theoretical depletion curve for world oil reserves of 274 billion tonnes (2000 billion barrels). In this curve,

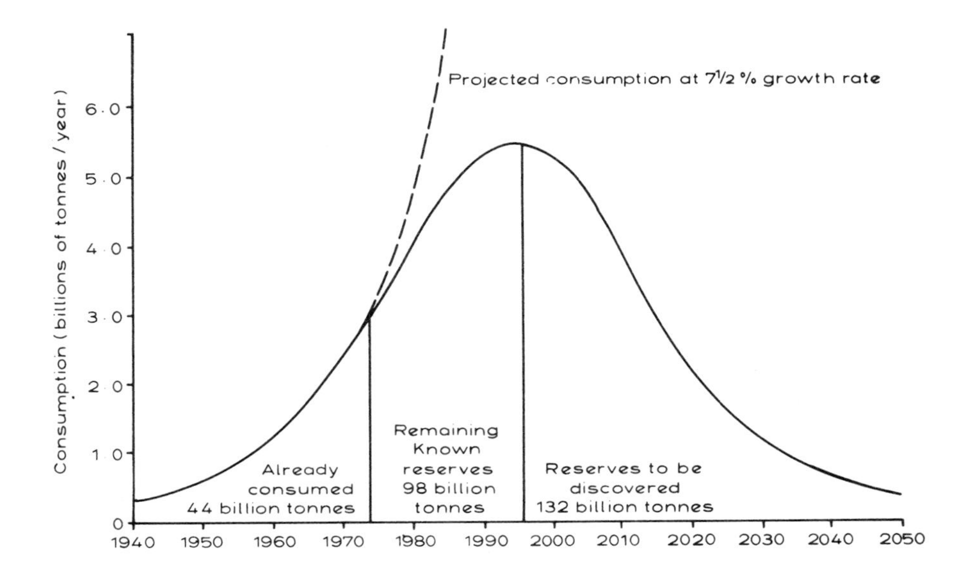

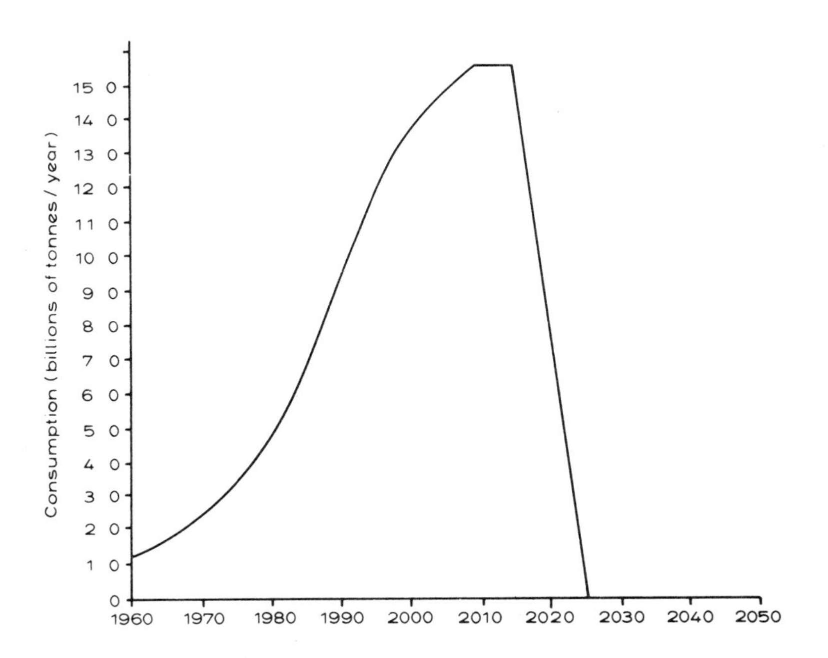

Figure 15–5 Theoretical depletion curves for world oil reserves of (a) 274 billion tonnes and (b) 548 billion tonnes. (After Hubbert and Warman, redrawn from Gerald Foley, *The Energy Question*. Harmondsworth: Penguin, 1976, p. 141, p. 143.)

world oil production peaks in the 1990s, followed by a gradual decline as the world accustoms itself to doing with less oil. The dotted line is a projection of the rate of growth prevalent in the world up to the Middle East war and the OPEC price increases of the early 1970s. This depletion curve suggests, therefore, that 'if the total recoverable reserves of 274 billion tonnes is anywhere near the correct figure, the end of the era of steady growth in oil consumption was very close in 1973, even without the action taken by the oil producing countries' (Gerald Foley, *The Energy Question*, Harmondsworth: Penguin, 1976, p. 140).

In Figure 15–5(b), however, we have quite a different explanation of the growth and depletion question, as it has been examined by the economic geographer Peter Odell (P. R. Odell, 'The future of oil: a rejoinder', *Geographical Journal*, June 1973, pp. 436–454). Here, Odell takes 548 billion tonnes (4000 billion barrels) as his estimate for total ultimately recoverable reserves. His depletion curve is based on the assumption that the per capita consumption of energy will continue to grow at its historic rate in the developed world and that, overall, it will be possible to expand production 'well into the 21st century at rates of consumption which will be five or more times their present levels'. In this scenario, therefore, production reaches a peak in 2015, at a stage when 85 per cent of the recoverable total has been consumed. The important point to note about the Odell depletion curve is that after about the year 2015, production must then collapse to zero within about ten years, a time-span in which an orderly withdrawal from such heavy dependence on oil would be quite impossible.

Before too much is read from such theoretical curves as presented above, we must add that Odell would consider it possible for conventional hydrocarbon resources to meet the demands for energy until perhaps the second quarter of the 21st century, but only given real rising prices and more stringent government controls. But the main point remains. Like many other basic requirements for an advancing world economy, there is simply not enough energy resource to continue at the mix and pace we have established, and the Western or industrial world's example and experience is not one that can be followed by the underdeveloped nations. Increasingly, it is not simply a matter of finding new sources, but of finding sufficient sources to serve the world's needs at the present. Ecumenopolis will be energy-poor unless we develop new alternatives and in time to meet the demand. Scarcities could only create conflicts between those who possess the energy and those without.

Given the inequality in energy potential, particularly of fossil fuels, this scarcity provides potential for geographical disorder of the most severe kind. Yet there are increasing adversities associated with an accelerating energy consumption (air pollution and various other environmental problems). A first-order dilemma is created: trouble is ahead whether we retreat or accelerate in the energy area.

THE WORLD FOOD CRISIS IN SPATIAL PERSPECTIVE

There is probably no problem of such long-standing and controversial nature, particularly in terms of solution, than the matter of food. In ancient times nations stored up vast quantities of grain for times of famine or shortage. It was common to have six or seven years of foodstuffs available for such unexpected crop conditions. Today each fall before wheat harvest the world commonly has less than one month of grain staples on hand. And even this is an exaggeration inasmuch as a good portion of the world's population is perennially undernourished, malnourished, or literally starving. The world today is no food cornucopia. The world is in a crisis situation with respect to food no less than it is with respect to energy. In point of fact, energy resources are a necessity to maintain the present level of food production in the world. Food production uses energy in many ways, and the conversion is not always managed economically. Since the natural productivity of the soil, unsupported by machinery, pesticides, irrigation and chemical fertilizers would not feed a tenth of the world's population, agriculture must make heavy demands on technology. Thus the energy crisis and the food crisis are interconnected. Made 'efficient' through technology, for example, British agriculture produces about half of Britain's food supply, but in the process also accounts for 5 per cent of the country's energy consumption.

Historically, the problem of deficit food has been solved by the four horses of the apocalypse: war, famine, pestilence, and death. These four horses still stalk the world and remain the primary base for population control in deficit areas. In the face of the problems, both past, present and future, many scientists, economists and geographers have speculated about the relationship between populations and their food supplies. Whatever their theories or viewpoints, however, they all refer back to Thomas Robert Malthus, who wrote his celebrated 'Essay on the Principle of Population' in 1798. Malthus argued that there must be a strong and constantly operating check on population from the difficulty of subsistence. His theory was produced deductively, and revised many times to become inductive. A basic characteristic of the Malthusian theory or doctrine is that the numerator, land or resources, remains relatively fixed while the denominator, population, continues to expand. Hence the ratio of land to man becomes ever smaller.

The appeal of Malthus in a finite world with rapidly-rising populations is obvious. Probably Malthus himself took a gloomy view of the future agricultural capacity of the world, but his arguments (often summarized in a sentence that while population increases geometrically, food supply only increases arithmetically) were illustrated by numerous countries and colonies in his own time. Nevertheless, by briefly quoting Malthus we can see how the basis of his arguments can have a relevance to our own times: 'Man is necessarily confined in a room. When acre has been added to acre till all the fertile land is occupied, the yearly increase of food must depend upon the melioration of the land already in possession. This is a fund, which, from the nature of all soils, instead of increasing, must be gradually diminishing'. He believed that there was '. . . the constant tendency in all animated life to increase beyond the nourishment prepared for it' . . . and that 'population invariably increases where the means of subsistence increase'.

The general flavour of these statements has great appeal to neo-Malthusians. Conscious of present-day famine and malnutrition, they, and others, can see no alternative to a future world in which these things will be always with us so long as human numbers increase at current rates (see P. V. S. Sukhatme, 'The World's Hunger and Future Needs in Food Supplies', in Ian Burton and Robert W. Kates, eds. *Readings in Resource Management*, Chicago and London: University of Chicago Press, 1965, pp. 38–94). The general approach to a solution has been to declare that there are simply too many people, or soon will be, and food is insufficient to go around. Not many people have been proclaiming recently that all we need is a better distribution of food. It has been calculated that if all the food of the world were divided equally we would all be undernourished.

Of course, such food scarcity is based on the food actually produced in any one year, not what could be produced. But the amount of food produced is largely related to prices for commodities, various crops under soil bank, and other economic strategies all of which tend to reduce potential production. Physically speaking, the earth is quite able, even under a more primitive technology, to feed adequately all the world's inhabitants. But because there are economic and social systems that circumvent full food production, it is perhaps wisest to turn to the remedies as they are now proposed.

The standard remedy focuses on not producing more food but on reducing the world's population. Some technologically advanced nations are pursuing 'zero population growth'. Related developments have occurred in the course of achieving zero population growth: (1) the availability of the birth control pill and other birth control devices, (2) reduction in status for having children and for the family relationships, and (3) deliberate attempts to abort unwanted children. In the United States alone some 1 million children are aborted each year. If it were not for the high percentage of women in child-bearing years, the United States would be losing population. Much the same situation has been repeated in Japan, Great Britain and in many urban-industrial nations of Europe.

It is ironic, therefore, that those nations most capable of feeding themselves have been doing the most in the way of population reduction. By contrast, the less developed areas of the world have as yet hardly been affected by family planning schemes. In terms of any solution to the imbalance between food and population, the zero population growth philosophy of Western nations has of late contributed nothing to the overall problem. In view of the ironic geographical response, such a philosophy may actively have caused greater problems for the future.

A no-growth population is also causing difficulties that were entirely unexpected initially. With a situation of stagnancy, new persons entering the job market are particularly limited. In many cases they must wait for someone already employed in any given activity to die or retire before an opening is available. Thus such stagnation tends to create rigidity and solidity of people to particular positions. This situation comes at a time in the Western world when labour unions have become extremely strong and are in a position to protect their present memberships.

Population control does not appear to be a meaningful solution to the food-population crisis for the urban industrial nations, although it clearly has relevance for many countries, particularly in South-east Asia.

The other alternative is to do much more to improve food production. In the short term, the main reliance clearly has to be placed on those countries that can still produce very large grain surpluses for export. The USA, for example, with only 8 million people in agriculture, currently produces about 240 million tonnes of grain per year compared with India's 105 million tonnes from 364 million people in agriculture. North America's stored surplus has frequently bailed out India and many other, smaller, countries in time of crop failure. *Ad hoc* crisis response is often too little and too late and the United Nations has frequently discussed its replacement by a permanent system of global food reserves to meet emergencies.

In the longer term, however, the point is made that this sort of transfer is both inadequate and inefficient. It cannot prevent such Third World shortages from occurring, and countries like the USA are unlikely to accept the permanent financial burden of production, storage and transportation. It is also evident, however, that the developing countries most in need, a 'hard core' of twenty-five or thirty of the very poorest nations, have little hope of acquiring adequate foreign exchange resources to buy more and more food on the world market.

In the medium term, the emphasis must be on improving the agricultural productivity of the developing countries. Enormous efforts have been made to do this during the last decades. But with few dramatic success stories, the underdeveloped countries have been able to do very little in the way of introducing new crops to their environments. In some countries, the 'Green Revolution' has not produced long-lasting effects, and even where crop yields have dramatically risen, the increases have been swallowed up by the greater number of (Malthusian) mouths. Perhaps it is still too soon to assess the effects, or establish a balanced picture, although the debating points are provided by Johnson:

'The enthusiastic supporters of the Green Revolution seem to imply that there is now an obvious path along which all or most developing countries can tread and in a few years have all the food they need and want. The gloom and doom

contingent really consists of two main groups. One argues that the impact of the Green Revolution on food production has been and will continue to be minor since the new seeds and production methods are really applicable to only a small part of the cultivated area of the developing nations. The other group argues that the Green Revolution, while successful in increasing food production, has resulted in so many dislocations that it is likely to be followed by a Red Revolution. The Green Revolution, so it is argued, has resulted in the rich getting richer and the poor getting poorer. The regions that find new seeds and production methods profitable are the recipients of much higher incomes while other regions that cannot adopt the Revolution are losing out because of lower food prices. And within the favoured regions, it is alleged that it is large farmers who are gaining; the smaller tenants are being pushed off their farms and the landless labourers are losing their jobs because of mechanization that has become profitable because of the higher yields.

As is so often the case, the truth lies somewhere in between.' (D. Gale Johnson, *World Agriculture in Disarray*, Fontana World Economic Issues, London: Fontana, 1973, p. 173.)

Another suggested approach has been that the world should change its diet (and its crops), including the elimination of meat or at the very least use grass-fed rather than grain- or corn-fed beef. These suggestions likewise, are fraught with problems. Much of the food now given to grain-fed animals is unsuitable in original form as food for human beings. Also, if grazing land were not used, it might simply go to waste because much of it is unsuitable for crops. Moreover, meat is the primary source of protein for many, and elimination of meat would have to be offset with other items such as fish or soybeans to avoid severe malnutrition problems. Finally, there is the difficult problem of changing diets that are in large part an important aspect of culture.

It has been suggested that the further domestication of animals might be an ineffective system of using the natural vegetation. Domesticated animals all eat generally from the same vegetative ecology. By contrast, wild animals, as in the savannah areas of Africa, survive in greater numbers than they would if domesticated. This is because each animal tends to eat from a different part of the natural vegetation, for example, some eat from certain grasses, some from low bush leaves, others from middle range leaves, and still others from the high branches of trees and shrubs. The question becomes one of obtaining the most food from a given area. For some areas natural growth and wild animals may prove to be a superior food source than the raising of domestic animals.

In other areas, massive irrigation diversions would seem most appropriate. Projects such as those that occur in much of the western United States and in Israel would appear to be model cases for agricultural improvement. 'Watering the deserts', indeed, has great appeal, but we should also be aware of its costs and limitations. Oil-rich Saudi Arabia, for example, a country described as the largest land body on the face of the earth where there is not even a stream, has embarked on agricultural development schemes relying upon the very expensive desalination of water. It has even contemplated, quite seriously, the feasibility of towing icebergs to Saudi Arabia from the Antarctic. In Saudi Arabia, agricultural areas developed in this way could produce fresh vegetables, but only for relatively small numbers of people who can afford them. In Egypt, by contrast, it will become a necessity to sustain millions of poor. It is planned to provide 800 000 hectares with irrigation from desalinated water by the year 2000, but such is its population growth that by the year 2100 Egypt will need the water equivalent of three River Niles.

Overall, however, there seems no reason why, in the middle term, the earth cannot provide enough food for many more millions of people. Many of the ways and means are evident: better seed, proper fertilizer, meaningful watering cycles, more productive crops, and the development of new lands would appear to be in order as a beginning. Elimination of many non-food crops such as tobacco would also seem appropriate. Combining fields and farms and eliminating many fences and roads that now take up potentially valuable land in rural areas would be other measures. And, of course, there are possibilities of developing food products from the seas. Moreover, many vitamins and nutrients can be manufactured synthetically from coal, petroleum, and other ingredients.

The food population crisis we might conclude, is not a physical crisis but a psychological, economic, and political crisis. The starvation of people about the world today is primarily a monument to human selfishness, not to human inability to physically provide food. But what about the longer-term? If we compare the present distribution of the world's agricultural land with the spreading, 'world city' envisaged by Doxiadis, we can see that much of the better agricultural lands of the world would be engulfed in concrete. Currently, we estimate that not much more than ten per cent of the earth's land surface is suitable for the growing of crops. There are environmental hazards of intensive cultivation, and there have already been some disastrous experiments made in the attempt to grow ground-nuts or grain by extending the cultivated area. Alteration of water balance, changes in micro-climate, soil erosion and deterioration of soil structure might well be the result of over-enthusiastic cultivation or inadequate husbandry.

It is considerations like these which bring back thoughts of the Malthusian equation.

ECONOMIC GROWTH PAINS: THE SHAPE OF THINGS TO COME?

According to so many authorities, the basic problem for the future of the world is the enormous increase in human numbers which, if proceeded with at the known rates of increase experienced in the last 200 years, will swamp the resources available to sustain them. We have seen that the population growth rate of the world has been increasing, with the doubling time shortened from 140 years by 1900, to 85 years by 1950, and to 39 years in 1960. It is basic to any comprehension of the world's future population numbers to learn more than we know already about the causes of this increase. And as the world proceeds to industrialize and urbanize further, it is critical to know what will happen to the rate of increase in the present non-industrial and developing countries. Admitting an exponential world population growth as described above, with concomitant urban and industrial growth, raises spatial problems at almost every level of magnitude: Will populations crowd in cities? Some will say that man has reached his highest estate in cities in 'conditions of controlled overcrowding', others that future levels of crowding can only reproduce the social and economic problems of present-day Calcutta many times over. Will megalopolis link up with megalopolis as Doxiadis (Figure 15–3) envisages?

Some authorities can see only Malthusian relationships between future world population and the food it will need. Since food is critical to survival, its scarcity may become a serious constraint to growth. The continual loss of available land to urban and industrial uses, they suggest, must mean that the productivity of the remainder must rise dramatically. Yet even if the total world supply could be maintained at its present level by augmentation (by irrigating deserts for example), the amount available is finite, and all of it subject to decreasing returns in relation to the inputs necessary to raise production.

Some of the pitfalls of estimating the stock of non-renewable resources have been already introduced in considering the world's energy crisis. We have difficulty in estimating reserves of non-renewable resources still available because as new discoveries are made, reserves go on rising. Yet still they are finite, and ever-rising demand for them brings fears that, whichever nation now or later still possesses them in abundance, the global stock is being depleted, perhaps exponentially. But few agree how close to depletion we are becoming. Oil and natural gas, for example, are today acquired from depths of only a few thousand metres, and we have only just begun to explore likely areas of this depth in the continental shelves of the world. One American geologist suggests that, in areas other than those where we now look for natural gas and at much greater depths, it may be possible to tap vast quantities of methane of inorganic or primaeval origin. While this is unproved, and may await the development of a new technology anyway, the sea is another area where cost, rather than technology, presently prevents the recovery of a wide range of chemicals and minerals.

If people and their numbers, the world's potential agricultural land and other material resources are considered separately, we find that each raises challenging questions. Yet if we try to consider them together they offer the most difficult problems for the future social, economic and political health of mankind. It is sometimes too glibly held that 'technology will provide in the future, just as it has brought us successfully to the present'. This would be a very 'Western' or 'Eurocentric' view, hardly tenable by three-quarters of the world's nations of peoples. Moreover, it neglects to account for the pollution which arises from our present industrial growth. If we consider this aspect along with the other matters already raised, we may gain further insights since pollution is further increased by mankind's growing dependence on a single world technology.

A thorough assessment of the effects of the pollution we have already brought into the world is an imperative, as MIT studies have amply revealed (Massachusetts Institute of Technology. *Man's Impact on the Global Environment: Report of the Study of Critical Environmental Problems* (*SCEP*). Cambridge, Mass. and London: MIT Press, 1970). Its presence on land, sea, and in the air often lies undetected until difficult or even impossible to reverse. Historically, it has attracted remedial action only in the face of local or regional disaster. We now know of many of the dangers, but we have very inadequate knowledge of the rate at which pollution is increasing or about the capacity of the earth to absorb it. Yet again, there must be an upper limit to the build-up of CO_2 in the atmosphere (caused by fossil fuel combustion) which may warm up the planet so much as to raise the level of the sea dangerously; there must also be a limit to the use of herbicides and pesticides, and a limit to the extent to which we can upset oceanic and terrestrial ecosystems by discharging industrial wastes and toxic materials (like heavy metals, oil, and radioactive substances), or nutrients (such as phosphorus) which can over-enrich lakes and coastal waters.

In 1972, the 'Club of Rome' (founded 1968 in Italy by Aurelio Peccei and seventy-five industrialists, scientists, and economists from around the world) performed a systems analysis of the world which linked the factors of population, resources, land, pollution and capital generation. Their findings were published in a famous report, *The Limits to Growth* (Meadows, D. et al, *The Limits to Growth*, Cambridge, Mass.: MIT, 1972). At the macro- or world-scale, its authors were concerned to explore the 'possible future' relations between the many sectoral activities we have considered in this book. The time-scale considered was from 1900 to 2100 AD, and their basic world simulation model was based on parameters set out previously by Forrester (Jay Forrester, *World Dynamics*. Cambridge, Mass.: MIT, 1971).

Stark assumptions were fundamental to this total-system or world model. They were, indeed, those we have already considered; there is a finite stock of exploitable, non-renewable resources; that there is a finite capacity for the environment to absorb pollutants; that there is a finite amount of potentially arable land; and that there is a finite yield of food obtainable from each sector of arable land. The Meadows team ran several models which varied the impact of expected development and trends. Thus the 'standard run' model (Figure 15–6) assumed that there will not be any great future

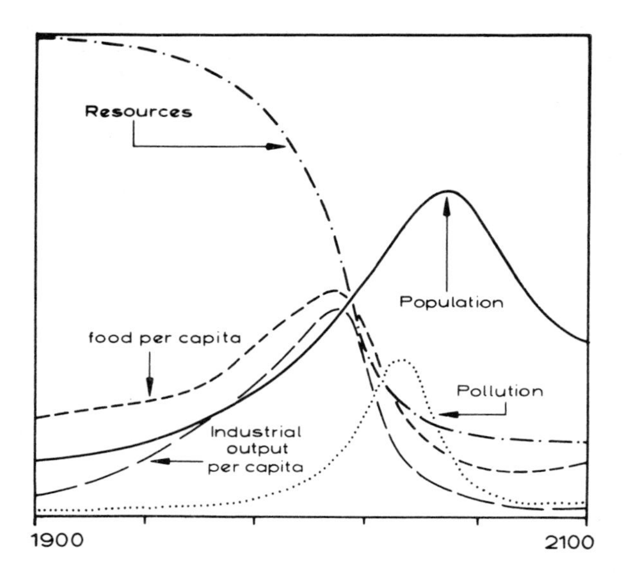

Figure 15–6 The 'Limits to Growth' world model standard run (Donella Meadows et al. *The Limits to Growth*, New York: Universal Books, 1972, p. 97).

changes either in human values or in the functioning of the global population system from the way it has operated over the last 100 years. This model shows the change in population, industrial output per capita, food per capita, pollution, non-renewable resources, crude birth and death rates, and services per capita from 1900 to 2100 AD.

It is vital to understand that each variable in Figure 15–6 is plotted on a different vertical scale and that the graph was intended to emphasize only general behaviour. Starting with 1900 sets the values of the variables to agree with their historic values 1900–1970. Essentially, the 'standard run' illustrated suggests that although world birth rate declines steadily, the death rate falls more quickly as the population rate soars. Industrial output, food and services per capita increase exponentially, producing a dramatic decline in the resource base. Food per capita and industrial output per capita then decline rapidly, paralleling the downward trend of resources as closely as they previously paralleled the upward curve of population. World population continues to climb for a time, until finally the birth rate curve intersects the death rate curve and there is a rapid 'collapse' of world population.

It is clear in the 'standard run' model illustrated that population collapse occurs because of the non-renewable resource depletion. But further runs of the model then seek to establish other possibilities. One of these assumes that advances in technology can double the amount of resources economically available. But this only offers a few more years of exponential growth: population 'collapse' is caused by pollution coupled with the depletion of non-renewable resources. With pollution a leading problem, a third run of the model assumes strict pollution control (quartering the problem). With resource policies the same as before, it was nevertheless a new assumption that unlimited nuclear power would unlock resource reserves and, as well, permit considerable resource-substitution. These changes allowed world population and industry to increase to the food-producing limit of agricultural land. The result is that while population and industrial output rose well beyond their previous peak values, in the later stages much capital had to be diverted to food production: in this third variation, therefore, it is food shortages which lead to 'collapse'.

A further Meadows world model then assumes that increased agricultural productivity will solve food shortages within a context of 'unlimited resources' plus pollution controls. As may be expected, in these circumstances world population and industry reach very high levels. But although each unit of industrial production generates less pollution, total production rises high enough to bring back the spectre of pollution; controls are overwhelmed and there is, again, an end to growth. Indeed, in this case, the model suggests that exponential population growth outstrips all kinds of capacity, bringing the world back to 1900 levels.

Yet another Meadows model includes assumptions of 'unlimited resources', pollution controls and increased agricultural productivity, and yet another adds 'perfect' birth control. Still, however, the world is led into catastrophe. Hence the 'limits to growth' in the prosperity of the world as seen by the Meadows team.

It is important to understand that in the background of both the *SCEP Environment Report* and *The Limits to Growth* study lies a paucity of data related to problems of global concern. This is true for all types of data, whether scientific, technical, economic, industrial, or social, while very few projections exist for rates of growth in the vital industrial, agricultural, and energy sectors. Even in the most fundamental area of demography we can note again the failures of population projection in the past and in the present. The sole determination of birth rates, for example, in *The Limits to Growth* models are the levels of industrialization and the food supply, it being assumed (on the basis of past experience) that birth rates will gradually fall as industrialization proceeds and material standards rise. Otherwise it is assumed that they will not fall unless the food supply becomes inadequate.

This, and many other criticisms, have been levelled at the Meadows' team assumptions (see, for example, Marie Jahoda, *Thinking About the Future: A Critique of 'Limits to Growth'*, London: Chatto & Windus, 1973). In a very perceptive review, three American geographers suggest that the assumption of a 250-year supply of minerals to current usage rates is too pessimistic. They would expect new technologies to win much more time than this and point out that 'a forecast made in 1870 would have omitted the principal source of energy in 1970, oil, and the fastest growing new source of energy, nuclear power. It would probably have excluded not only all the synthetic materials, fibers, and rubbers but probably also aluminium and other metals' (Berry, B. J. L., Conkling, E. C., and Ray, D. M. *The Geography of Economic Systems*, Englewood Cliffs, New Jersey: Prentice-Hall, 1976, p. 500). On food and arable land, they suggest the need for a clearer distinction between physical limits and the political and economic limits to the production and distribution of the world's food and they challenge the inevitability of decreasing returns from agricultural inputs. Finally, they question the assumption that pollution is directly or linearly associated with increasing world industrial output and agriculture.

Since the Meadows' world models were launched, several scientists have rerun them with new sets of assumptions. One of these (Figure 15–7) shows that 'new technology' assumptions can increase world productivity, which in turn increases the standard of living, which eventually slows down birth rates to a steady state. The result is a world which, even with steeply rising populations and reducing natural resources, can successfully avoid the Malthusian catastrophes of the 'Meadows world'. But the Club of Rome and its associates remain pessimistic. Their revamped models were demonstrated in Houston, Texas, in 1977: 'The computer was asked when the world would run out of oil if known reserves were twice as large as they are now. It quickly produced a plunging graph unable to come up with any likely scenario that would prevent massive starvation in India by 2025 or widespread death from starvation there in the 1990s' (Donald Q. Innis, 'Planning the death of growth', *The Geographical Review*, April 1978, pp. 232–233).

The Meadows models and their present derivatives are, of course, not predictions of the future realities, they are more of a reminder that our urban-industrial civilization faces critical choices. Should we try to live within self-imposed limits, or can we pursue economic growth in the belief that successive technological breakthroughs can over-

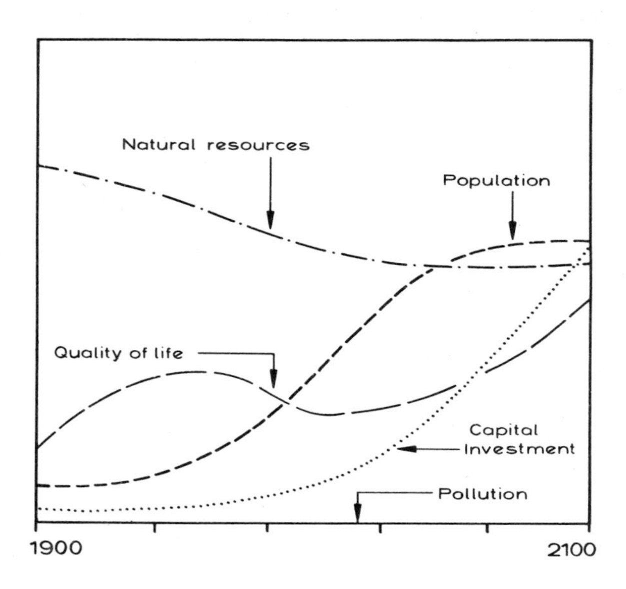

Figure 15–7 A technological and optimistic alternative world model (Robert Boyd, 'World dynamics, a note'. *Science*, August 1972, pp. 516–519).

come the potential natural limits? Standing back from such models, one lesson might be that, formally or informally, we should make assumptions about the future of the world which are neither heavily burdened with portents of doom, nor so euphoric that we close our eyes to problems. Principally, however, they present us with the case for developing some kind of new and sustainable world society before the one in which we live becomes untenable. For the spatial and economic systems that have evolved in the world, although they are increasingly comprehended by economic geographers, must also lead us to consider their future.

Perhaps the situation was aptly described by Lewis Carroll in *Alice in Wonderland*: Alice was asking the cat, 'Would you tell me please which way I ought to go?' 'That depends a great deal on where you want to get', said the cat. 'I do not care, so long as I get somewhere,' Alice said. 'Oh, you are sure to do that,' replied the cat, 'if only you walk long enough.'

But there is the rub: we do not have time to wander aimlessly in our search for a better world for all its peoples. We may leave Malthus behind in his 'room', but we are still bound to the life-support systems of 'spaceship Earth'. A spaceship, we might conclude, which is more like a greasy football which needs better looking after.

BIBLIOGRAPHY

Bauer, P. T. 'Western guilt and third world poverty'. *Commentary*, January 1976, pp. 31–38.
Bernarde, M.A. *Our Precarious Habitat*. New York: W. W. Norton, 1970.
Berry, B. J. L. *The Human Consequences of Urbanization*. London: Macmillan, 1974.
Berry, B. J. L., Conkling, E. C., and Ray, D. M. *The Geography of Economic Systems*. Englewood Cliffs, New Jersey: Prentice-Hall, 1976.
Bunge, W. W. 'The geography of human survival'. *Annals of the Association of American Geographers*, September 1973, pp. 275–295.

Blaut, J. M. 'The theory of development'. In *Radical Geography: Alternative Viewpoints on Contemporary Social Issues*, Ed. Peet, R., pp. 309–314. London: Methuen, 1978.

Blowers, A., Hamnett, C., and Sarre, P. *The Future of Cities* (Open University Set Book). London: Hutchinson, 1976.

Council of Europe *Population Decline in Europe*. London: Arnold, 1978.

Dassman, R. F. *Environmental Conservation*. Chichester: Wiley, 4th edition, 1976.

Detwyler, T. R., and Marcus, M. G. *Urbanization and Environment*. Belmont, California: Duxbury Press, 1972.

Doxiadis, C. A. 'Ecumenopolis: the coming world-city'. In *Cities of Destiny*, Ed. Toynbee, A. pp. 336–358. London: Thames & Hudson, 1967.

Falcon, W. 'The green revolution: generation of problems'. *American Journal of Agricultural Economics*, December 1970, pp. 698–710.

Foley, G. *The Energy Question*, Harmondsworth: Penguin, 1976.

Forrester, J. *World Dynamics*. Cambridge: Wright-Allen Press, 1971.

Grigg, D. *The Harsh Lands*. London: Macmillan, 1970.

Guinness, P. 'Nuclear Power and the American Energy Crisis'. *Geography*, Vol. 64, Part I, January 1979, pp. 12–16.

Hardin, G., and Baden, J., eds. *Managing the Commons*. London: Freeman, 1977.

Holdren, J. P., and Ehrlich, P. R., eds. *Global Ecology*. New York: Harcourt Brace Jovanovich, 1971.

Hoyle, B. S., ed. *Spatial Aspects of Development*. Chichester: Wiley, 1974.

Jahoda, M. et al. *Thinking About the Future: A Critique of 'Limits to Growth'*. London: Chatto & Windus, 1973.

Johnson, D. G. *World Agriculture in Disarray*. London: Macmillan, 1973.

Johnson, S. *The Population Problem, Sources for Contemporary Issues Series*. Newton Abbot: David & Charles, 1975.

Johnson, B. L. C. *India, Resources and Development*. London: Heinemann, 1977.

Jones, E. and Van Zandt, E. *The City: Yesterday and Tomorrow*. London: Aldus Jupiter, 1975.

Keeble, D. E. 'Models of economic development'. In *Models in Geography*. Ed. Chorley, R. J., and Haggett, P. London: Methuen, 1967.

MacNeill, J. W. *Environmental Management*. Ottawa: Government of Canada, 1971.

Meadows, D. H., Meadows, D. L., Randers, J., and Behrens, W. W. *The Limits to Growth*. New York: Universe Books, 1972.

MIT *Man's Impact on the Global Environment: Report of the Study of Critical Environmental Problems (SCEP)*. Cambridge, Mass. and London: MIT Press, 1970.

Mountjoy, A. B. *Industrialization and Developing Countries*. London: Hutchinson, 1975.

Novick, D. *A World of Scarcities: Critical Issues in Public Policy*. New York and Toronto: Halsted Press, 1976.

Odell, P. R. 'The future of oil: a rejoinder'. *Geographical Journal*, Vol. 139, Pt. 3, 1973, pp. 436–454.

Odell, P. R. 'World energy in balance'. *The Geographical Magazine*, May 1974, pp. 378–380.

Odell, P. R., and Vallenilla, L. *The Pressures of Oil: A Strategy for Economic Revival*. London: Harper and Row, 1978.

Prescott, J. R. V. *The Political Geography of the Oceans*. New York: John Wiley & Sons, 1975.

Sheskin, I. M., and Osleeb, J. P. 'Natural gas: a geographic perspective', *Geographical Review*, January 1977, pp. 71–85.

Schultz, T. W. *Transforming Traditional Agriculture*. New Haven: Yale University Press, 1964.

Simmons, I. G. 'Conservation'. In *Evaluating the Human Environment: Essays in Applied Geography*, Ed. Dawson, J. A., and Doornkamp, J. C. Chapter 11, pp. 249–275. London: Arnold, 1973.

Stockholm Conference on the Human Environment. *Human Settlements: The Environmental Challenge* (compendium of United Nations Papers). London: Macmillan, 1973.

The Geographical Magazine 'World Map of the Energy Market'. *The Geographical Magazine*, May 1974, pp. 408–410.

Thomas, T. M. 'World energy resources: survey and review'. *Geographical Review*, April 1973, pp. 246–258.

Wagstaff, H. R. *A Geography of Energy*. Dubuque, Ia.: William C. Brown, 1974.

Ward, B. *Human Settlements: Crisis and Opportunity*. Ottawa: Government of Canada, 1974.

Warman, H. R. 'The future of oil'. *Geographical Journal*, Vol. 138, Pt. 3, 1972, pp. 287–297.

Williams, A. F. 'Transport strategy for a world without oil'. *The Geographical Magazine*, September 1978, pp. 825–833.

Zelinsky, W. *A Prologue to Population Geography*. Englewood Cliffs, N.J.: Prentice-Hall, 1966.

Glossary

Accessibility The relative degree of ease with which a location may be reached from other locations.

Acculturation The borrowing of items from one culture by another such that the cultures coming into contact with one another are affected.

Actual range The distance between two competing communities or business centres where the market area is split between them. The actual range can only be determined with reference to a competing centre. Another name for the actual range is *breaking point.*

Administrative principle In central place theory, the principle which states that a higher-order service area wholly includes seven service areas of the next lower order.

Affinity Affinity is the spatial attraction of one type of activity to another. Clustering often occurs as a result.

Agglomeration Spatial grouping together of activities or people for mutual benefit. Particular savings, or economies, accrue to such groupings of retailers and industries.

Agglomerative economies Benefits that are derived from a grouping of activities.

Amenities Features of the environment—natural or human—which are perceived as pleasant and attractive.

Ancillary linkages Connections, or linkages, between unlike functions such that both benefit. Such linkages lead to the mixing of unlike functions in an area, for example, office buildings and restaurants.

Assembly costs A manufacturer's transport costs on raw material inputs.

Back haul The return-trip shipment. Back haul is extremely important in ocean shipping because most ships are unstable when empty and therefore require ballast.

Basic activity Any activity that contributes directly to the economic base of a community by bringing in money from 'outside' the community. A basic activity is in contrast to a service activity, which serves only the people in the local community.

Basic workers Workers who bring money into a community from outside. They are in contrast to service workers, who generate money only within the community.

Beneficiation A process by which low-grade ores are enriched on-site by a process that removes some of the tares.

Bioculture The general name for the husbandry of animals. The most common form is ranching.

Boreal forest Another name for the taiga, or northern forest. This is a forest of softwoods, usually spruce, characterized by rather thin stands of trees having small trunks.

Break-of-bulk A shipment's division into parts —typically at ports—upon transfer from water to land transport.

By-products A useful secondary product from a manufacturing process.

Central area The area situated between a core demand area and an outlying area.

Central business district The older historic retail and office hub of the city.

Central city In the United States, the municipality in a metropolitan area which contains the central business district. (See also *Inner City*.)

Central place function Activity offered from a place to the surrounding hinterland, both urban and rural.

Central place theory A theory developed by Walter Christaller pertaining to the size and spacing of urban places. The primary thesis is that such places are centrally situated within an hierarchical hexagonal lattice of complementary market areas. Related subconcepts of central place theory include threshold, actual range, and outer range.

Centrality A state of high accessibility; the quality of being at the centre of a transportation system.

Centrifugal forces Those outward forces within a city which cause activities to move farther and farther outward toward the periphery. These forces are counterbalanced by centripetal forces which pull things toward the centre.

Ceteris paribus Literally means to let everything be placed as it is. In popular use, it means to hold other things constant while only the thing to be examined is allowed to change—hence a conceptual device.

Clear cutting The practice of removing all trees and vegetation from an area as part of the timber operation. Clear cutting is highly controversial because of its negative environmental, aesthetic, recreational, and wildlife features.

Cluster A close spatial proximity of settlements.

Cluster economies Similar to *Agglomerative economies*.

Colonial Often used in the limited sense of dependence of a less developed area upon a more developed area.

Commodity The item shipped from the primary to the secondary sector. Commodities are to be distinguished from *products*, which are items moving from the secondary to the tertiary sectors, and goods, which move from the tertiary sector to the consumer.

Comparative advantage The variation of an area's or location's suitability for different activities; based on the idea that all locations have certain activities which are more profitable than others.

Competition Refers to more than one place or enterprise seeking the same resources or customers; it results in a division of territory.

Complementarity A state that exists if the varying advantages of two or more locations or areas permit a mutually beneficial linkage, usually by trade.

Concentration The tendency of people or activities to congregate, or cluster, in space.

Congestion On a transport link, the condition of retarded flow, resulting in increased costs.

Connectivity The degree of direct linkage from one location to other locations on a transport network.

Consumer services Services provided directly to the consumer. Consumer services are to be distinguished from producer services, which are services provided to primary, secondary, and tertiary sectors.

Containerization The process of placing all manner of freight in large containers for shipment. This process allows efficient loading and unloading as well as direct transfer from one mode to another.

Convenience goods Goods used very frequently, such as groceries or petrol.

Core area The nucleus area in the core, central, and outlying area concept of regional growth and development. The core area of the United States generally conforms to the industrial belt.

Corner area An area situated off the main flows of transportation or geographically on the edge of a region. Basically the same as a peripheral area.

Cut-over areas Formerly forested areas that were once highly profitable for timber production but were abused in the extraction process.

Demand in space The value of goods and/or services desired by customers; may be expressed for unit areas or per capita.

Dependent variable As the Y variable in regression analysis, it refers to that variable which receives the response from the independent, or X, variable. In geographical analysis, the dependent variable is the pattern or location for which understanding is desired.

Disaffinity A general repulsion between two land uses, which usually locate in different parts of the city.

Diseconomies The diminishing returns or profitability that sometimes results from greater size (as of a city) or output (as of a plant).

Dispersed city A term some authors apply to a cluster of towns with some specialization of function. Other authors apply it to a built-up urban area which sprawls into the surrounding countryside.

Distribution costs See *Procurement costs.*

Diversification For firms, refers to a variety of outputs, usually in more than one industrial sector. For cities, it indicates that there is no unusual dependence on or specialization in particular industries.

Dual society A society in which there are primarily two types of classes: the very poor and the very rich. A dual society is characteristic of many underdeveloped areas.

Economic margin In agricultural location theory, refers to the farthest locations from which goods can be profitably shipped to commercial markets.

Economics of Scale. See *Scale economies.*

Efficiency Best use of territory; inefficiency in space means less-than-optimal use of territory.

Entail Literally, to cut up or carve an estate for inheritance purposes in a prescribed way. A principle of ownership in which the ownership of property is restricted to the owner's lineal descendent or to a particular inheritance class.

Environment The natural and cultural setting within which people and firms exist; used here mainly to treat variations in natural conditions, such as landforms, or climate.

Environmental determinism The idea that the physical environment determines or critically influences the way in which people behave economically in any given place. Environmental possibilism is a milder form of this concept.

Equilibrium A theoretical state of stability. Any deviation from this state would decrease efficiency or profitability; equilibrium prices are values (for labour, land, or capital) corresponding to these conditions.

Extensive Refers to a relatively low level of inputs or outputs per unit area: thus 'extensive agriculture'.

External diseconomies Outside factors and influences that tend to cause inefficiencies in manufacturing and other operations. External diseconomies characteristically result from too-large concentrations of competing activities in a particular area.

External economies Outside factors and influences that tend to benefits (higher sales, profits, lower costs) in manufacturing and other operations. The primary location of external economy is the large city, particularly the infrastructure that is provided for the accommodation of industrial production.

Exurban The zone beyond the built-up area of the city, but within which commuters to the urban area are dominant.

Fallowing The practice of letting certain fields remain without cropping for a season or more. In semiarid areas fallowing is commonly practised to allow the land to build up in moisture.

Forelands Diverse areas shared by other ports. All ports have hinterlands (the area served on the land side, or back, of the port) and forelands (the area served on the water side, or front, of the port).

Footloose industry Manufacturing industry which is not based upon resource constraints such as coalfields, but which has the ability to choose a wide range of locations.

Freeways A common name for interstate highways in the United States which are usually four or more lanes in width with separated traffic lanes, The word 'freeway' was coined to differentiate such routes from similar-looking tollways, throughways, and tolled parkways.

Fringe Another name for peripheral area or corner area. Fringe means on the edge of economic action.

Functional region Area under the economic and social domination of a centre; nodal region is the more usual term.

Geographical inertia The tendency of older industrial regions to survive by the contraction and adaptation of old, heavy industries in situ, and by the development of new light industry.

Ghetto An area of the city distinguished by ethnical, racial, or religious character; usually, but not necessarily, a low income area.

Gravity model A mathematical description based on Newtonian physics applied to social phenomena. The principle is that interaction, or pull force, is directly related to the size of a place and inversely related to the distance one is from that place. Both size and distance are commonly modified as to weight by various exponents.

Gross domestic product (GDP) The total value of goods and services produced inside the country.

Hierarchy The concept that urban places, with their trade areas, may be grouped into distinctive levels of functional importance, and that the individual consumer will travel to smaller, closer places for everyday purchases and to larger, more distant places for less-demanded and perhaps more durable goods.

Horizontal integration A system of obtaining efficiencies in an overall manufacturing operation by controlling backward and forward linkages in the manufacturing process.

Independent variable See *Dependent variable*.

Indigenous Refers to the native or natural place of people or commodities. For example, potatoes and corn are indigenous to South America.

Industrial complex A set of specific industries which are closely related, usually because each industry makes significant purchases from the others.

Inelasticity of demand A demand for a commodity, product, good, or service which is presumed to be little affected by price. Salt is such a commodity.

Information Knowledge about the environment, technology, and other conditions that would be necessary for optimal decisions. Thus an information field is the geographical distribution (around an individual group) of knowledge about other people or areas.

Infrastructure That system of services and facilities within a city or elsewhere which aids industry and other economic activities; for example, roads, governmental services, and marketing mechanisms.

Inner city The central portion of a large city or metropolis. The term is increasingly used in the United States as a synonym for the central city. In some communities the inner city refers to a loosely defined area within the central city.

Inner range The minimal radial distance from a business which contains sufficient purchasing power to provide a threshold. The geographical equivalent of the marketing concept of threshold.

Inner zone The innermost component of a large city which commonly contains the central business district as well as, perhaps, an inner city ethnic minority.

Innovation Using an idea to lead to change, often beneficial, in individual behaviour or in a production process. The innovation process may take place in waves over time and space.

Input-output Refers to the pattern of purchases and sales among sectors of the economy; especially useful in tracing the effects of change in one sector on the behaviour of other sectors.

Intensive Refers to a relatively high level of inputs and/or outputs per unit area. Used most commonly in reference to agriculture.

Interdependence Indicates that because of specialization and trade, what occurs at one location affects what happens at many other locations.

Intervening opportunity In the movement of goods, services, people, indicates the presence of closer, better opportunities which greatly diminish the attractiveness of even slightly farther ones.

Isodapane In industrial location theory, a line or contour of constant total cost (assembly, production, and distribution).

Kennedy round A major agreement in 1967 on tariff reductions amongst the industrial nations.

Labour productivity The relative cost or labour per unit of output.

Landscape Refers either to the systematic human pattern of occupance, or to the visual make-up of natural or man-made territory. Thus 'townscape' may refer to the visual patterning of the city, with special reference to roads, buildings, street 'furniture'.

Light industry A class of industry within cities which may be made on the basis of space-use intensity or its inputs, processes, or outputs. To be contrasted with heavy industry which commonly occupies considerable land and engages in primary, rather than secondary, manufacturing.

Linkages The pattern of interdependence among industries. May take many forms such as forward linkage or backward linkage.

Location freedom The notion that an activity is free to locate anywhere; probably only a theoretical freedom.

Location rent See *Rent gradient*.

Location triangle In location theory, a simple diagram of optimum location in the case of three markets and/or material sources.

Marginal cost The actual cost of producing one more item. If a business is operating under scale economies, marginal costs decrease with increasing diseconomies and increase with each increasing output.

Marginal farmer The farm entrepreneur with very low net return per man/hour.

Market A place or location where goods and services are demanded and exchanged.

Market principle In central place theory, that arrangement of the hierarchy of places which will minimize aggregate distance travelled to centres; service areas of larger, higher-order places include one-third of the service areas of each of the six neighbouring lower-order places.

Mark-up Mark-up is simply the difference in price between seller and buyer. In retailing, it refers to the difference between the wholesale price and retail price. In general the higher the mark-up the lower the volume.

Milkshed The zone of an urban market's fluid milk supply.

Multiple nuclei theory One of the three major theories of urban morphology, along with the concentric zone theory by Park and Burgess and the sector theory by Hoyt. The multiple nuclei theory, by Harris and Ullman, postulates that there are multiple concentrations of residential, industrial, and commercial activity within any large city and that such is caused by logical economies of necessity and convenience.

Multiplier The ratio between basic and service workers. The smaller the number of basic workers as compared with service workers (B/S ratio), the greater the multiplier effect.

Multipurpose trip Trip whereby more than one establishment is visited or more than one function is accomplished on such a circuit.

Nesting In central place theory, refers to the tendency for service areas of lower-order places to be wholly included in the service areas of larger, higher-order places.

Network, transportation The actual physical system of links and nodes on which movement can take place.

Nodal region The area which is dependent on or is dominated by a nodal centre at the heart of a transport network.

Nonoptimal behaviour Any decision making, whether intentional or by default, which results in a less-than-maximum profit outcome. Perhaps a very common circumstance even in the world of business and industry as other than profit-motive considerations are held to be important.

Oligopoly, spatial Refers to a fairly stable shared regional market, usually with a common price structure.

On-site costs Costs of production which are not related to transportation of materials to or from the farm or factory but which occur at the site. Such costs commonly include labour, rent, and those directly attributable to scale economies and to environmental features.

Orientation The tendency for various kinds of industries to locate at markets (market orientation), at resources (resource or raw-material orientation), or because transport costs are decisive (transport orientation).

Outer range The farthest distance from which customers will frequent a central place.

Outlying area The outermost region from a central core of production, market, and population. Outlying areas are at a great disadvantage because of the probability of intermediate opportunities.

Penthouse effect The phenomena of high rents on the uppermost floors of high buildings. The high rents are the result of both such site amenities as view and certain psychological values attributed to being 'on the top'.

Peripherality The state of being at the edge of a community and its interacting systems, far from the controlling centres of the culture or economy.

Precession wave A wave of urban land preparation preceding the actual built-up edge of the metropolis. Here land changes ownership, otherwise made ready for urban occupancy.

Primacy An unusually high proportion of population and economic activity in the single largest city of a country, usually the capital.

Primary activities Commodities produced in the first or primary sector. These include forestry, fishing, mining, and agriculture in which basic processing or conversion of raw materials take place (as distinguished from secondary activities, which utilize the outputs of primary manufacturers).

Processing or production costs The actual costs —mainly capital and labour—incurred in the conversion process, not including transportation costs.

Procurement costs The costs of transporting to the place of production the items required in the production process. In manufacturing these costs represent the cost of shipping the raw materials to the plant. The costs of shipping the finished products to market are called distribution costs.

Producer services Services provided in the quaternary sector which are designed to facilitate the production and flow of commodities, products, and goods from the primary, secondary, and tertiary sectors.

Products The finished products of an industry in the secondary sector. Products are to be distinguished from output of the primary sector, in this text called commodities, and those of the tertiary sector, called goods.

Quaternary sector The fourth, or service, sector of the economy. This sector consists of both producer services and consumer services.

Randomness Indicates that locational uncertainty or imprecision which may be expected to result from many small and unknown factors.

Range In central place theory, the maximum distance over which a seller will offer a good or service; or from which a purchaser will travel for it. See specific types of range, such as *Actual, Inner* and *Outer*.

Rank-size rule Notes the empirical tendency for the product obtained by multiplying a city's rank times its size to equal a constant, the population of the country's largest city.

Refraction As used in geography, the bending of routes such that shipment is made for the minimum cost. Lower cost routes are favoured in length, whereas the distance shipped along higher cost routes are minimized.

Region A portion of space which, according to specified criteria, possesses meaningful unity; see *Uniform region* and *Nodal region*.

Regional convergence Implies that as an economy tends to reach an equilibrium state, prices, incomes, and other measures of value will tend to equalize.

Regional planning The conscious attempt of government to influence the course of economic and social development. Usually it is a result of welfare considerations, more rational use of resources, or alleviation of poverty.

Relative location The advantages or disadvantages of a particular location measured with reference to all competing locations.

Rent gradient The decline in rent (value) with distance from a market of centre. Such decline occurs because the farther any given place is from the centre, the higher must be the transportation cost to that centre. As used in agricultural and urban structure theory, it is therefore a measure of the value to the landowner of the relative accessibility of a certain location. Because of competition for the more accessible locations, a distance–decay gradient in location rent may be expected to develop outward from the central point of greatest accessibility.

Resources In an economic sense, any valued aspect of the environment; as used in this book, demanded natural materials, such as water, minerals, or soil. We may also speak of human resources, valuing numbers, skills, etcetera.

Retail gravitation The notion that the attractiveness of a seller to a customer varies inversely with the distance between them.

Scale economies The principle that as production volume increases, the average cost per unit of output decreases because of certain efficiencies.

Secondary sector The second sector of production, or manufacturing.

Sector theory The tendency for sectors or wedges projecting outward from a city's centre to be devoted to different uses and social classes. One of several major theories of urban morphology or land use.

Self-sufficiency The attempt of the local economy to provide by itself for all its needs and demands.

Service area (or trade area) The territory from which most of a seller's customers originate.

Shifting cultivation The type of cultivation most characteristic of primitive economies in tropical areas. Fields are shifted from place to place about a central village over time, a cyclical process made necessary due to exhaustion of soil fertility.

Spatial adjustment In the sense of the business firm, implies changes in the location of a firm or of its suppliers' markets in the face of external change (perhaps change in resources or markets).

Spatial diffusion The process of the gradual spread over space of people or ideas from critical centres of origin.

Spatial interaction The interrelation of locations usually in terms of movement of people or communications; the level of interaction varies inversely with distance between locations.

Spatial organization The aggregate pattern of use of space by a society.

Spatial parameters The degree of spatial choice, or freedom, in the location of any particular activity. Today there are usually more acceptable choices in location than earlier because of improvements in transportation.

Spatial relations The ways in which space and distance influence behaviour and location decisions.

Sporadic activities Activities that are distributed unevenly and so as not to coincide with the distribution of the market or the population.

Stranded areas An area that was once economically equal with areas about it but which has fallen behind so much that it appears to have been stranded, or by-passed, in development and growth.

Subsistence The condition of a local economy's ability to provide only for its basic food and shelter needs without significant surplus.

Substitution In seeking maximum profitability at a location or for a firm, it may be possible to exchange more labour for less capital, or in agriculture, more fertilizer for less land, depending on which is scarcer.

Surburbia That urbanized portion of the metropolis which lies beyond the municipal boundaries of the central city in the United States. This term also connotes a fairly low density of occupance and a particular lifestyle.

Terminal costs The costs associated with the unloading of freight at the point of origin or destination. Commonly applied to any cost of loading or unloading.

Tertiary sector The third production sector. This consists of wholesaling and retailing establishments.

Threshold The minimum number of sales necessary to support in viable fashion a particular establishment or function. The marketing equivalent of inner range.

Transferability The extent of connections between places, or the degree to which a good or service may be transported.

Transportation principle In central place theory, that arrangement of the hierarchy of places which results in the most efficient transport network; see *Network, transportation.*

Trickle-down process The process by which housing of the higher income group is passed down to the next income group.

Ubiquitous industries A pattern of industry which is in direct proportion to the distribution of population. Market-type industries are commonly ubiquitous.

Underemployment A case of individuals working either part time of part of the year, but seeking full-time work. In some activities like agriculture, it can refer to full-time occupation which inefficiently uses time and labour.

Uniform region A territory or space for which the internal variation of specified criteria is appreciably less than the variation between this area and other areas.

Urban rent theory The theory of bid-rents, whereby different activities compete to pay rent for the same sites, which provides an explanation for the land-use pattern within cities.

Vertical integration The control of forward and backward linkages of production in the manufacturing process.

Weight-loss ratio The difference between raw material volume and finished product volume for a manufacturing firm as expressed by a general ratio formula.

Wholesaling Process by which products from the manufacturing sector are moved to the retailing sector. Part of the tertiary sector.

Conversion Factors

LENGTH

1 inch = 2·5400 centimetres
1 foot = 0·3048 metres
1 statute mile = 1·6093 kilometres

1 centimetre = 0·3937 inches
1 metre = 3·3281 feet
1 kilometre = 0·6214 miles

AREA

1 acre = 0·4047 hectares
1 hectare = 10 000 square metres
1 square mile = 2·5899 square kilometres
1 square kilometre = 0·3861 square miles

1 hectare = 2·4710 acres
1 square kilometre = 100 hectares

WEIGHT

1 lb = 0·4536 kilogram
1 tonne = 1000 kilograms = 2204·6 lb
1 Imperial ton = 2240 lb = 1·0161 tonnes
1 short (US) ton = 2000 lb = 0·9072 tonnes

1 kilogram = 2·2046 lb

VOLUME

1 cubic foot = 0·0283 cubic metres
1 Imperial pint = 0·5682 litres

1 cubic metre = 35·3147 cubic feet
1 litre = 1·7600 Imperial pints

1 Imperial gallon = 1·2009 US gallons = 4·5461 litres
1 US gallon = 0·8327 Imperial gallons = 3·7854 litres
1 barrel = 42 US gallons = 34·97 Imperial gallons = 159·0 litres

Index